Lecture Notes in Mathematics

Edited by A. Dold and B. Eckmann

Subseries: Department of Mathematics, University of Maryland
Adviser: M. Zedek

1276

Carlos A. Berenstein (Ed.)

Complex Analysis II

Proceedings of the Special Year
held at the University of Maryland, College Park, 1985–86

Springer-Verlag
Berlin Heidelberg New York London Paris Tokyo

Editor

Carlos A. Berenstein
Department of Mathematics, University of Maryland
College Park, MD 20742, USA

Mathematics Subject Classification (1980): 32-06

ISBN 3-540-18357-4 Springer-Verlag Berlin Heidelberg New York
ISBN 0-387-18357-4 Springer-Verlag New York Berlin Heidelberg

Printed in Germany

Printing and binding: Druckhaus Beltz, Hemsbach/Bergstr.
2146/3140-543210

INTRODUCTION

The past several years have witnessed a striking number of important developments in complex analysis of both one and several variables. Through these advances the essential unity of these two previously rather separate branches of function theory has become increasingly apparent. More and more, ideas and constructs that first arose in function theory of one variable are playing an important role in the several variables theory. At the same time, techniques developed originally for use in several variables have found fruitful applications to problems in classical function theory. Examples of the former phenomenon include the development of a capacity theory for the Monge-Ampère operator and recent extensions of the Henkin-Ramirez representation formulas and their application to interpolation problems in $\mathbb{C}^n$. In the second category, the systematic use of the inhomogeneous Cauchy-Riemann equation has led to important developments in the theory of H^∞, as well as other Banach algebras on the unit disk. It has also inspired the consideration of many new questions about open Riemann surfaces. Finally the brilliant solution of the Bieberbach Conjecture by Louis de Branges offers irrefutable testimony (as if any were needed) to the continued vitality of classical ideas and approaches.

It is in this context that the Department of Mathematics of the University of Maryland decided to dedicate its sixteenth Special Year to the subject of Complex Analysis. The objective of these Special Years has been to provide a forum for the exchange of ideas between the members of the department, long-term visitors and the other participants. (In spite of the electronic mail and other gadgets, a lot of mathematics is still done on a face to face basis!) This time we have had over one hundred and fifty mathematicians from several different countries, who participated in the Special Year from July, 1985 to December, 1986. This participation has been made possible by a substantial contribution from our department as well as the very generous support of the National Science Foundation and the Argonne Universities Association. The Organizing Committee of the Special Year, D. Hamilton, J.A. Hummel, L. Zalcman and myself, would like to express their appreciation for this support to Dr. J. Polking, Dr. A.I. Thaler and Dr. J.V. Ryff of the NSF, Chancellor George A. Russell of the AUA and Professors J. Osborn and N. Markley, past and present chairmen of our department.

It is clear that this Special Year has been a very successful one and the three volumes of Proceedings are just a token proof of it. On behalf of the Organizing Committee, I would also like to thank all the participants who made this organizational effort worthwhile. Last, but not least, our heartfelt thanks to the administrative personnel of our department, among them D. Kennedy, D. Forbes, S. Matthews, and B.L. Smith, and to the very skillful S. Smith, V. Sauber, J. Nagendra, K. Aho, and I. Grove who did a magnificent job at typing these Proceedings. Finally, there is one person who, in our opinion and the opinion expressed in many letters of our visitors, merits a special round of applause, since without her immense dedication and generosity we seriously doubt we would have fared so well, we refer to Mrs. N. Lindley.

To conclude, a word about these three volumes. The reader will find here both surveys of different areas of Complex Analysis as well as many new results and insights. We had asked for manuscripts that were accessible not only to specialists in the subject but to a broader audience; we think that the authors have responded to that request and these Proceedings will prove to be a very useful source of reference and problems in Complex Analysis.

Carlos A. Berenstein

Table of Contents

INTRODUCTION III

Table of Contents Vol. II V
Table of Contents Vol. I (Lect. Notes in Math. Vol. 1275) VII
Table of Contents Vol. III (Lect. Notes in Math. Vol. 1277) IX

List of Participants (see Vol. I Lect. Notes in Math. Vol. 1275)

SPECIAL YEAR PAPERS -- VOLUME II

H. Alexander
Polynomial Hulls and Linear Measures 1

E. Amar
Estimations, par Noyaux, des Solutions de l'Equation $\bar{\partial}u = f$ . . . 12

Richard Beals and Nancy K. Stanton
The Heat Equation and Geometry for the $\bar{\partial}$-Neumann Problem . . . 25

S. Bell
Extendibility of the Bergman Kernel Function 33

Bo Berndtsson
$\bar{\partial}_b$ and Carleson Type Inequalities 42

Joaquim Bruna
Boundary Absolute Continuity of Functions in the Ball Algebra 55

Urban Cegrell
Subharmonic Functions and Minimal Surfaces 60

Anne-Marie Chollet
Propriétés de Recouvrement des Sous-Ensembles de la Frontière d'un Domaine Strictement Pseudo-Convexe 67

S. Coen
Some Properties of the Canonical Mapping of a Complex Space into Its Spectrum 79

C. Robin Graham
Scalar Boundary-Invariants and the Bergman Kernel 108

Robert E. Greene and Steven G. Krantz
Biholomorphic Self-Maps of Domains 136

J. Korevaar
Some $\mathbb{C}^N$ Capacities and Applications 208

R. Meise, S. Momm and B. A. Taylor
Splitting of Slowly Decreasing Ideals in
Weighted Algebras of Enire Functions 229

A. Meril and D. C. Struppa
Convolutors in Spaces of Holomorphic Functions 253

C. A. Berenstein and Daniele C. Struppa
A Remark on "Convolutors in Spaces of Holomorphic Functions" . . 276

R. Michael Range
Integral Representations in the Theory of the
$\bar{\partial}$-Neumann Problem 281

Pascal J. Thomas
Tents and Interpolating Sequences in the Unit Ball 291

John Wermer
Balayage and Polynomial Hulls 303

J. Wiegerinck
Convergence of Formal Power Series and Analytic Extension . . . 313

SPECIAL YEAR PAPERS -- VOLUME I

Don Aharonov
Bazilevič Theorem and the Growth of Univalent Functions. 1

J. Arazy, S. Fisher and J. Peetre
Möbius Invariant Spaces of Analytic Functions 10

Albert Baernstein, II
Dubinin's Symmetrization Theorem 23

James E. Brennan
Functions With Rapidly Decreasing Negative Fourier Coefficients 31

Chi-tai Chuang
On the Theory and Applications of Q_m-normal Families of Meromorphic Functions 44

J. A. Cima and T. H. MacGregor
Cauchy Transforms of Measures and Univalent Functions 78

Matts Essén
Harmonic Majorization, Harmonic Measure and Minimal Thinness . . . 89

H. M. Farkas
Unramified Coverings of Hyperelliptic Riemann Surfaces 113

Carl H. FitzGerald
Arcs on Which a Univalent Function Has Small Magnitude 131

Paul M. Gauthier
Uniform and Better-Than-Uniform Elliptic Approximations On Unbounded Sets 149

F. W. Gehring and G. J. Martin
Discrete Convergence Groups 158

F. Haslinger and M. Smejkal
Representation and Duality in Weighted Frechet Spaces of Entire Functions 168

Simon Hellerstein
Real Zeros of Derivatives of Meromorphic Functions and Solutions of Differential Equations - A Survey with Problems and Conjectures 197

James A. Jenkins
Some Estimates for Harmonic Measures 210

Robert Kaufmann
Hausdorff Measures and Removable Sets 215

John L. Lewis
Approximations of Sobolev Functions and Related Topics 223

David Minda
Inequalities for the Hyperbolic Metric and Applications to Geometric Function Theory 235

P. J. Rippon
Subharmonic Functions With Unilateral Growth Conditions 253

Glenn E. Schober
Kirwan's Conjecture 266

A. A. Shaginyan
On Potential Approximation of Vector Fields 272
On Tangential Harmonic Approximation and Some Related Problems . . 280

Harold S. Shapiro
Unbounded Quadrature Domains 287

SPECIAL YEAR PAPERS -- VOLUME III

D. Drasin
Some Recent Successes in Value-Distribution Theory 1

R. Dwilewicz
Bergman-Szegö Type Theory for CR Structures. 15

G. Gentili
Regular Complex Geodesics for the Domain
$D_n = \{(z_1,\ldots,z_n) \in \mathbb{C}^n: |z_1| + \ldots + |z_n| < 1\}$ 35

D. Jerison and A. Sánchez-Calle
Subelliptic, Second Order Differential Operators 46

Karl Oeljeklaus and Wolfgang Richthofer
Recent Results on Homogeneous Complex Manifolds 78

Y-T. Siu
Kähler-Einstein Metrics for the Case of Positive
First Chern Class. 120

W. Stoll
Algebroid Reduction of Nevanlinna Theory 131

T. V. Tonev
An Infinite-Dimensional Generalization of the Shilov Boundary
and Infinite Dimensional Analytic Structures in the Spectrum
of a Uniform Algebra 242

T. V. Tonev
New Relations Between Sibony-Basener Boundaries. 256

M. Tretkoff
Picard-Fuchs Differential Equations for the Quadratic Periods
of Abelian Integrals of the First Kind 263

E. Vesentini
Holomorphic Families of the Holomorphic Isometries 290

P-M. Wong
Complex Monge-Ampère Equation and Related Problems 303

H. Wu
Liouville Theorems 331

A FINAL WORD 350

POLYNOMIAL HULLS AND LINEAR MEASURE

H. Alexander

1. Let X be a compact subset of $\mathbb{C}^n$. We denote the polynomially convex hull of X by $\hat{X}$. By definition $\hat{X} = \{x \in \mathbb{C}^n : |f(x)| \le \|f\|_X$ for all polynomials $f\}$. We shall consider cases of the following generic result.

<u>THEOREM</u>. Let X be a "thin" subset of $\mathbb{C}^n$. Then $\hat{X}\setminus X$ is a (possibly empty) pure one-dimensional analytic subvariety of $\mathbb{C}^n\setminus X$.

The first results of this type were due to J. Wermer [W]. A "thin" set meant a real analytic closed curve. Later, Stolzenberg [S] and Bishop [B] treated the case of a finite family of smooth curves. In [A1], the theorem was proved for a <u>connected</u> set of finite linear measure.

1.1. <u>The Stolzenberg-Wermer picture</u>. Each of these proofs has an associated figure in the plane. I shall try to convey the main geometric construction of the proof without going into all of the details. Take X to be a smooth curve in $\mathbb{C}^n$. Suppose that x is a point of $\hat{X}\setminus X$. One must show that $\hat{X}$ is a one-variety near x. Choose a polynomial f such that $f(x) = 0 \notin f(X)$. Then by Sard's theorem, there are two rays from the origin such that the wedge between the two rays meets $f(X)$ in a finite number of disjoint arcs on each of which the angular polar coordinate is monotonic and such that the inverse image of $f|X$ over each of these arcs γ is itself a finite union of arcs in X each of which maps homeomorphically by f to γ. Label the components $\{\Omega_k\}$ of the complement of $f(X)$ in the wedge and let Ω_o be the unbounded component of the complement of $f(X)$. See figure 1. The idea of the proof is to show that X is analytic "over" the $\{\Omega_k\}$ in the wedge by starting with Ω_o where the result is obvious because $\hat{X}$ is empty over Ω_o, and then showing that if $\hat{X}$ is analytic over Ω_k, then X is analytic over Ω_{k+1}. This is called "crossing over the edges". Without going into the details, I want to mention one of basic facts which is used; it is a consequence of the F. and M. Riesz theorem.

<u>Proposition</u>. Let Ω be a Jordan domain in the plane with rectifiable

---

Supported in part by a grant from the National Science Foundation.

boundary; let μ be a finite Borel measure on $\partial\Omega$ which annihilates polynomials in z. Let E be a Borel subset of $\partial\Omega$ such that $|\mu|(E) = 0$, then the linear measure of E is zero.

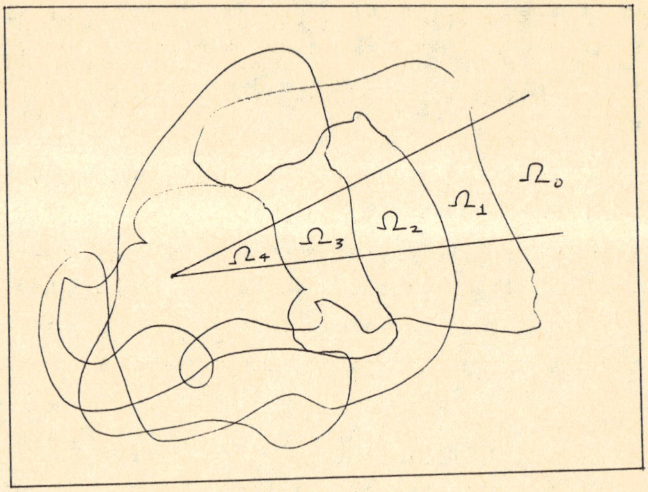

Figure 1

1.2. <u>The finite linear measure picture</u>. Now consider the case of a connected set X of finite linear measure. We would like to cross over the edges as before, but, in general, the picture of figure 1 cannot be achieved and we must cross over a set of positive measure.

Let L be a compact plane set of finite linear measure. We shall say that two components Ω_1 and Ω_2 of the complement of L are <u>amply adjacent</u> if there exists a rectangle $R = [a,b] \times [c,d]$ in the plane such that the bottom edge lies in Ω_1 and the top edge lies in Ω_2, and such that there exists a compact subset K_1 of $[a,b]$ of positive measure such that vertical lines $x = t$ for t in K_1 meet $R \cap L$ in exactly one point. As t varies over K_1 these points of $R \cap L$ give a compact subset K of R which is mapped homeomorphically to K_1 by projection to the x-axis.

The set $[a,b]\backslash K_1$ consists of open intervals. Fix one of them (u,v) and consider the intersection of L with $(u,v) \times [c,d]$. Consider the smallest rectangle R' of the form $(u,v) \times [s,t]$ which contains this intersection. By using the fact that L is <u>connected</u> one can show that the length of the perimeter of R' is no more than four times the linear measure of this intersection. The perimeter of R' is a disjoint union of: two points of K, a "bottom" open arc which lies in Ω_1 and a "top" open arc which lies in Ω_2. Carry out the same construction in all of the complementary interals of K_1. Next, form two Jordan domains J_1 and J_2 in R with J_j in Ω_j as follows. For J_1 take the bottoms of all of the perimeters of the rectangles R' together with K. It is easy to show that this forms a rectifiable arc in R. Complete this with the bottom half of the perimeter of R to form a rectifiable Jordan curve which bounds a Jordan domain J_1 lying in Ω_1. Obtain J_2 in Ω_2 in the same way. See figure 2. Now, with $L = f(X)$ for a polynomial f, the crossing over the edge argument can be carried out from J_1 to J_2. The main point being that the boundaries of J_1 and J_2 intersect in the set K of positive linear measure. We refer to [A1] for the complete details.

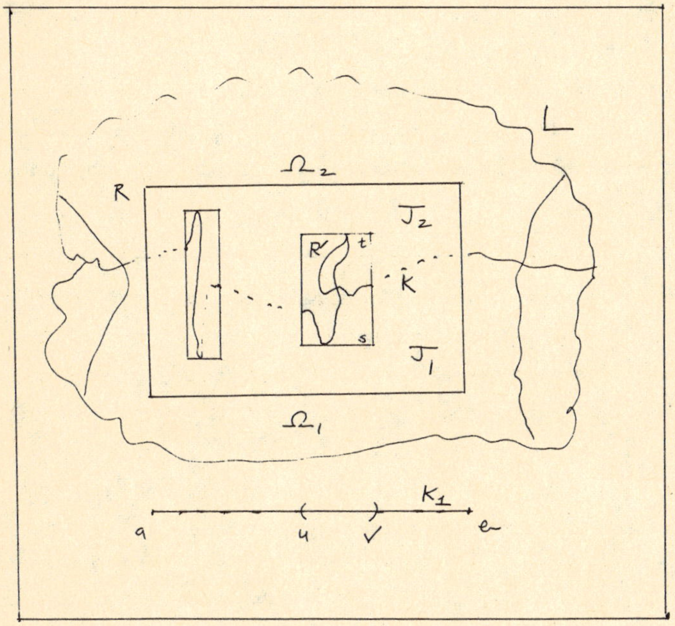

Figure 2

2. The connectedness assumption on X is clearly needed in the construction of 1.2 and it was an open question whether the assumption could be dropped. In [A2] the following was proved.

<u>THEOREM</u>. There exists a compact subset X of $\mathbb{C}^2$ of finite linear measure such that $\hat{X}\backslash X$ is non-empty and is not a pure one-dimensional subvariety of $\mathbb{C}^2\backslash X$.

The set X is not highly pathological; it consists of a countaable disjoint union of real analytic simple closed curves. Also $Z := \hat{X}\backslash X$ is a countable union of analytic subsets of $\mathbb{C}^2\backslash X$. The reason that Z is not an analytic subset of $C^2\backslash X$ is that countably many of these subvarieties cluster at some points of Z, where of course Z does not then have the structure of a one-dimensional analytic subset.

The set X in $\mathbb{C}^2$ is constructed using certain functions of Beurling [Be]. We refer to [A2] for full details but here we shall describe the z_1-projection of $\hat{X}$. This projection, denote it by Q, looks like the closed unit disk with a countable number of disjoint disks removed. These disks cluster at every point of the unit circle and the sum of their radii is convergent. See figure 3. The z_1-projection of X is the boundary of Q. X itself consists of the unit circle $\{(z,0) : |z| = 1\}$ together with real analytic curves which project to the inner boundary curves of Q. The set Z is infinitely sheeted over Q and has infinite area. Z is the union of graphs of functions f which are smooth on Q and analytic on its interior and identically zero on the unit circle. Such functions were produced by Beurling.

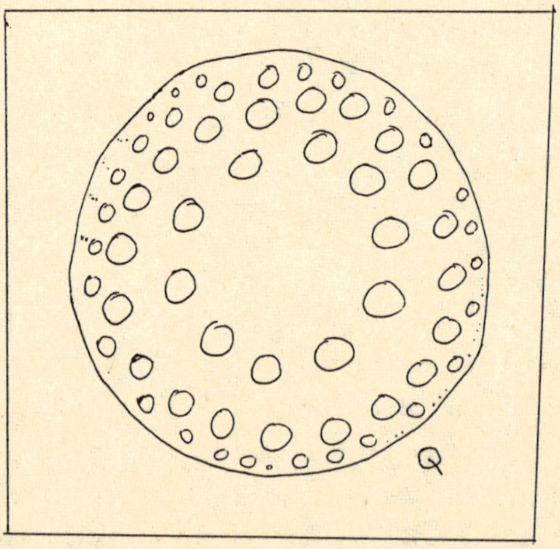

Figure 3

2.1. <u>Crossing over edges</u>. This example was found by looking for a situation in which crossing over edges and also the conclusion of the proposition of 1.1 do not hold. If "crossing over edges" were valid in this case, we could cross over from the exterior of the unit disk to the interior of Q and conclude that $\hat{X}$ is analytic, and in particular finite sheeted, over the interior of Q, which is not true. This means that the interior of Q is not amply adjacent to the exterior of the unit disk, even though the boundary of Q has finite linear measure.

Likewise the conclusion of the proposition of 1.1 is not valid for the domain W which is the interior of Q. To see this, note that the existence of the function f of the last paragraph implies that there is a Jensen representing measure m supported on the boundary of Q for the origin (which we assume lies in W) for the algebra $A(Q)$ (functions continuous on Q, analytic on W) which has no mass on the unit circle. Then the measure $z \cdot dm$ annihilates polynomials and has no mass on the unit circle.

3. We shall consider the hull of a rectifiable Jordan curve. In this case we can be more precise about the structure of the hull.

<u>THEOREM</u> Let X be a rectifiable Jordan curve in $\mathbb{C}^n$ which is not polynomially convex. Then $V := \hat{X}\backslash X$ is an irreducible analytic one-dimensional subvariety of $\mathbb{C}^n\backslash/x$ and has finite area. In fact,

$$4\pi \text{ area}(V) \leq L^2,$$

where L is the length of X.

We shall give a brief sketch of the proof; the full details will appear in [A3]. We know already that V is an analytic subvariety of $\mathbb{C}^n\backslash X$.

(a) First we show directly that V consists of a single analytic branch. The main idea is to take a polynomial f and to consider a component Ω of $\mathbb{C}\backslash f(X)$ which is amply adjacent to the unbounded component of $\mathbb{C}\backslash f(X)$. One shows that the multiplicity of any branch of V over Ω is the same as the multiplicity of V itself over Ω. Here one uses the fact that any branch of V clusters at all of X, because any proper subset of X is polynomially convex.

(b) Let f be a polynomial. Let $\{\Omega_k\}$ be the components of $\mathbb{C}\backslash f(X)$. Then since X is a curve in $\mathbb{C}^n$, we can view $f \circ X$ as a plane curve; it has a winding number (index) n_k about points of Ω_k; n_k may be positive, negative or zero. We also have a multiplicity m_k of f on V over Ω_k.

<u>Proposition</u>. There exists an orientation of X such that for all

polynomials f and all k, we have $m_k = n_k$.

Remark. This would be obvious if we knew that Stokes' theorem were valid in this setting. For then we could integrate $d(f-c)/(f-c)$ over X and the argument principle would give the proposition. The point is that we do not know if Stokes' theorem holds, in fact, it is implicit in the statement of Stokes' theorem that the area is finite - the fact we are trying to prove. We shall return to this question below.

(c) The isoperimetric inequality. Let γ be a rectifiable closed (not necessarily simple) plane curve. Let $\{\Omega_k\}$ be the components of $\mathbb{C}\setminus\gamma$ and let n_k be the index of γ about points of Ω_k. Then the isoperimetric inequality asserts that

$$4\pi\Sigma n_k \text{area}(\Omega_k) \le L^2,$$

where L is the length of γ. This version will suffice for our purposes but it may be of interest that this remains valid if n_k is replaced by its absolute value or even its square.

(d) We can now complete the proof that V has finite area. We apply (b) for $f = z_k$ and also (c) for $\gamma = z_k \circ X$ and conclude that the area of $z_k(V)$ taken with multiplicity is finite. (We may assume that V has zero area over $z_k(X)$ itself, otherwise V would lie in a lower dimensional subspace.) Since the area of V is the sum of the areas with multiplicity of the coordinate projections, we are through.

4. Globevnik and Stout [GS] have considered holomorphic mappings from the unit disk into $\mathbb{C}^n$ whose global cluster sets have finite linear measure. The methods and results of Section 3 can also be applied to such questions. However, rather than going into this, I shall consider a one-variable result which was conjectured by Globevnik and Stout.

THEOREM. Let f be an analytic function on the unit disk whose global cluster set $C(f)$ has finite linear measure. Then f extends to be continuous on the closed unit disk.

At the Maryland conference I learned from Lee Stout that this theorem had also been proved by C. Pommerenke. As my proof is relatively short, it may be of interest to have it written down and I shall present it here.

Proof. Since $C(f)$ is connected and of finite linear measure, it is bounded. This implies that f is a bounded function. To prove that f extends to be continuous at points of the unit circle, it suffices to show that the cluster set of f at each point of the unit circle reduces to a singleton. We argue by contradiction and suppose not. That is, suppose that s is a point of the unit circle and that $C(f,s)$, the cluster set of f at s, is not a singleton. Then $C(f,s)$ is a continuum and so its projection to the x-axis or y-axis is a non-degenerate interval. Without loss of generality, we may assume that the x-projection of $C(f,s)$ is the interval $[a,b]$ with $a < b$.

Let $N(t)$ be the number of points of $C(f)$ hit by the vertical line $x = t$. Then, because $C(f)$ has finite linear measure, it follows that N is integrable and, in particular, finite almost everywhere. Choose c with $a < c < b$ such that $N(c)$ is finite and also such that the line $x = c$ (which we shall denote by l) contains none of the critical values of f (which are countable in number).

Let $l \cap C(f) = \{p_1,p_2,\ldots,p_d\}$ where p_i have increasing y-coordinates. See figure 4. I claim that $f^{-1}(p_i)$ is finite for each i. First observe that, by the argument principle, f has constant finite multiplicity for points in the same component of the complement of $C(f)$. The line l is divided by the points p_i into open line segments each of which lies in a single component of the complement of $C(f)$. Let m be the larger of the two multiplicities of f over the two components which contain the two open segments of l which have p_i as an endpoint. Then $f^{-1}(p_i)$ has at most m points. Otherwise, if $f^{-1}(p_i)$ had at least $m + 1$ points then there would be arcs through these $m + 1$ points which would map by f to cover parts of the two segments at p_i more than m times, a contradiction.

Thus $Q := f^{-1}(\{p_1,p_2,\ldots,p_d\})$ is a finite subset of the unit disk.

Let σ be one of the open line segments on l having p_i and p_{i+1} as endpoints for some i. Let τ be a connected component of $f^{-1}(\sigma)$. There are only a finite number of such components, by the finite multiplicity of f. Then, because l contains no critical values of f, τ is an open arc in the open unit disk. Each "end" of τ has a connected cluster set in the closed unit disk.

Let z be a cluster point of τ in the open unit disk. I claim that z is a point in Q. If not, then clearly $f(z)$ lies in σ and so $f^{-1}(\sigma)$ near z is just an open arc through z and so z cannot be a cluster point of τ.

Since a cluster set of τ must be connected, we can now say that these cluster sets are either a single point of Q or a closed subarc of the unit circle.

I claim that these subarcs γ on the unit circle must reduce to a single point. Suppose not. Consider the radial limits of f along γ. By Fatou's theorem, these exist almost everywhere. As these radii must hit τ infinitely often, since f maps τ into σ, we conclude that these limits are either p_i or p_{i+1}. This implies that the boundary values of the non-constant function f assume a single value on a set of positive measure. This is a contradiction. We now know that τ has limits at both "ends" and that these limits are either a point of Q or a point of the unit circle.

I claim that <u>some</u> τ has s as a limit. Suppose not. Then since there are only a finite number of τ's (a finite number correspond to each of a finite number σ's) and each of them is bounded away from s, there exists a small disk centered at s such that its intersection W with the open unit disk is disjoint from all of the τ's. Consider the open set $f(W)$. It is disjoint from all of the σ's; i.e., it can meet 1 only at a p_i. Since it is open it cannot meet 1 at any point. Since $f(W)$ is connected it must lie on one side of 1. Therefore the closure of $f(W)$ lies on one side of 1. But this closure contains $C(f,s)$ and this set is supposed to project to the interval $[a,b]$, which is divided by the line 1. Contradiction.

We now have an arc τ that approaches s. I claim that the limit of f on this arc exists and is a point of 1. In fact, the cluster set of f on this arc is connected and is contained in 1, since f maps τ into 1. Also this cluster set is contained in $C(f)$. Since $C(f) \cap 1 = \{p_1, p_2, \ldots, p_d\}$ is finite, we conclude that the limit exists and equals one of the p_i.

Thus there exists an arc τ in the unit disk which approaches s and such that the limit of f exists on this arc and the limit has real part c. Now repeat the same argument with a different c' with $a < c' < b$ and get an arc τ' approaching s on which f has a different limit. This contradicts the theorem of Lindelof. Q.E.D.

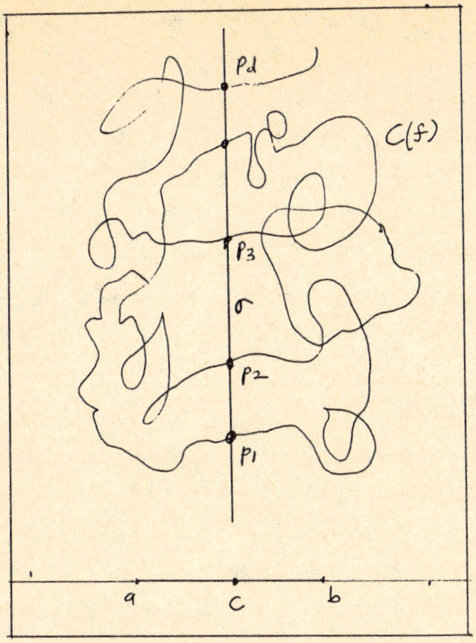

Figure 4

5. Open Questions.

5.1. <u>Finite area</u>. Let X be a <u>connected</u> compact set in $\mathbb{C}^n$ of finite linear measure. We know that $\hat{X}\setminus X$ is a one-dimensional analytic subset of $\mathbb{C}^n\setminus X$. In general, $\hat{X}\setminus X$ can have a countable number of analytic branches. Does $\hat{X}\setminus X$ have finite area? Does each component of $\hat{X}\setminus X$ have finite area? It seems quite likely that the answer to both questions is yes. It may be that the methods of [A3] can be extended to cover these cases. By the example of Section 2, we know that the connectedness assumption on X cannot be dropped.

5.2. <u>Hulls</u>. Let X be a compact subset of $\mathbb{C}^n$ which is totally disconnected and has finite linear measure. Is X polynomially convex? This would follow by a standard application of the argument principle [S] if one knew that $\hat{X}\setminus X$ were a one-dimensional analytic subset of $\mathbb{C}^n\setminus X$ if it were not empty. To answer this in the negative one might try to find an example like that in Section 2 but with X totally disconnected - this appears to be difficult.

5.3. <u>Stokes' Theorem</u>. Let X be a rectifiable Jordan curve in $\mathbb{C}^n$ which is not polynomially convex. Then as indicated in Section 3,

$V := \hat{X}\setminus X$ is an irreducible one-dimensional analytic subset of $\mathbb{C}^n\setminus X$ of finite area. The finite area condition implies that integration over V defines a global current which we shall denote as $[V]$. It is a positive current of type $(1,1)$. Since X is rectifiable, integration of 1-forms over X (oriented as in the proposition of Section 3, part (b)) is well-defined and this gives a global current $[X]$. Now one can ask whether Stokes' theorem holds in this context; i.e. does

$$b[V] = [X]$$

hold?

This is closely related to the "moment condition" introduced by Harvey and Lawson [HL]. By definition, the moment condition holds for X if $[X](\omega) = 0$ for every global holomorphic $(1,0)$-form ω. If Stokes' theorem holds, then applying it to a holomorphic $(1,0)$-form ω, we conclude that the moment condition holds. In the case of smooth X (here smooth will mean C^1) Harvey and Lawson proved the converse. This suggests the following problems for X a rectifiable Jordan curve. Show that the moment condition holds for X if X is not polynomially convex. The converse is easy to see. If X is polynomially convex, then polynomials are uniformly dense in the space of continous functions on X. From this it follows that the moment condition cannot hold.

For the case of smooth X, it appears that the equivalence of the two conditions as well as the validity of Stokes' theorem are true and easily obtained from results already in the literature. Namely: if X satisfies the moment condition, then Harvey and Lawson proved that $b[W] = [X]$ for a one-variety defined in $\mathbb{C}^n\setminus X$. It follows that W is $\hat{X}\setminus X$ and so X is not polynomially convex.

Now for the converse, we assume that X is not polynomially convex. We want to verify Stokes' theorem for $V := \hat{X}\setminus X$. N. Sibony suggested to me that his work [Si] on positive currents would be useful in doing this. In fact by the results of [Si] (or using the fact that V has finite area by Section 3 and appealing to the theory of locally flat currents of [F]) one concludes that $b[V] = f[X]$ with f a real integrable function on X where the meaning of the right-hand side is: $f[X](\varphi) = [X](f\varphi)$ for global 1-forms φ. Since $0 = b(b[V]) = b(f[X])$, it follows that f is a constant. Now because Figure 1 holds in this context we know that there are points of X at which V is attached as a complex one-manifold with boundary X (locally). This is enough to conclude that f is identically one and so Stokes' theorem holds.

References

[A1] H. Alexander, Polynomial approximation and hulls in sets of finite linear measure in $\mathbb{C}^n$, Amer. J. Math 93 (1971),65-74.

[A2] ———————, The polynomial hull of a set of finite linear measure in $\mathbb{C}^n$, to appear in J. d'Analyse Math.

[A3] ———————, The polynomial hull of a rectifiable curve in $\mathbb{C}^n$, to appear in Amer. J. Math.

[Be] A. Beurling, Sur les fonctions limites quasi analytiques des fractions rationnelles, VIII Congres des Mathematicians Scandinaves, Stockholm, 1934, 199-210.

[B] E. Bishop, Analyticity in certain function algebras, Trans. Amer. Math. Soc. 102 (1962), 507-544.

[F] H. Federer, Geometric measure theory, Springer Verlag, New York, 1969.

[GS] J. Globevnik and E.L. Stout, Boundary regularity for holomorphic maps from the disc to the ball, preprint.

[HL] R. Harvey and B. Lawson, On boundaries of complex analytic varieties, Ann. Math. 102 (1975), 233-290.

[Si] N. Sibony, Quelques problemes de prolongement de courants en analyse complexe, Duke Math. J. 52 (1985) 157-197.

[S] G. Stolzenberg, Uniform approximation on smooth curves, Acta Math. 115 (1966) 185-198.

[W] J. Wermer, The hull of a curve in $\mathbb{C}^n$, Ann. Math. 62 (1958), 550-561.

Department of Mathematics
University of Illinois at Chicago
Chicago, Illinois 60680

ESTIMATIONS, PAR NOYAUX, DES SOLUTIONS DE L'EQUATION $\bar\partial u = f$

Eric Amar

1. Introduction.

Very recently, B. Berndtsson showed that if f is a $(0,1)$ form, $\bar\partial$ closed in the unit ball $\mathbb{B}$ of $\mathbb{C}^n$, with coefficients in $L^2(\mathbb{B})$, there is a solution u to the problem $\bar\partial u=f$ with u in $L^2(\partial\mathbb{B})$.

Precisely we get:

<u>Theorem</u>: Let f be a $(0,1)$ form $\bar\partial$ closed in $\mathbb{B}$ and such that its coefficients are in $L^p(\mathbb{B})$ for $2\leq p\leq 2n+2$, then the solution u to $\bar\partial u = f$ given by the Skoda's kernels or by Charpentier's ones of exponent $k\geq 1$ is in $L^r(\partial\mathbb{B})$ with

if $2\leq p<2n+2$, $\qquad r = \dfrac{2np}{2n+2-p}$

if $p = 2n+2$, $\qquad u$ is in $B.M.O.(\partial\mathbb{B})$.

In the first part of this work we use the results of [3] to show the estimates $L^p(\mathbb{B})-L^r(\partial\mathbb{B})$ for $2<p\leq 2n+2$ with the kernels. (For $p>2n+2$ the solution is in a Lipschitz class [3].)

In the second part we want the limit estimate $p=2$ and we must choose regularized Skoda's kernels or Charpentier's kernels of order $k>0$.

The estimate in this part is more delicate because we must show that an operator is a singular integral operator on the sphere, and this is done the same way that in [1].

I want to thank H. Skoda for a very interesting conversation on this subject.

I. Estimations L^p.

Soit f une $(0,1)$ forme $\bar\partial$ fermée dans $\mathbb{B}$, on dira que $f\in L^p_{(0,1)}(\mathbb{B})$ si ses coefficients sont dans $L^p(\mathbb{B})$ pour lamesure de Lebesgue.

On a alors la

Proposition 1.1: Si f est dans $L^p_{(0,1)}(\mathbb{B})$ pour $2<p<2n+2$ et est $\bar{\partial}$ fermeé dans $\mathbb{B}$ alaors les noyaux admissibles fournissent une solution dans $L^r(\partial\mathbb{B})$ avec

$$r = \frac{2np}{2(n+1)-p}$$

Preuve: on va appliquer les résultats de [3]. Pour cela il nous faut voir à quelle classe $V^\infty(0,1)$ appartient f. On fait la preuve dans $\mathbb{C}^2$ pour alléger les notations. Dans ce cas étudions le "mauvais" coefficients:

$$\bar{\xi} f_3 = \bar{\xi}_1 f_2 - \bar{\xi}_1 f_2 = \text{coeff. de } f\wedge\bar{\partial}\rho \tag{1.1}$$

Soit $Q(z,h)$ une pseudo-boule:

$$Q(z,h) = \left\{\eta \in \mathbb{B} \;/\; \left|1 - \frac{z\bar{\eta}}{|\eta|}\right| < h \text{ et } 1-|\eta| < h\right\} \tag{1.2}$$

et posons

$$T = \bigcup_{i=1}^{N} Q(\xi_i,h_i) = \bigcup_{i=1}^{N} Q_i \tag{1.3}$$

où les pseudo-boules Q_1 sont disjointes.
ll nous faut estimer l'intégrale:

$$I = \int_T |f_3| \frac{d\sigma}{\sqrt{1-|\xi|^2}} \tag{1.4}$$

Passons en polaire, et posons $g=f_3 \in L^p(\mathbb{B})$:

$$I \le \int_0^1 \left\{\left|\int_{(T\cap S_r)} g(r\xi)d\sigma(\xi)\right|\right\} \frac{dr}{\sqrt{1-r^2}} \tag{1.5}$$

Soit $h = \max_i (h_i)$, par Hölder il vient:

$$\left|\int_{(T\cap S_r)} g(r\xi)d\eta(\xi)\right| \le \|g(r.)\|_p |(T\cap S_r)|^{1/q} \tag{1.6}$$

avec q conjugué de p. Mais on a:

(1.7) $|(T\cap S_r)| \leq |\overline{T}\cap S_1)|\chi_{[1-h,1[}$, avec $S_1 = \partial\mathbb{B} = \mathbb{S}$

d'où

(1.8) $$I \leq \int_{1-h}^{1} \|g(r.)\| \frac{dr}{\sqrt{1-r^2}} |(\overline{T}\cap S)|^{q^{1/q}} \leq \|g\|_p |(\overline{T}\cap S)|^{1/q} h^{1/q-1/2}$$

grâce encore à Hölder.

D'où, car $h^n \leq |(\overline{T}\cap S)|$ il vient:

(1.9) $$I \leq \|g\|_p |(\overline{T}\cap S)|^{1/q+1/nq-1/2n}$$

donc on a bien:

(1.10) $$f \in V^{\alpha}_{(0,1)}(B) \quad \text{avec} \quad \alpha = \frac{(2n+1)p-2(n+1)}{2np}.$$

Rappelons alors les resultats de [3] avec T l'opérateur résolvant $\overline{\partial}$ associé aux noyaux admissibles il vient:

(1.11) $$\begin{cases} T \text{ est borné de } V^{\alpha}(\mathbb{B}) \text{ dans } L^{r,\infty}(\partial\mathbb{B}) \text{ avec } r = \frac{1}{1-\alpha} \text{ pour } \alpha<1 \\ T \text{ est borné de } V^1(\mathbb{B}) \text{ dans } BMO\ (\partial\mathbb{B}) \end{cases}$$

où $L^{r,\infty}$ est l'espace faible de Lorentz.
On en deduit donc ici:

$\forall$ p > 2 il vient:

(1.12) $$\begin{cases} T \text{ est borné de } L^{p}(\mathbb{B}) \text{ dans } L^{r,\infty}(\partial\mathbb{B}) \text{ avec } r = \frac{4p}{6-p};\ 2<p<6 \\ T \text{ est borné de } L^6(\mathbb{B}) \text{ dans } BMO\ (\partial\mathbb{B}) \end{cases}$$

Soit alors $6 < p_0 < 2$; il existe p tel que $2 < p < p_0$ et on a donc (1.12); appliquant à (1.12) le théorème d'interpolation de Marcinkiewicz on en déduit que

(1.13) T est borné de $L^{p_0}(\mathbb{B})$ dans $L^r(\partial\mathbb{B})$ avec $r_0 = \frac{4p}{6-p}$

et la proposition 1.1.

Le cas p = 2 apparait donc ici comme un cas limite où la mesure $\dfrac{f_3}{\sqrt{1-|\xi|^2}}$ n'est même pas une mesure bornée dans $\mathbb{B}$ en général.

Toutefois le résultat reste valable en restreignant un peu la classe de noyaux.

II. Estimations L^2.

§1 - Réduction du problème

Soit donc $K_t(z,\xi)$ le noyau résolution ∂ ; on a: si $f = f_1 d\bar{\xi}_1 + f_2 d\bar{\xi}_2$ une (0,1) forme, posons

$$f_3 d\bar{\xi}_1 \wedge d\bar{\xi}_2 = f \wedge \bar{\partial}\rho = (\xi_1 f_2 - \xi_2 f_3) d\bar{\xi}_1 \wedge d\bar{\xi}_2$$

La solution est alors donnée par:

$$u(z) = \sum_{i=1}^{3} \int_{\mathbb{S}} \int_0 K_t^i(z,\xi)\ f_i(t\xi)\ t^3 dt d\sigma(\xi)$$

où on a paramétré la measure volume dans $\mathbb{B}$ par $t^3 dt d\sigma$ et où on a les noyaux:

$$K_t^i(z,\xi) = \frac{(1-t^2)^2 \bar{z}_i}{D_t(z,\xi)} \quad i=1,2$$

et

$$K_t^3(z,\xi) = \frac{(\bar{z}_1\bar{\xi}_2 - \bar{z}_2\bar{\xi}_1)(1-t^2)}{D_t(z,\xi)}$$

avec

$$D_t(z,\xi) = (1-tz\cdot\bar{\xi})^3(1-t\bar{z}\cdot\xi).$$

On ne va s'intéresser qu'au 3° noyau, les autres étant plus simples à estimer. On pose donc:

$$K_t(z,\xi) = K_t^3(z,\xi).$$

Soit alors $f \in L^2(0,1)(\mathbb{B})$, f_1 et f_2 sont dans $L^2(\mathbb{B})$ donc f_3 aussi avec $g = f_3$ on veut montrer que

$$(1.1) \qquad G(z) = \int_{\mathbb{B}} K_t(z,\xi) g(r\xi) d\sigma(t\xi)$$

est dans $L^2(S)$, ce qui équivaut à:

$$(1.2) \qquad \forall H \in C^\infty(\mathbb{S}), \quad \left| \int_{\mathbb{S}} G(z)\ H(z)\ d\sigma(z) \right| \le c \|H\|_{L^2}$$

par dualité, donc encore, en remplacant G par son expression, avec

(1.3) $$\forall H\in\mathcal{C}^{\infty}(\mathbb{S}),\ h(t,\xi) = \int_{\mathbb{S}} K_t(z,\xi)\ H(z)\ d\sigma(z)$$

(1.4) $$\left|\int_{\mathbb{B}} (h(t,\xi)\ g(t\xi)\ d\sigma(t\xi))\right| \le c\|H\|_{L^2}$$

c'est à dire que:

(1.5) $$\|h(t,\xi)\|_{L^2(\mathbb{B})} \le \|H\|_{L^2(\mathbb{S})}$$

où

(1.6)

$$\int_{\mathbb{B}} |h(t,\xi)|^2 d\sigma(t\xi) = \int_0^1\int_{\mathbb{S}}\int_{\mathbb{S}\times\mathbb{S}} K_t(z,\xi)\overline{K_t(w,\xi)}\ H(z)\ \overline{H(w)}d\sigma(z)d\sigma(w)d\sigma(\xi)dt$$

aurement dit, il suffit de montrer que l'opérateur S associé au noyau:

(1.7) $$S(z,w) = \int_0^1\int_{\mathbb{S}} K_t(z,\xi)\overline{K_t(w,\xi)}d\sigma(\xi)dt$$

est borné de $L^2(\mathbb{S})$ dans $L^2(\mathbb{S})$ pour conclure.

Imitant fidèlement la méthode (et les calculs) de [1] on va montrer que S admet une décomposition presque othogonale et donc S sera un opérateur associé à une intégrale singulière donc bornée sur $L^2(\mathbb{S})$[5]. On aura besoin des rappels suivants:

2. <u>Rappels et notations</u>.

σ désigne la mesure de Lebesgue sur $\mathbb{S} = \partial\mathbb{B}$. On notera $\delta(z,w)$ la pseudo distance

(2.1) $$\delta(z,w) = |1-z.\overline{w}|$$

On sait qu'alors, [6], $\delta^{1/2}$ est une distance.
Si Q(z,h) dénotes la pseudo boule de centre z et de rayon h on a:

(2.2) $$\sigma(Q(z,h)) = \sigma\{\xi\in\mathbb{S}\ \text{ t.q.}|1-z.\overline{\xi}| < h\} \simeq h^2.$$

Soit $t\in[0,1[$, z, ξ dans $\mathbb{S}$ alors : avec $u = 1 - t \in]0,1]$

(2.3) $\quad |1-tz.\bar{\xi}|=|u+(1-u)(1-z.\bar{\xi})| \geq |u+\delta(z,\xi)|$ pour $u \leq 1/2$.

En effet on a que $\mathrm{Re}(1-z.\bar{\xi}) \geq 0$ donc:

(2.4) $$|u+(1-u)(1-z.\bar{\xi})|^2=|u+(1-u)\mathrm{Re}(1-z.\bar{\xi})|^2+|\mathrm{Im}(1-z.\bar{\xi})|^2$$

$$\geq u^2+(1-u)^2|1-z.\bar{\xi}|^2 \geq \tfrac{1}{4}\ |u+\delta(z,\xi)|^2$$

On a aussi [6] :

(2.5) $$\delta(z,\xi) \leq |z-\xi| \leq \delta(z,\xi)^{1/2}.$$

Posons

(2.6) $$\Omega(z,\xi) = z_1\xi_2 - z_2\xi_1$$

il vient

(2.7) $$|\Omega(z,\xi)| \leq \delta(z,\xi)^{1/2}$$

en effet:

(2.8) $$|\Omega(z,\xi)|=|(z_1-\xi_1)\xi_2+\xi_1(\xi_2-z_2)| \leq 2|z-\xi|$$

on a aussi

(2.9) $$|\Omega(z,\xi)-\Omega(w,\xi)=|(z_1-w_1)\xi_2-(z_2-w_2)\xi_1| \leq \delta(z,w)^{1/2}$$

et

(2.10) $$|(z-w).\bar{\xi}| \leq \delta(z,w)+2\delta(z,w)^{1/2}\delta(z,\xi)^{1/2}$$

car

$$(z-w).\bar{\xi}=(z-w).\bar{z}+(z-w).\bar{\xi}-(z-w).\bar{z}=(1-w.\bar{z})+(z-w).(\bar{\xi}-\bar{z})$$

d'où (2.10) en utilisant (2.5).
On utilisera aussi le lemme simple suivant:

<u>Lemme 2.1</u>: Soient z,w,ξ trois points sur $\mathbb{S}$ tels que $\delta(z,\xi) \geq 4\delta(z,w)$ alors si v est un point quelconque du segment réel $[z,w]$ on a :

$$\delta(v,\xi) = |1-\xi.\bar{v}| \geq \tfrac{1}{4}\ \delta(z,\xi).$$

<u>Preuve</u>: comme $\delta^{1/2}$ est une distance il vient:

(2.11) $$\delta(z,\xi)^{1/2} \leq \delta(z,v)^{1/2}+\delta(v,\xi)^{1/2}$$

mais les pseudo-boules dans $\bar{\mathbb{B}}$, $\{\eta \in \bar{\mathbb{B}}$ t.q. $|1-z.\bar{\eta}| < h\}$ sont convexes donc

(2.12) $$\delta(z,v) \leq \delta(z,w) \quad \text{car} \quad v \in [z,w]$$

d'où

2.13) $\delta(v,\xi)^{1/2} \geq \delta(z,\xi)^{1/2} - \delta(z,w)^{1/2} \geq \frac{1}{2}\,\delta(z,\xi)^{1/2}$

d'où le lemme.

3. Estimations.

Reprenant la méthode de [1], on pose u=1-t et

(3.1) $$a_r(z,\xi) = \int_r^{2r} \int_{\mathbb{S}} K_t(z,\xi)\,\overline{K_t(w,\xi)}\,d\sigma(\xi)\,dt \quad \text{pour} \quad r \in \,]0,1].$$

Ces a_r vont constituer la décomposition presque orthogonal de $S(z,\xi)$ au sens de M. Cotlar et si on montre que:

A) $$\int_{\mathbb{S}} a_r(z,\xi)\,d\sigma(z) = 0$$

B) $$\int_{\mathbb{S}} |a_r(z,\xi)| \left[1 + \frac{\delta^{1/2}(z,\xi)}{r^{1/2}}\right] d\sigma(z) \leq C \qquad \forall r \in \,]0,1]$$

C) $$\int_{\mathbb{S}} |a_r(z,\eta) - a_r(w,\eta)|\,d\sigma(\eta) \leq C\,\frac{\delta^{1/2}(z,w)}{r^{1/2}} \qquad \forall r \in \,]0,1]$$

alors S sera un opérateur borné de $L^2(\mathbb{S})$ dans $L^2(\mathbb{S})$ [5].

Voyons A.

(3.2) $$I = \int_{\mathbb{S}} K_t(z,\xi)\,d\sigma(z) = \int_{\mathbb{S}} \frac{\Omega(z,\xi)(1-r^2)}{D_t(z,\xi)}\,d\sigma(z)$$

Le changement de variables, pour ξ sur $\mathbb{S}$:

(3.3) $$z_1' = z.\overline{\xi} \quad \text{et} \quad z_2' = \Omega(z,\xi)$$

est un changement unitaire, donc laisse invariant $d\sigma$ et S d'où

(3.4) $$I = (1-t^2) \int_{\mathbb{S}} \frac{z_2'}{(1-tz_1')^3(1-t\overline{z}_1')}\,d\sigma(z)$$

et par Fubini (par invariance par rapport a z_2')

(3.5) $$I = (1-t^2) \int_{\mathbb{D}} \frac{1}{(1-tz_1)^3(1-t\overline{z}_1)} \left\{\int_{|z_2|^2=1-|z_1|^2} z_2\,d\theta\right\} d\lambda(z_1) = 0$$

où on a posé $d\theta$ la mesure de Lebesgue normlisée sur $|z_2| = \sqrt{1-|z_1|^2}$ et $d\lambda$ la mesure de Lebesgue sur $\mathbb{D}$.

On a alors:

$$(3.6)\qquad \int_{\mathbb{S}}\left\{\int_{\mathbb{S}} K_t(z,\xi)\,\overline{K_t(w,\xi)}\,d\sigma(z,w)\right\}d\sigma(z) = 0$$

par Fubini en énchangeant l'ordre d'intégration.
On en déduit A par intégration entre r et 2r.

Voyons B. On pose

$$(3.7)\qquad I = \int_{\mathbb{S}} |a_r(z,w)|\left[1+\frac{\delta^{1/2}(z,w)}{r^{1/2}}\right]d\sigma(w)$$

Remplaçant a_r par son expression et majorant en mettant les modules dans les intégrales il vient:

$$(3.8)\qquad I \le \int_r^{2r}\int_{\mathbb{S}\times\mathbb{S}} |K_{1-u}(z,\xi)|\,|K_{1-u}(w,\xi)|\left[1+\frac{\delta^{1/2}(z,w)}{r^{1/2}}\right]d\sigma(z)\,d\sigma(w)\,du$$

mais on a:

$$(3.9)\qquad K_{1-u}(z,\xi) = \frac{|\Omega(z,\xi)|\left[1-(1-u)^2\right]}{|u+(1-u)(1-z.\bar{\xi})|^4} \le \frac{u\,\delta(z,\xi)^{1/2}}{(u+\delta(z,\xi))^4}$$

grace á: (2.3) et (2.7).

D'où:

$$(3.10)\qquad I \le \int_r^{2r}\int_{\mathbb{S}\times\mathbb{S}}\left[1+\frac{\delta^{1/2}(z,w)}{r^{1/2}}\right]u^2\,\frac{\delta^{1/2}(z,\xi)\;\delta^{1/2}(w,\xi)}{(u+\delta(z,\xi))^4(u+\delta(w,\xi))^4}\,d\sigma(\xi)\,d\sigma(w)\,du$$

(3.11) Décomposons $\mathbb{S}\times\mathbb{S}$ ainsi:

pour z fixé sur $\mathbb{S}$, ξ varie dans la "couronne":

$$C_n(z) = \{\xi\in\mathbb{S} \text{ t.q. } \delta(z,\xi) \in [2^{n+1}u,\ 2^n u]$$

pour z et ξ fixés, w varie dans la "couronne";

$$C_m(z,\xi) = \{w\in\mathbb{S} \text{ t.q. } \delta(\xi,w) \in [2^{m+1}u,\ 2^m u]$$

où n et m sont dans $\mathbb{Z}$.

Mais on a:

$$(3.12)\qquad \delta^{1/2}(z,w) \le \delta^{1/2}(z,\xi) + \delta^{1/2}(w,\xi)$$

car $\delta^{1/2}$ est une distance, d'où de façon claire:

(3.13) $$I \le \int_r^{2r} \Big[\sum_{n,m\in\mathbb{Z}} \Big(1+ \frac{2^{\frac{n+1}{2}} u^{1/2} + 2^{\frac{m+1}{2}} u^{1/2}}{r^{1/2}}\Big) \frac{u^2 2^{\frac{n+1}{2}} u^{1/2} \; 2^{\frac{m+1}{2}} u^{1/2}}{u^4 [1+2^4]^4 [1+2^m]^4 u^4} \times$$

$$\times \int_{C_n(z)} \Big\{ \int_{C_m(z,\xi)} d\sigma(w) \Big\} d\sigma(\xi) \Big] du$$

utilisant alors que $u \le 2r$ et que

(3.14) $$\Big(1+2^{\frac{n}{2}+1} + 2^{\frac{m}{2}+1}\Big) \le \Big(1+2^{\frac{m}{2}+1}\Big)\Big(1+2^{\frac{n}{2}+1}\Big)$$

il vient, grace a (2.2), car $\sigma(C_n(z)) \le 2^{n+2}u^2$

(3.15) $$I \le \int_r^{2r} \Big[\sum \frac{\Big(1+2^{\frac{n}{2}+1}\Big) 2^{\frac{n+1}{2}}}{r^{1/2}} 2^{2n+2} \Big]^2 \frac{du}{u}$$

La série converge visiblement d'où finalement : $I \le c$, $\forall r \in]0,1]$.

Voyons C. On pose:

(3.16) $$I = \int_{S} |a_r(z,\eta) - a_r(w,\eta)| \, d\sigma(\eta)$$

d'où en faisant passer les valeurs absolues dans les intégrales:

(3.17) $$I \le \int_r^{2r} \int_{S\times S} |K_{1-u}(\eta,\xi)| \, |K_{1-u}(z,\xi) - K_{1-u}(w,\xi)| \, d\sigma(\xi) d\sigma(\eta)$$

Il nous faut évaluer la différence:

(3.18) $$|K_{1-u}(z,\xi) - K_{1-u}(w,\xi)| = u \left| \frac{\overline{\Omega}(z,\xi)}{D_{1-u}(z,\xi)} - \frac{\overline{\Omega}(w,\xi)}{D_{1-u}(w,\xi)} \right| \le$$

$$\le |\Omega(z,\xi)| \left| \frac{1}{D_{1-u}(z,\xi)} \frac{1}{D_{1-u}(w,\xi)} \right| + \frac{u}{|D_{1-u}(w,\xi)|} |\Omega(z,\xi) - \Omega(w,\xi)|.$$

Posons alors:

(3.19) $$F(\lambda) = \frac{1}{\lambda^3 \bar{\lambda}} \text{ pour } \lambda \in \mathbb{C}$$

On a alors:

(3.20) $$|F(\lambda) - F(\lambda')| \le |\lambda - \lambda'| \frac{1}{|\lambda''|^5}$$

par Taylor, où λ'' est dans le segment réel $[\lambda,\lambda']$

Mais $\dfrac{1}{D_t(z,\xi)} = F(1-t\, z.\bar{\xi})$ d'où

(3.21)
$$\left|\frac{1}{D_t(z,\xi)} - \frac{1}{D_t(w,\xi)}\right| \le \frac{t|(z-\bar{w}).\xi|}{|1-t\nu.\xi)|^5}$$

où ν est un point de segment réel $[z,w]$ (on remarque que ν n'est pas sur $\mathbb{S}$ en général mais dans $\bar{\mathbb{B}}$)

Examinons deux cas:

1er cas: $\delta(z,\xi) \ge 4\delta(z,w)$ et $\delta(w,\xi) \ge 4\delta(z,w)$.

Grâce au lemme 2.1 on a alors:

(3.22)
$$\delta(\nu,\xi) \ge \delta(z,\xi)$$

On majore alors ainsi, grâce encore à (2.3) et à (2.10)

(3.23)
$$\left|\frac{1}{D_t(z,\xi)} - \frac{1}{D_t(w,\xi)}\right| \le \frac{|(z-w).\bar{\xi}|}{|u+\delta(z,\xi)|^5} \le \delta^{1/2}(z,w)\ \frac{[\delta(z,w)^{1/2}+2\delta(z,\xi)^{1/2}]}{u+\delta(z,\xi))^5}$$

2° cas: $\delta(z,\xi) < 4\delta(z,w)$ ou $\delta(w,\xi) < 4\delta(z,w)$

On majore brutalement:

(3.24)
$$\left|\frac{1}{D_{1-u}(z,\xi)} - \frac{1}{D_{1-u}(w,\xi)}\right| \le \frac{1}{(u+\delta(z,\xi))^4} + \frac{1}{(u+\delta(w,\xi))^4}$$

Majorons encore le 2° terme de (3.18): grâce à (2.9):

(3.25)
$$\frac{u}{|D_{1-u}(w,\xi)|}\,|\Omega(z,\xi)-\Omega(w,\xi)| \le \frac{u\ \delta(z,w)^{1/2}}{(u+\delta(w,\xi))^4}$$

On peut maintenant revenir à (3.17)

(3.26)
$$I \le J+L$$

avec, grâce à (3.18)

(3.27)
$$J = \int_r^{2r}\int_{\mathbb{S}\times\mathbb{S}} K_{1-u}(z,\xi)\,|u|\Omega(z,\xi)|\left|\frac{1}{D_{1-u}(z,\xi)} - \frac{1}{D_{1-u}(w,\xi)}\right| d\sigma(\eta)\,d\sigma(\xi)\,du$$

et

(3.28)

$$L = \int_r^{2r} \int_{\$\times\$} |K_{1-u}(\eta,\xi)| \left| \frac{1}{|D_{1-u}(w,\xi)|} \right| |\Omega(z,\xi)-\Omega(w,\xi)| d\sigma(\eta) d\sigma(\xi) du$$

On va décomposer J en trois parties suivant les cas examinés précédemment

(3.29)

$$J = \int_r^{2r} \int_{\delta(z,\xi)<4\delta(z,w)} \int_{\$} + \int_r^{2r} \int_{\delta(w,\xi)<4\delta(z,w)} \int_{\$} + \int_r^{2r} \int_{\substack{\delta(w,\xi)\geq 4\delta(z,w) \\ \delta(z,\xi)\geq 4\delta(z,w)}} \int_{\$}$$

i.e. $J = J_1+J_2+J_3$.

Voyons J_1: on y majore : $|\theta(z,\xi)| \leq \delta(z,\xi)^{1/2} \leq 2\delta(z,w)^{1/2}$ d'où

(3.30)

$$J_1 \leq \delta(z,w)^{1/2} \int_r^{2r} \frac{\delta(\eta,\xi)^{1/2}}{[u+\delta(\eta,\xi)]^4} u^2 \left\{ \frac{1}{[u+\delta(z,\xi)]^4} + \frac{1}{[u+\delta(w,\xi)]^4} \right\} d\sigma(\eta) d\sigma(\xi) du$$

J_1 se décompose encore en 2 dans les accolades; voyons le premier terme, J_{11}.

On introduit le découpage de $\$\times\$$ déjà vu au (3.11) pour B:

$$C_n(z) = \{\xi\in\$ \quad \text{t.q.} \quad \delta(z,\xi)\in[2^n_u,\ 2^{n+1}_u[\,\}$$

$$C_m(z,\xi) = \{\eta\in\$ \quad \text{t.q.} \quad \delta(\xi,\eta)\in[2^m_u,\ 2^{m+1}_u]\}$$

On trouve alors par des calculs identiques:

(3.31) $$J_{11} \leq \delta(z,w)^{1/2} \int_r^{2r} \sum_{n,m} \frac{2^{\frac{m+1}{2}}\, 2^{2m+2}}{[1+2^m]^4} \frac{2^{2n+2}}{[1+2^n]^4} \frac{du}{u^{3/2}}$$

La série double converge et donc il vient:

(3.32) $$J_{11} \leq \frac{\delta(z,w)^{1/2}}{r^{1/2}}$$

Le terme J_{12} est analogue avec les couronnes $C_n(w)$, $C_m(w,\xi)$ d'où

(3.33) $$J_{12} \leq \frac{\delta(z,w)^{1/2}}{r^{1/2}}$$

Voyons J_2.

Cette fois on a $\delta(w,\xi) < 4\delta(z,w)$ mais

(3.34) $$\delta(z,\xi)^{1/2} \le \delta(z,w)^{1/2} + \delta(w,\xi)^{1/2} \le 3\delta(z,w)^{1/2}$$

Ce qui permet d'obtenir le même résultat que pour J_1

Voyons J_3.

Utilisant les majorations précédentes (3.23) il vient

(3.35) $$J_3 \le \int_r^{2r}\int_{\$\times\$} \frac{|\Omega(\eta,\xi)|u^2|\Omega(z,\xi)|}{(u+\delta(\eta,\xi))^4}\,\frac{|(z-w)\cdot\bar{\xi}|}{(u+\delta(z,\xi))^5}\,d\sigma(\xi)d\sigma(\eta)du$$

d'où, par (2.10) encore: et sur $\delta(z,\xi) \ge 4\delta(z,w)$ il vient:

(3.36)

$$|(z-w)\cdot\bar{\xi}| \le \delta(z,w)^{1/2}[\delta(z,w)^{1/2}+2\delta(z,\xi)^{1/2}] \le \delta(z,w)^{1/2}\delta(z,\xi)^{1/2}$$

d'où portant dans J_3

(3.37) $$J_3 \le \delta(z,w)^{1/2}\int_r^{2r} \frac{\delta(\eta,\xi)^{1/2}u^2\delta(z,\xi)^{3/2}}{[u+\delta(\eta,\xi)]^4\,[u+\delta(z,\xi)]^4}\,d\sigma(\xi)d\sigma(\eta)du$$

Reprenant encore (3.11) on obtient:

(3.38) $$J_3 \le \frac{\delta(z,w)^{1/2}}{r^{1/2}}$$

Il nous reste a màjorer L:

(3.39) $$L=\int_r^{2r}\int_{\$\times\$}|K_{1-u}(\eta,\xi)|\left|\frac{u}{|D_{1-u}(w,\xi)|}\right||\Omega(z,\xi)-\Omega(w,\xi)|d\sigma(\eta)d\sigma(\xi)$$

mais (2.9) donne $|\Omega(z,\xi)-\Omega(w,\xi)| \le \delta(z,w)^{1/2}$

et le découpage avec les couronnes en w permet d'obtenir

(3.40) $$L \le \frac{\delta(z,w)^{1/2}}{r^{1/2}}$$

L'opérateur S est donc bien une intégrale singulière sur S et donc est borné de $L^2(\$)$ dans $L^2(\$)$.

REFERENCES

[1] Amar, E., Extension de formes $\bar{\partial}_b$ fermées et solutions de l'équation $\bar{\partial}_b u = f$. Ann. Scuola Norm. Sup. Pisa, Vol. VII, n°1, (1980), p.155-179.

[2] Amar, E., Extension de fonctions holomorphes et courant. Bull. Sc. Math., 107, (1983), p.25-48.

[3] Amar, E. and Bonami, A., Measures de Carleson d'ordre α et solutions au bord de l'equation $\bar{\partial}$. Bull. Soc. Math. France, 107, (1979), p.23-48.

[4] Charpentier, P., Formules explicites pour les solutions minimales de l'équation $\bar{\partial}u=f$ dans la boule et le polydisque. Ann. Inst. Fourier, 30-4 (1980), p.121-154.

[5] Coifman, R. and G. Weiss, Analyse harmonique non commutative sur certains espaces homogènes. Lect. Notes in Math., n° 242, Springer-Verlag.

[6] Rudin, W., Function theory in the unit ball of $\mathbb{C}^n$. Grundelehren der Math., Springer-Verlag (1983).

[7] Skoda, H., Valeur au bord pour les solutions de l'opérateur d" et caractérisation des zéros de la classe de Nevanlinna Bull. Soc. Math. France, 104 (1976), p.225-229.

Université de Bordeaux I
Mathematiques et Informatique
351, Cours de la Libération
33405 Talence Cedex
France

THE HEAT EQUATION AND GEOMETRY FOR THE $\bar{\partial}$-NEUMANN PROBLEM

Richard Beals*

and

Nancy K. Stanton**

1. Background.

The heat equation for the $\bar{\partial}$-Neumann problem is a complex analogue of a classical problem in Riemannian geometry. In this section, we describe some of the Riemannian results.

Let M be a compact oriented n dimensional Riemannian manifold with smooth boundary bM. One powerful method of relating geometry and analysis on M is through the heat equation. Let Δ denote the Neumann realization of the Laplace-Beltrami operator on M, i.e., $\Delta = \delta d$ on functions f satisfying $Nf = 0$ where N is the inward unit normal vector field on bM and δ is the formal adjoint of d. A function $f \in C^2(\bar{M}\times\mathbb{R}^+)$ solves the <u>heat equation</u> with Neumann boundary conditions if

$$\begin{aligned} &Nf = 0 \quad \text{on} \quad bM\times\mathbb{R}^+ \\ &\left(\frac{\partial}{\partial t} + \Delta\right) f = 0 \quad \text{on} \quad \bar{M}\times\mathbb{R}^+ . \end{aligned} \tag{1.1}$$

The <u>initial value problem</u> for the heat equation is: given $f_0 \in C^0(\bar{M})$, find a solution f of the heat equation with

$$\lim_{t\to 0^+} f(\cdot,t) = f_0 . \tag{1.2}$$

The solution is $f(x,t) = (e^{-t\Delta}f_0)(x)$, $(x,t) \in \bar{M}\times\mathbb{R}^+$, where $e^{-t\Delta}$ is the semigroup generated by $-\Delta$.

Let $0 = \lambda_1 < \lambda_2 \le \lambda_3 \le \ldots$ be the spectrum of Δ. Then

$$\operatorname{tr} e^{-t\Delta} = \sum e^{-t\lambda_i} . \tag{1.3}$$

The trace has an asymptotic expansion as $t \longrightarrow 0^+$,

---

*Research partially supported by NSF grant DMS-8402637

**Research partially supported by NSF grant DMS-8200442-01

(1.4) $$\operatorname{tr} e^{-t\Delta} \sim (4\pi t)^{-n/2} \sum_{j=0}^{\infty} c_j t^{j/2}$$

where c_0 is the volume of M. McKean and Singer [10] proved that c_j is the sum of integrals over M and bM of universal polynomials (depending only on n and j) in geometric invariants of M and bM. The heat semigroup is given by integration against the heat kernel $p(x,y,t) \in C^{\infty}(\bar{M}\times\bar{M}\times\mathbb{R}^{+})$ and

(1.5) $$\operatorname{tr} e^{-t\Delta} = \int_M p(x,x,t)\,dV(x),$$

where dV is the volume element on M. The proof of (1.4) and the geometric interpretation of the coefficients come down to a detailed construction and analysis of a good enough approximation to p.

The calculation of c_0 is equivalent, via an Abelian theorem and Karamata's Tauberian Theorem, to a famous result of Weyl [15].

<u>Theorem 1.6 (Weyl)</u>. Let $N(\lambda)$ denote the number of eigenvalues (counted with multiplicities) of Δ which are less than λ. Then, as $\lambda \longrightarrow \infty$,

(1.7) $$N(\lambda) \sim \frac{\text{Volume } M}{(4\pi)^{n/2}\Gamma(n/2+1)}\,\lambda^{n/2} .$$

The work of McKean and Singer was generalized to elliptic boundary value problems by Greiner [8] and Seeley [12].

2. The heat equation for the $\bar{\partial}$-Neumann problem.

One analogous problem in several complex variables is the heat equation for the $\bar{\partial}$-Neumann problem. Let M be a compact complex $n+1$ dimensional Hermitian manifold with smooth boundary bM, $n \geq 1$. Let N denote the inward unit normal vector field on bM and J_0 the almost complex structure on $\bar{M}$. Set $N'' = \frac{1}{\sqrt{2}}(N+iJ_0N)$ and let ν denote contraction with N''. Define

(2.1) $$\Lambda_b^{p,q} = \left\{u \in \Lambda^{p,q}\big|_{bM} : \nu u = 0\right\} .$$

Then

(2.2) $$\nu : \Lambda^{p,q}\big|_{bM} \longrightarrow \Lambda_b^{p,q-1} .$$

A form $u \in C^1(\Lambda^{p,q}(\bar{M}))$ satisfies <u>$\bar{\partial}$-Neumann boundary conditions</u> if

(2.3) $$\nu u = 0 \quad \text{and} \quad \nu\bar{\partial}u = 0 .$$

The $\bar\partial$-Laplacian $\square$ is defined by

$$\text{Dom}\,\square = \left\{u \in C^2(\Lambda^{p,q}(\bar M)) : \nu u = 0,\ \nu\bar\partial u = 0\right\}$$

(2.4)

$$\square = \bar\partial\bar\partial^* + \bar\partial^*\bar\partial\ .$$

A form $F(z,t) \in C^2(\Lambda^{p,q}(\bar M \times \mathbb{R}^+))$ solves <u>the heat equation for the $\bar\partial$-Neumann problem</u> if

(2.5)

$$\text{for fixed } t,\ F(\cdot,t) \in \text{Dom}\,\square$$

$$\left(\frac{\partial}{\partial t} + \square\right)F = 0\ .$$

The <u>initial value problem</u> for the heat equation for the $\bar\partial$-Neumann problem is: given $f \in C^0(\Lambda^{p,q}(\bar M))$, find a solution F of the heat equation (2.5) satisfying

(2.6)
$$\lim_{t \to 0^+} F(\cdot,t) = f\ .$$

Suppose $\bar M$ satisfies condition $Z(q)$: at each point of bM, the Levi form has at least $n + 1 - q$ positive eigenvalues or at least $q + 1$ negative eigenvalues. Then the initial value problem has a solution, and it is given by $F(\cdot,t) = e^{-t\square}f$, where $e^{-t\square}$ is the semigroup generated by $-\square$. This semigroup is given by integration against a smooth kernel $p(z,w,t) \in C^\infty(\Lambda^{p,q}\otimes\Lambda^{q,p}(\bar M \times \bar M \times \mathbb{R}^+))$,

(2.7)
$$F(z,t) = \int_M p(z,w,t) \wedge *f(w)\ .$$

As in the Riemannian case, we have

(2.8)

$$\operatorname{tr} e^{-t\square} = \sum e^{-t\lambda}$$

$$= \int_{\bar M} \operatorname{tr} p(z,z,t)\ .$$

The sum in the first line of (2.8) is over all eigenvalues of $\square$, counted with multiplicity. In the second line, the local trace $\operatorname{tr} : \Lambda^{p,q}\otimes\Lambda^{q,p} \longrightarrow \Lambda^{n+1,n+1}$ is the linear map satisfying $\operatorname{tr} u\otimes v = u \wedge *v$.

To study the asymptotic behavior of $\operatorname{tr} e^{-t\square}$ we need a good description of the heat kernel p. We sketch our construction. The details are in [2]. The difficulty is that the $\bar\partial$-Neumann boundary conditions are non-elliptic, so the classical methods and elliptic pseudodifferential calculus used by McKean-Singer, Greiner and Seeley do not apply. We use Calderon's method of reduction to the boundary. This reduces the construction of p to the inversion of a first order classical parabolic pseudo-differential operator on bM. Because the boundary condition is non-elliptic, this pseudodifferential operator is not invertible in the classical parabolic calculus.

Our description of the reduction to the boundary follows that of Greiner and Stein [9] in their study of the $\bar{\partial}$-Neumann problem itself. We write

$$e^{-t\Box} = G + H \tag{2.9}$$

where G is the Green's operator for the heat equation, i.e., the fundamental solution of the initial value problem for the Dirichlet realization of $\frac{\partial}{\partial t} + \Box$, and H is a correction term. For $f \in C^0(\Lambda^{p,q}(\bar{M}))$, Hf solves the heat equation with initial value 0 and hence

$$Hf = J(Hf|_{bM}) \tag{2.10}$$

where J is the Poisson operator for the heat equation. Now, $\nu e^{-t\Box}f = 0 = \nu Gf$, so $\nu Hf = 0$, and thus $Hf \in \Lambda_b^{p,q}$. By the second $\bar{\partial}$-Neumann boundary condition,

$$\nu\bar{\partial}J(Hf|_{bM}) = -\nu\bar{\partial}Gf \; . \tag{2.11}$$

Let

$$\Box_h^+ = \nu\bar{\partial}J \tag{2.12}$$

acting on sections of $\Lambda_b^{p,q}(bM\times\mathbb{R}^+)$. Then we can rewrite (2.11) as

$$\Box_h^+(Hf|_{bM}) = -\nu\bar{\partial}Gf \; . \tag{2.13}$$

The operator $\Box_h^+$ is a classical first order parabolic pseudodifferential operator on $\Lambda_b^{p,q}(bM\times\mathbb{R}^+)$. Condition $Z(q)$ allows us to invert $\Box_h^+$ in a mixed pseudodifferential operator algebra which includes both the classical operators of parabolic type and the class of operators introduced in [1] to study the heat equation for $\Box_b$. We have a full calculus for this algebra. By (2.9), (2.10), and (2.13),

$$e^{-t\Box} = G - J\Box_h^{+-1}\nu\bar{\partial}G \; . \tag{2.14}$$

Because the second term vanishes to infinite order in M as $t\longrightarrow 0^+$, standard methods show

$$\operatorname{tr} e^{-t\Box} = \operatorname{tr} G + \operatorname{tr} H_0 + r \tag{2.15}$$

where H_0 is a pseudodifferential operator on bM in the mixed algebra described above and r vanishes to infinite order as $t\longrightarrow 0^+$. Now, $\operatorname{tr} G$ is well known [8], [10], [12]. Our pseudo-differential operator calculus implies that the kernel of H_0 is a convergent sum of progressively smoother kernels vanishing for $t < 0$, so the asymptotic behavior of $\operatorname{tr} H_0$ is determined by the traces of these summands. Each summand is, locally, a matrix each of whose entries is a convolution of two functions, one with classical parabolic homogeneity and one with the non-isotropic parabolic homogeneity required in the calculus of [1]. Thus, to obtain an asymptotic expansion for $\operatorname{tr} e^{-t\Box}$, we need to analyze the asymptotic behavior in time of the convolution of two functions, each with a different kind of homogeneity. The trace of such a convolution kernel has a complete asymptotic expansion for small time. This leads to the following theorem.

<u>Theorem 2.16</u>. Suppose $\bar{M}$ satisfies condition $Z(q)$. Let $\Box_{p,q}$ denote the $\bar{\partial}$-Laplacian on (p,q) forms on $\bar{M}$ satisfying $\bar{\partial}$-Neumann boundary conditions. Then, as $t \longrightarrow 0^+$

$$\operatorname{tr} \exp(-t\Box_{p,q}) \sim t^{-n-1}(c_0 + \sum_{j\geq 1} (c_j + c'_j \log t) t^{j/2}). \tag{2.17}$$

If $q = n + 1$, then $c'_j = 0$ all j, since the $\bar{\partial}$-Neumann conditions are then just the Neumann condition. We have not yet worked out examples, but our proof of Theorem 2.16 leads us to believe that $\log t$ terms do occur. If so, this is a feature not present in the Riemannian case, (1.4). Logarithmic terms have been encountered in problems with singularities, for example, in the work of Brüning and Heintze [4], Brüning and Seeley [5], Callias and Uhlmann [6], and Cheeger [7]. Another difference between the complex and Riemannian cases is in the leading coefficient c_0. Unlike the Riemannian case, if $q \leq n$, the boundary contributes to c_0 : c_0 is the sum of a multiple of the volume of M and the integral over bM of a positive function of the Levi form (see [2, Theorem 11.7]). In the case of $(0,q)$ forms on a bounded strictly pseudoconvex domain in $\mathbb{C}^{n+1}$ equipped with the Euclidean metric, this is equivalent to Métivier's Weyl-type theorem [11].

<u>3. A geometric interpretation</u>.

To obtain a geometric interpretation of the coefficients in (2.17), we want a canonical connection on bM which preserves both the metric and the CR structure. Assume that the Levi form is definite on M. Then, if the metric is a Levi metric, the Webster-

C.M. Stanton connection [14], [13] has these properties. It is also convenient to assume that the metric is Kähler in a neighborhood of bM, so that parallel transport along geodesics normal to bM preserves $T^{1,0}(\bar{M})$ in a neighborhood of bM. Under these geometric assumptions, we have the following result [3].

Theorem 3.1. Let $\bar{M}$ be a compact complex $n + 1$ dimensional manifold with smooth boundary bM. Suppose the Levi form is definite on bM, that $\bar{M}$ is equipped with a Hermitian metric which is Kähler in a neighborhood of bM and induces a Levi metric on bM and that $Z(q)$ is satisfied. Then the coefficients c_j in the asymptotic expansion (2.17) can be written as $c_j = c''_j + c'''_j$. The coefficients c''_{2j} are integrals over M of universal poly- nomials in the components of the curvature and torsion of the Hermitian connection and their covariant derivatives with respect to this connection. The coefficients c''_{2j+1} are zero. The coefficients c'_j and c'''_j are integrals over bM of universal polynomials in the components of the second fundamental form of bM, the curvature and torsion of the Webster-C.M Stanton connection, and their covariant derivatives with respect to this connection, as well as components of the Hermitian curvature of $\bar{M}$ and its Hermitian covariant derivatives.

The proof of Theorem 2.16 shows that the coefficients c_j and c'_j are integrals over M and bM of certain functions. If the induced metric on bM is a Levi metric, inspection of this proof shows that, in suitable local coordinates in a neighborhood of $x_0 \in \bar{M}$, these functions, evaluated at x_0, are universal polynomials in the derivatives of the coefficients of an appropriate local basis of vector fields, evaluated at x_0. To prove the theorem, we show that its hypotheses allow us to use the geometry to determine a basis and suitable coordinates such that the Taylor polynomials of the coefficients of these vector fields have the desired geometric interpretation.

We remark that, if $\bar{M}$ is embedded in an $n + 1$ dimensional complex manifold $\tilde{M}$ without boundary and the Levi form on bM is definite, there is a metric satisfying the hypotheses of the theorem. As the potential function for the Kähler metric near bM, one can take $e^{\pm u}$ where u is a defining function for M which satisfies the Monge-Ampère equation on bM; the sign is determined on each component of bM by the sign of the Levi form.

Bibliography

1. R. Beals, P.C. Greiner, and N.K. Stanton, The heat equation on a CR manifold, J. Differential Geometry, 20 (1984), 343-387.

2. R. Beals and N.K. Stanton, The heat equation for the $\bar{\partial}$-Neumann problem, I, preprint.

3. R. Beals and N.K. Stanton, The heat equation for the $\bar{\partial}$-Neumann problem, II, preprint.

4. J. Brüning and E. Heintze, The asymptotic expansion of Minakshisundaram-Pleijel in the equivariant case, Duke Math. J. 51 (1984), 959-980.

5. J. Brüning and R. Seeley, Regular Singular Asymptotics, Advances in Math. 58 (1985), 133-148.

6. C.J. Callias and G.A. Uhlmann, Singular asymptotic approach to partial differential equations with isolated singularities in the coefficients, Bull. A.M.S. 11 (1984), 172-176.

7. J. Cheeger, Spectral geometry of singular Riemannian spaces, J. Differential Geometry 18 (1983), 575-657.

8. P.C. Greiner, An asymptotic expansion for the heat equation, Arch. Rational Mech. Anal. 41 (1971), 163-218.

9. P.C. Greiner and E.M. Stein, Estimates for the $\bar{\partial}$-Neumann Problem, Math. Notes 19, Princeton Univ. Press, Princeton, N.J., 1977.

10. H.P. McKean, Jr., and I.M. Singer, Curvature and the eigenvalues of the Laplacian, J. Differential Geometry 1 (1967), 43-69.

11. G. Métivier, Spectral asymptotics of the $\bar{\partial}$-Neumann problem, Duke Math. J. 48 (1981), 779-806.

12. R. Seeley, Analytic extension of the trace associated with elliptic boundary value problems, Amer. J. Math. 91 (1969), 963-983.

13. C.M. Stanton, Intrinsic connections for Levi metrics, in preparation.

14. S.M. Webster, Pseudo-hermitian structures on a real hypersurface, J. Differential Geometry 13 (1978), 25-41.

15. H. Weyl, Das asymptotische Verteilungsgesetz der Eigenwerte linearer partieller Differentialgleichungen (mit einer Anwendung auf die Theorie der Hohlraumstrahlung), Math. Ann. 71 (1912), 441-479.

Richard Beals
Department of Mathematics
Yale University
New Haven, CT 06520

Nancy K. Stanton
Department of Mathematics
University of Notre Dame
Notre Dame, IN 46556

Extendibility of the Bergman kernel function

S. Bell*
Purdue University
W. Lafayette, IN 47907

1. Introduction. If Ω is a bounded pseudoconvex domain in $\mathbf{C}^n$ of finite type (in the sense of D'Angelo [6]), then Kerzman's theorem [7] asserts that the Bergman kernel function $K(z,w)$ associated to Ω extends to be in $C^\infty((\bar{\Omega}\times\bar{\Omega})-\Delta)$ where $\Delta = \{(z,z): z \in b\Omega\}$ (see also [2]). In this note, I shall prove that the Bergman kernel $K(z,w)$ associated to a bounded strictly pseudoconvex domain Ω with real analytic boundary enjoys the follwing property. If **a** and **b** are distinct points in $\bar{\Omega}$, then there exist balls $B_{\mathbf{a}}$ and $B_{\mathbf{b}}$ centered at **a** and **b**, respectively, such that $K(z,w)$ extends to be holomorphic in z and antiholomorphic in w on the set $B_{\mathbf{a}} \times B_{\mathbf{b}}$. I shall also prove local versions of this result which apply when only one of the points **a** or **b** is strictly pseudoconvex or when the boundary is real analytic near only one of the points.

2. Preliminaries. We must first recall some facts from the theory of the $\bar{\partial}$-Neumann problem. The reader who is not familiar with this theory should consult the work of Kohn ([8,9]). In this section, we assume that Ω is a bounded pseudoconvex domain in $\mathbf{C}^n$. The boundary of Ω is not assumed to be smooth. The $\bar{\partial}$-Neumann operator N associated to Ω exists and maps the space $L^2_{(0,1)}(\Omega)$ of $(0,1)$ forms on Ω with coefficients in $L^2(\Omega)$ into itself (see Catlin [5]). Furthermore, N satisfies the estimate $\|N\alpha\|_0 \le C\,\|\alpha\|_0$ where C is a positive constant which does not depend on α

---

* Sloan Fellow, research partially supported by NSF grant DMS-8420754.

and where the norm is the norm on $L^2_{(0,1)}(\Omega)$ defined via $\|\Sigma\alpha_i d\bar{z}_i\|_0{}^2 = \int_\Omega \Sigma|\alpha_i|^2 dV$. Here dV denotes Lebesque measure on $\mathbf{C}^n$. The operator $\bar\partial^* N$ is also defined on $L^2_{(0,1)}(\Omega)$ and maps this space into $L^2(\Omega)$ as a bounded operator (see [5]). If s is a positive integer, the space of $(0,1)$ forms on Ω with coefficients in the Sobolev space $W^s(\Omega)$ will be denoted by $W^s{}_{(0,1)}(\Omega)$. The obvious norm on this Hilbert space will be written $\|\alpha\|_s$. If it is necessary to take this norm on a domain D which is unequal to Ω, the notation will be modified thus: $\|\alpha\|_s{}^D$.

If the boundary of Ω is real analytic near a strictly pseudoconvex boundary point z, then it follows from the local analytic hypoellipticy of the $\bar\partial$-Neumann problem proved by Tartakoff [10] and Trèves [11] that $N\alpha$ extends to be real analytic near z whenever α is a $(0,1)$ form in $L^2_{(0,1)}(\Omega)$ whose support is contained in a subdomain D which is a positive distance away from z. More generally, we shall say that the $\bar\partial$-Neumann problem on Ω satisfies local condition Q at a boundary point z if $b\Omega$ is real analytic near z and if $N\alpha$ extends to be real analytic at z whenever α is a form in $L^2_{(0,1)}(\Omega)$ which is supported away from z. (It is proved in [4] that a similar condition on the Bergman projction is equivalent to local analytic hypoellipticity.) Thus the $\bar\partial$-problem on a bounded pseudoconvex domain satisfies local conditon Q at all its strictly pseudocovnex real analytic boundary points.

The Bergman projection P associated to Ω is the orthogonal projection of $L^2(\Omega)$ onto its subspace $H(\Omega)$ consisting of holomorphic functions. The Bergman projection is related to the $\bar\partial$-Neuamann operator via Kohn's formula $P = I - \bar\partial^* N\bar\partial$. The Bergman kernel $K(z,w)$ associated to Ω is uniquely determined by the property that

$$P\varphi(z) = \int_\Omega K(z,w)\,\varphi(w)\,dV_w$$

for all φ in $L^2(\Omega)$. The kernel $K(z,w)$ is holomorphic in z and antiholomorphic in w, and $K(z,w)$ is equal to the conjugate of $K(w,z)$.

and where the norm is the norm on $L^2_{(0,1)}(\Omega)$ defined via $\|\Sigma\alpha_i d\bar{z}_i\|_0{}^2 = \int_\Omega \Sigma|\alpha_i|^2 dV$. Here dV denotes Lebesque measure on $\mathbf{C}^n$. The operator $\bar{\partial}^* N$ is also defined on $L^2_{(0,1)}(\Omega)$ and maps this space into $L^2(\Omega)$ as a bounded operator (see [5]). If s is a positive integer, the space of (0,1) forms on Ω with coefficients in the Sobolev space $W^s(\Omega)$ will be denoted by $W^s_{(0,1)}(\Omega)$. The obvious norm on this Hilbert space will be written $\|\alpha\|_s$. If it is necessary to take this norm on a domain D which is unequal to Ω, the notation will be modified thus: $\|\alpha\|_s{}^D$.

If the boundary of Ω is real analytic near a strictly pseudoconvex boundary point z, then it follows from the local analytic hypoellipticy of the $\bar{\partial}$-Neumann problem proved by Tartakoff [10] and Trèves [11] that $N\alpha$ extends to be real analytic near z whenever α is a (0,1) form in $L^2_{(0,1)}(\Omega)$ whose support is contained in a subdomain D which is a positive distance away from z. More generally, we shall say that the $\bar{\partial}$-Neumann problem on Ω satisfies local condition Q at a boundary point z if $b\Omega$ is real analytic near z and if $N\alpha$ extends to be real analytic at z whenever α is a form in $L^2_{(0,1)}(\Omega)$ which is supported away from z. (It is proved in [4] that a similar condition on the Bergman projction is equivalent to local analytic hypoellipticity.) Thus the $\bar{\partial}$-problem on a bounded pseudoconvex domain satisfies local conditon Q at all its strictly pseudocovnex real analytic boundary points.

The Bergman projection P associated to Ω is the orthogonal projection of $L^2(\Omega)$ onto its subspace $H(\Omega)$ consisting of holomorphic functions. The Bergman projection is related to the $\bar{\partial}$-Neuamann operator via Kohn's formula $P = I - \bar{\partial}^* N\bar{\partial}$. The Bergman kernel K(z,w) associated to Ω is uniquely determined by the property that

$$P\varphi(z) = \int_\Omega K(z,w)\, \varphi(w)\, dV_w$$

for all φ in $L^2(\Omega)$. The kernel K(z,w) is holomorphic in z and antiholomorphic in w, and K(z,w) is equal to the conjugate of K(w,z).

3. Statement of results. Suppose Ω is a bounded pseudoconvex domain in $\mathbf{C}^n$. Suppose further that **a** and **b** are distinct boundary points of Ω and that the boundary of Ω is C^∞ smooth near the points **a** and **b**. Let $K(z,w)$ denote the Bergman kernel associated to Ω. We shall prove

Theorem 1. If **b** is a point of finite type (in the sense of D'Angelo), if the boundary of Ω is real analytic near **a**, and if the $\bar\partial$-Neumann problem on Ω satisfies local condition Q at **a**, then there exist balls B_a and B_b centered at **a** and **b**, respectively, such that $K(z,w)$ extends to be in $C^\infty(B_a \times (\bar\Omega \cap B_b))$ as a function which is holomorphic in z and antiholomorphic in w on $B_a \times (\Omega \cap B_b)$.

Theorem 2. If the boundary of Ω is real analytic near both points **a** and **b**, and if the $\bar\partial$-Neumann problem on Ω satisfies local condition Q at both points **a** and **b**, then there exist balls B_a and B_b centered at **a** and **b**, respectively, such that $K(z,w)$ extends to be holomorphic in z and antiholomorphic in w on $B_a \times B_b$.

Corollary 3. If the boundary of Ω is real analytic and strictly pseudoconvex near the points **a** and **b**, then the kernel function $K(z,w)$ extends to be holomorphic in z and antiholomorphic in w in a neighborhood of the point $(\mathbf{a},\mathbf{b})$ in $\mathbf{C}^n \times \mathbf{C}^n$.

4. The proofs. To prove Theorem 1, we shall need to use a fact which is proved in [2]. Since **b** is a smooth boundary point of finte type, it is possible to construct a small subdomain D of Ω whose boundary coincides with that of Ω near **b** with the following properties.

1) $D \subset \Omega$, and D is a pseudoconvex domain with C^∞ smooth boundary which is of finite type in the sense of D'Angelo,
2) the boundary of D does not contain **a**,
3) there is a ball $W = B(\mathbf{b};\eta)$ of radius η centered at **b** in $\mathbf{C}^n$ such that

$(\Omega \cap W) = (D \cap W)$,

4) the Bergman kernel $K_D(z,w)$ associated to D extends to be in $C^\infty((\bar{D} \times \bar{D}) - \Delta_D)$ where $\Delta_D = \{(z,z) : z \in bD\}$.

Let χ_1 be a real valued C^∞ function on $\mathbf{C}^n$ which is compactly supported in $B(\mathbf{b};\eta/2)$ and which is equal to one on $B(\mathbf{b};\eta/3)$. Let $\chi_2 = 1 - \chi_1$. If s is a positive integer, an operator Φ^s can be constructed with the following properties (see [3]):

a) Φ^s is a bounded operator in $W^s(D)$ norm from the space of holomorphic functions on D into the space $W^s{}_0(D)$, the closure of $C^\infty{}_0(D)$ in $W^s(D)$,

b) Φ^s is a linear differential operator with coefficients in $C^\infty(\bar{D})$,

c) $P_D \Phi^s = P_D$ on $C^\infty(\bar{D})$ where P_D denotes the Bergman projection associated to D.

The argument used in [3] to construct Φ^s and to prove property (a) also yields

d) the operator which maps a holomorphic function h in $W^s(D)$ to $\Phi^s(\chi_2 h)$ is bounded in the $W^s(D)$ norm and $\Phi^s(\chi_2 h) \in W^s{}_0(D)$.

We now define a function $\psi(z,w)$ which is in $L^2(\Omega)$ as a function of z for each fixed value of $w \in D$ via

$$\psi(z,w) = \chi_1(z) K_D(z,w) + \Phi^1(\chi_2(z) K_D(z,w)).$$

Here, the opereator Φ^1 acts in the z variable only and the parameter w is restricted to be in D. Note that because $W^1{}_0(D)$ can be viewed as a subspace of $W^1{}_0(\Omega)$ via extension by zero, we may think of $\psi(z,w)$ as a function of z defined on all of Ω supported in D which is in $W^1{}_0(\Omega)$. The properties (1)-(4) and (a)-(d) imply that for w restricted to the set $Y := B(\mathbf{b};\eta/4) \cap \Omega$, the family of functions $\{v(z) = \psi(z,w) : w \in Y\}$ is uniformly bounded in W^1 norm on the set $z \in \Omega - B(\mathbf{b};\eta/3)$. Furthermore, for any fixed $w \in \bar{D}$, $\psi(z,w)$ is holomorphic in z on the open set $B(\mathbf{b};\eta/3) \cap \Omega$.

We now claim that for $w \in D$ and $z \in \Omega$, we have $K(z,w) = (P\psi(\zeta,w))(z)$ where the Bergman projection P associated to Ω is understood to act in the ζ variable only.

To be more precise, we have

$$K(z,w) = \int_\Omega K(z,\zeta)\, \psi(\zeta,w)\, dV_\zeta .$$

Indeed, if g is in $H(\Omega)$, then by applying property (c) of Φ^1, we obtain

$$\int_\Omega \psi(\zeta,w)\, \overline{g(\zeta)}\, dV_\zeta = \int_D (\chi_1 K_D(\zeta,w) + \chi_2 K_D(\zeta,w))\, \overline{g(\zeta)}\, dV_\zeta$$

$$= \int_D K_D(\zeta,w)\, \overline{g(\zeta)}\, dV_\zeta = \overline{g(w)}$$

and hence the function $v(\zeta) = \psi(\zeta,w)$ projects to a holomorphic function with the reproducing property at w. This property is uniquely possessed by the function $h(\zeta)=K(\zeta,w)$.

To finish the proof of Theorem 1, we must use a Baire category argument. I wish to thank M. S. Baouendi for suggesting the idea of this argument to me. Define a set of functions E on Ω to be the set of all functions $u \in L^2(\Omega)$ such that $u = \bar\partial^* N\alpha$ for some $\bar\partial$-closed (0,1) form α in $W^1{}_{(0,1)}(\Omega)$ which is supported in D and which satisfies $\|\alpha\|_1 \le 1$. Note that if $u \in E$, then u extends to be real analytic in some ball centered at **a** because the $\bar\partial$-problem on Ω is assumed to satisfy condition Q at **a**. We now wish to prove that there is a fixed ball $B_{\mathbf{a}}$ centered at **a** such that <u>every</u> function in E extends to be real analytic on $B_{\mathbf{a}}$. First, note that E is a closed subset of $L^2(\Omega)$. Indeed, if $u_j = \bar\partial^* N\alpha_j$ is a sequence of functions in E where the α_j are as in the definition of E and if u_j converges to u in $L^2(\Omega)$, we may choose a subsequence of $\{\alpha_j\}$ which converges weakly in $W^1{}_{(0,1)}(\Omega)$ to a form α in $W^1{}_{(0,1)}(\Omega)$ such that $\bar\partial\alpha=0$ and $\|\alpha\|_1 \le 1$. Furthermore, α is supported in D. This subsequence converges strongly in $L^2{}_{(0,1)}(\Omega)$ to α by Rellich's Lemma. Thus, because the operator $\bar\partial^* N$ is bounded in L^2 norms, we conclude that $u = \bar\partial^* N\alpha$ and consequently, that $u \in E$. Hence E is a closed subset of a Hilbert space and is therefore of the second category.

Let S_{jk} denote the set of functions u in E such that u extends to be real analytic on $B(\mathbf{a};1/k)$ and such that the $L^2(B(\mathbf{a};1/k))$ norm of u restricted to $B(\mathbf{a};1/k)$ is less than j. It is clear that $E = \cup S_{jk}$ where the union ranges over all positive integers j and k. Therefore, at least one of the sets, say S_{JK}, has a non-empty interior in E. Since S_{JK} is convex and balanced, it must contain a neighborhood of zero in E. Hence, every function in E of sufficiently small $L^2(\Omega)$ norm extends to be holomorphic on $B(\mathbf{a};1/K)$. It is clear that the bound on the $L^2(\Omega)$ norm of functions in E is irrelevant because $\lambda E \subset E$ for any complex number λ with $|\lambda| \leq 1$. Thus, there is a ball $B_{\mathbf{a}}$ centered at $\mathbf{a}$ such that every function in E extends to be real analytic on $B_{\mathbf{a}}$. Furthermore, we have proved that there exists a constant c such that if α is a $\bar\partial$-closed (0,1) form in $W^1{}_{(0,1)}(\Omega)$ supported on D, then the function $u = \bar\partial^* N\alpha$ is real analytic on $B_{\mathbf{a}}$ and the $L^2(B_{\mathbf{a}})$ norm of u is bounded by $c\|\alpha\|_1$. Thus, we have that for such a form α, there exists a positive constant C which does not depend on α such that

$$\|\bar\partial^* N\alpha\|_0{}^G \leq C\,\|\alpha\|_1 \quad \text{where} \quad G = B_{\mathbf{a}} \cup \Omega. \tag{3.1}$$

By shrinking $B_{\mathbf{a}}$, if necessary, we may assume that the closure of $B_{\mathbf{a}}$ does not intersect the closure of D. We now claim that $K(z,w)$ extends holomorphically to $B_{\mathbf{a}}$ as a function of z for each fixed w in Y. Indeed, let $\alpha_W = \bar\partial_Z\, \psi(z,w)$ where the notation $\bar\partial_Z$ indicates that we are applying $\bar\partial$ in the z-variable only. Note that α_W is a $\bar\partial$-closed (0,1) form which is supported on $D - B(\mathbf{b};\eta/3)$ and which is uniformly bounded in $W^1{}_{(0,1)}(\Omega)$ as a function of z for w in Y. Because Kohn's formula yields that

$$K(z,w) = \psi(z,w) - (\bar\partial^* N\alpha_W)(z)\,, \tag{3.2}$$

we conclude that $K(z,w)$ extends to be holomorphic in z on $B_{\mathbf{a}}$ for w in Y.

Finally, the uniform smoothness of $K(z,w)$ mentioned in the statement of Theorem 1 follows easily from formula (3.2). Indeed, this identity can be differentiated under the operators with respect to $\bar{w}$ as many times as desired without changing the crucial properties of the functions involved. Estimate (3.1) and Sobolev's Lemma can be applied to prove that, after shrinking $B_{\mathbf{a}}$ if necessary, $K(z,w)$ belongs to $C^{\infty}(B_{\mathbf{a}}\times(\bar{\Omega}\cap B(\mathbf{b};\eta/4))$.

Proof of Theorem 2. We shall use the same sets and functions that we constructed in the proof of Theorem 1. The key formula is again (3.2). In the setting of the hypotheses of Theorem 2, however, we my apply Theorem 1 to the subdomain D at $\mathbf{b}$. Indeed, we now have that the function $\psi(z,w)$ extends to be antiholomorphic in w past the boundary of Ω near $\mathbf{b}$ when z is a fixed point in $\Omega-B(\mathbf{b},\eta/3)$. Thus, in addition to its other important properties, the form α_w now possesses the property that its coefficients extend past $b\Omega$ near $\mathbf{b}$ antiholomorphically in w for any fixed z in Ω. Now, if $B_{\mathbf{a}}$ and $B_{\mathbf{b}}$ are sufficiently small balls centered at $\mathbf{a}$ and $\mathbf{b}$, respectively, then the kernel function is given by $K(z,w)=\psi(z,w)-(\bar{\partial}^*N\alpha_w)(z) = (\bar{\partial}^*N\alpha_w)(z)$ when $(z,w)\in(B_{\mathbf{a}}\cap\Omega)\times(B_{\mathbf{b}}\cap\Omega)$. Furthermore, the function on the right hand side of this equation extends to be antiholomorphic on all of $B_{\mathbf{b}}$ as a function of w when z is restricted to be in $B_{\mathbf{a}}$. Thus $K(z,w)$ extends to be holomorphic in z and antiholomorphic in w on $B_{\mathbf{a}}\times B_{\mathbf{b}}$.

Remark. If $\mathbf{b}$ is not assumed to be of finite type in Theorem 1, a much weaker result can be obtained. In this case, reasoning similar to that above yields that $K(z,w)$ extends to be holomorphic in z and antiholomorphic in w on $B_{\mathbf{a}}\times(B_{\mathbf{b}}\cap\Omega)$, but no statement can be made about the regularity of $K(z,w)$ as w approaches the boundary of Ω in $B_{\mathbf{b}}$. The one subtle point in the argument concerns the subdomain D. This domain can be constructed to be an arbitrarily small smoothly bounded pseudoconvex domain which is as C^1 close to a ball as desired. Then a result of David Barrett [1] applies to yield that the Bergman projection associated to D maps $W^1(D)$ into itself boundedly. Thus, the Bergman kernel $K_D(z,w)$ associated to D is uniformly in $W^1(D)$ as a function of z for w restricted to be in a fixed compact subset of D. Now the reasoning used in the proof of Theorem 1 can be applied.

REFERENCES

1. D. Barrett, **Regularity of the Bergman projection and local geometry of domains**, Duke Math. J. **53** (1986), 333-343.
2. S. Bell, **Differentiability of the Bergman kernel and pseudo-local estimates**, Math. Zeit.**192** (1986), 467-472.
3. S. Bell, **Biholomorphic mappings and the $\bar\partial$-problem**, Ann. of Math. **114** (1981), 103-113.
4. S. Bell, **Boundary behavior of holomorphic mappings**, Several Complex Variables: Proceedings of the 1981 Hangzhou Conference, Birkhauser, 1984.
5. D. Catlin, **Necessary conditions for subellipticity of the $\bar\partial$-Neumann problem**, Ann. of Math. **117** (1983), 147-171.
6. J. P. D'Angelo, **Real hypersurfaces, orders of contact, and applications**, Ann. Math. **115** (1982), 615-637.
7. N. Kerzman, **The Bergman kernel function. Differentiability at the boundary**, Math. Ann. **195** (1972), 149-158.
8. J. J. Kohn, **Harmonic integrals on strongly pseudoconvex manifolds, I and II**, Ann. of Math. **78** (1963), 112-148 and Ann. of Math. **79** (1964), 450-472.
9. J. J. Kohn, **A survey of the $\bar\partial$-Neumann problem**, Proc. Symp. Pure Math. **41**, Amer. Math. Soc., Providence, 1984.
10. D. Tartakoff, **The local real analyticity of solutions to $\Box_b$ and the $\bar\partial$-Neumann problem**, Acta Math. **145** (1980), 177-204.
11. F. Trevès, **Analyitc hypo-ellipticity of a class of pseudodifferential operators with double characteristics and applications to the $\bar\partial$-Neumann problem**, Comm. Partial Differential Equations **3** (1978), 475-642.

$\bar\partial_b$ AND Carleson TYPE INEQUALITIES

Bo Berndtsson

Introduction.

Let $\mathcal{D}$ be a smooth strictly pseudoconvex domain in $\mathbb{C}^n$ and f a $\bar\partial$-closed (0,1) form in $\mathcal{D}$. If u is a function defined on $\partial\mathcal{D}$ that can be obtained as the boundary values of a solution to

(1) $$\bar\partial u = f$$

we write (following [6])

(2) $$\bar\partial_b u = f\ .$$

The equation (2) can be given a dual formulation: If h is a smooth $\bar\partial$-closed form of bidegree $(n,n-1)$ any solution to (2) must satisfy

(3) $$\int_{\partial\mathcal{D}} uh = \int_{\mathcal{D}} f \wedge h$$

Conversely, if this equation holds for all such h, then u satisfies (2) (in a suitable sense, see [6]). Since h is of total degree $2n-1$, on the boundary h can be written as

$$h^* dS$$

where dS is surface measure on $\partial\mathcal{D}$ and h^* is some density. Therefore, a necessary condition to have, e.g., $u \in L^p(dS)$ $(p > 1)$ is that

$$\exists C\ \forall h\ \left|\int_{\mathcal{D}} f \wedge h\right| \le C\, \|h^*\|_{L^q(dS)}$$

where, of course, q is the dual exponent. An argument involving the Hahn-Banach theorem shows that such an estimate is actually also sufficient to have a solution in $L^p(dS)$. So, formulated this way, the problem of solving (2), amounts to an estimate where, vaguely speaking, we want to control a closed $(n,n-1)$-form in $\mathcal{D}$ by its boundary values.

If $n = 1$ and $\mathcal{D}$ is the unit disc an optimal estimate of this kind is given by Carleson's inequality. In this case h is just a holomorphic function (times $d\zeta$), and h^* equals $i\zeta$ times its boundary values, and the inequality says that a positive measure μ in $\mathcal{D}$ satisfies

(4) $$\int_{\mathcal{D}} |h|\, d\mu \le C \int_{\partial\mathcal{D}} |h^*|\, dS$$

if and only if μ is a Carleson measure (see [3] and section 1 below).

The starting point for this paper is the simple observation that Carleson measures can be (almost) characterized as the measures that satisfy an estimate

$$\mu \leq (1-|\zeta|^2)\Delta\phi$$

where ϕ is bounded and subharmonic in the disc. This gives a simple proof of (4) by Green's formula, and also suggests a possible generalization of Carleson's condition to several variables. In general terms a form should satisfy a "Carleson condition" if it can be controlled by the hessian of a bounded plurisubharmonic function. This suggests a relation between (4) and the weighted L^2-estimates for $\bar\partial$ of Hörmander ([4]). In section 2 we give a variant of Hörmander's inequality, where we take into account the boundary values of the solution, and show that it can be thought of as a generalization of (4). We then use this to derive some (weighted) L^2-estimates for (2).

The main problem in the higher dimensional case is that those methods, so far, work only in the L^2-norms. As a matter of fact, all non-trivial estimates in other norms that we know of are obtained by constructive methods, i.e., integral formulas. In the final section we give some remarks on the case of L^∞-estimates, and show that a uniform estimate is equivalent to weighted L^2-estimates for a large class of weights. We then formulate a condition on a form f which could possibly give uniform estimates at least if, say, $\mathcal{D}$ is the ball. Expressed in invariant language the condition is that both f and ∂f should be uniformly bounded when measured in some Kähler metric with bounded potential. The fact that this condition implies the existence of a bounded solution when $\mathcal{D}$ = unit disc, is essentially Wolff's theorem ([3] p. 322), and by the same argument that Wolff used, e.g., the Corona theorem for two generators will follow in any domain where this theorem can be generalized.

Finally, we refer to the article by Amar and Bonami [1] and Skoda [6], for L^p-estimates by integral kernels.

The material in this paper was worked out when I visited the UCLA and the Université de Paris-Sud at Orsay. It's a pleasure to thank both those institutions for their hospitality.

Section 1: Carleson measures.

Let μ be a positive measure in the unit disc, Δ. One says that μ is a Carleson measure, or $\mu \in \mathcal{C}$, if

$$\exists A \ \forall \delta > 0 \ \forall z \in \partial\Delta \quad \int_{|\zeta - z| < \delta} d\mu(\zeta) < A\delta .$$

The reason for the introduction of this class of measures comes from the following theorem of Carleson (see [3] p. 63):

Theorem 1: $\exists C$ such that

$$\int_{\Delta} |h| d\mu \le C \int_{\partial\Delta} |h||d\zeta| \quad \forall h \in H(\bar{\Delta})$$

iff $\mu \in \mathcal{C}$.

We will now compare this to another type of condition. Let us say that $\mu \in \mathcal{C}^*$ if

(8) $\exists \phi \in L^{\infty}(\Delta)$, ϕ subharmonic, such that $\mu \le (1-|\zeta|^2)\Delta\phi$.

We then have the following

Proposition 2: a) $\mathcal{C}^* \subseteq \mathcal{C}$

b) Suppose $\mu \in \mathcal{C}$, and moreover that μ is absolutely continuous w.r.t. Lebesgue measure, $d\lambda$, with a density k that satisfies

$$|k(\zeta)| \le C/(1-|\zeta|^2)$$

Then $\mu \in \mathcal{C}^*$.

Remark: It is clear that, e.g., a pointmass is a Carleson measure which is not in $\mathcal{C}^*$, so the two classes are not identical. However, in most applications of Carleson measures the extra technical assumption in b) can be assumed to hold.

Proof: A proof has already been published in [2], but for convenience we give an outline here, too. By the Riesz decomposition, if ϕ is bounded and subharmonic

$$\phi = P[\phi] + G[\Delta\phi]$$

where P is the Poisson integral and G the Green potential. If $\phi \in L^{\infty}$, $P[\phi] \in L^{\infty}$, so $G[\Delta\phi] \in L^{\infty}$. Hence

$$G\left[\frac{\mu}{1-|\zeta|^2}\right] \in L^{\infty} \tag{1}$$

if $\mu \in \mathcal{C}^*$. It's even more clear that (1) implies that $\mu \in \mathcal{C}^*$, since we can take our ϕ to be the Green potential of $\mu/(1-|\zeta|^2)$. On the other hand, it is well known (see [3] p. 239), that $\mu \in \mathcal{C}$ iff

$$(2) \qquad \int \frac{1-|z|^2}{|1-\bar{\zeta}\cdot z|^2}\, d\mu(\zeta) \in L^\infty .$$

But

$$G\left[\frac{\mu}{1-|\zeta|^2}\right](z) = C \int_{|\zeta|<1} \log\left|\frac{\zeta-z}{1-\bar{\zeta}z}\right|^2 \frac{d\mu(\zeta)}{1-|\zeta|^2}$$

$$= C \int_{|\zeta|<1} \log\left[1-\frac{(1-|\zeta|^2)(1-|z|^2)}{|1-\bar{\zeta}z|^2}\right]\frac{d\mu(\zeta)}{1-|\zeta|^2} .$$

Therefore a) follows from $\log(1-x) \le -x$. To prove b) we decompose

$$G\left[\frac{\mu}{1-|\zeta|^2}\right](z) = \int_{\left|\frac{\zeta-z}{1-\bar{\zeta}z}\right|<\frac{1}{10}} + \int_{\left|\frac{\zeta-z}{1-\bar{\zeta}z}\right|\ge\frac{1}{10}}$$

The second term is estimated by using $\log(1-x) \sim -x$ if $x < \frac{9}{10}$, and the first term is handled by the extra technical assumption.

<u>Remark</u>: It is interesting to note that the Carleson measures that come up in practice are often given directly by an inequality (*). For instance, the basic point in Carleson's theorem on interpolation with bounded holomorphic functions is that if $\{a_n\}$ is a hyperbolically separated sequence of points in Δ then

$$\mu = \Sigma(1-|a_n|^2)\delta_{a_n} \in \mathcal{C} \quad \text{iff}$$

$$(3) \qquad \exists\delta > 0: \quad \forall k \quad \prod_{j\ne k}\left|\frac{a_j-a_k}{1-\bar{a}_k a_j}\right|^2 \ge \delta .$$

After taking logarithms in (3) this is essentially Proposition 2. The difference is that we consider only measures that are never too concentrated around one point, whereas in (3) one takes out the point at the singularity of the Green's kernel. To take another example, in Wolff's proof of the Corona theorem one uses that if $h \in H^\infty$ then

$$(1-|\zeta|^2)|h'|^2 \in \mathcal{C} .$$

This is also an example of Proposition 2 if we take $\phi = |h|^2$.

Next we will show how Theorem 1 can be proved from Proposition 2. Let ϕ and v be subharmonic in Δ. Then Green's formula implies

$$2\int_{\partial\Delta} e^{\phi+v}|d\zeta| = \int_\Delta (1-|\zeta|^2)\Delta e^{\phi+v} + e^{\phi+v}(\Delta(|\zeta|^2-1))d\lambda$$

$$\ge \int_\Delta (1-|\zeta|^2)[\Delta\phi+\Delta v+|\nabla(\phi+v)|^2]e^{\phi+v}d\lambda$$

$$\ge \int_\Delta e^{v+\phi}(1-|\zeta|^2)\Delta\phi\, d\lambda .$$

If ϕ is bounded and μ satisfies

$$\mu \le (1-|\zeta|^2)\Delta\phi$$

we get

$$\int_\Delta e^v d\mu \le C\int_{\partial\Delta} e^v |d\zeta|$$

and taking $v = \log h$, we get the inequality of Theorem 1, when $\mu \in \mathcal{C}^*$. From this it is not hard to get the general case by smoothing out a measure $\mu \in \mathcal{C}$. Rather than write this out in detail, let us consider a measure $\mu \in \mathcal{C}$ of the form

$$\mu = \Sigma\left(1-|a_n|^2\right)\delta_{a_n}$$

(i.e., something which in a way is as far as possible from satisfying our extra assumption). Around each point a_n we place a disc with center a_n and radius $(1-|a_n|^2)/10$. Let $\tilde{\mu}_n$ be the mass $1-|a_n|^2$, spread out uniformly over this disc and

$$\tilde{\mu} = \Sigma\tilde{\mu}_n .$$

Then $\tilde{\mu} \in \mathcal{C}^*$ and by the mean value property

$$\int_\Delta g d\tilde{\mu} \ge \int_\Delta g d\mu$$

if g is subharmonic. Hence in particular

$$\int_\Delta e^v d\mu \le \int_\Delta e^v d\tilde{\mu} \le C \int_{\partial\Delta} e^v |d\zeta|$$

if v is subharmonic. Clearly a similar argument works in general.

Before proceeding to the case of differential forms we consider holomorphic functions in the ball in $\mathbb{C}^n$. In [5] Hormander has determined the measures that correspond to Carleson measures in this setting. We say that $\mu \in \mathcal{C}_n$ if

$$\exists A\ \forall z \in \partial B\ \forall\delta > 0 \quad \int_{|1-\bar{\zeta}z|<\delta} d\mu(\zeta) \le A\delta^n.$$

Then, it is proved in [5], that $\mu \in \mathcal{C}_n$ iff

$$\exists C\ \forall h \in H(\bar{B}_n) \int_{B_n} |h|d\mu \le C \int_{\partial B_n} |h|dS \tag{4}$$

where B_n is the ball in $\mathbb{C}^n$. To give the statement corresponding to Proposition 2 we introduce the differential operator L by

$$Lu = \Sigma u_{k\bar{k}} - \Sigma u_{j\bar{k}}\zeta_j\bar{\zeta}_k .$$

If Δ_B is the Laplacian defined by the Bergman metric in the ball we have

$$Lu = \frac{1}{1-|\zeta|^2} \Delta_B u .$$

As before, we say that a positive measure in B lies in $\mathcal{C}_n^*$ if

$$\exists \phi \in L^\infty(B_n) \text{ such that } \mu \le L\phi$$

(hence it is assumed in particular that ϕ is L-subharmonic).

Proposition 3: a) $\mathcal{C}_n^* \subseteq \mathcal{C}_n$

b) If $\mu \in \mathcal{C}_n$ and $\mu = kd\lambda$ with

$$k(\zeta) \le \frac{C}{1-|\zeta|^2} \text{ then } \mu \in \mathcal{C}_n^* .$$

The proof of this Proposition is completely parallel to the proof of Proposition 2, one just has to use Green's function w.r.t. Δ_B instead of the euclidean one. One can also copy the argument in the disc to prove (4) from Proposition 3. Indeed, it seems that for almost any elliptic operator one can introduce a class of "Carleson measures" associated with the operator, and obtain inequalities like (4) for exponentials of subsolutions of the operator (a slight refinement of the estimate shows actually that one can replace e^v by v^α for any $\alpha > 1$, in the inequalities). What is a little bit surprising is the fact that (4), which basically is an inequality for plurisubharmonic functions, is associated to one single operator.

Section 2: Inequalities for harmonic forms and $\bar\partial_b$.

As we have seen in the introduction, the $\bar\partial_b$ problem is related to estimates of $\bar\partial$-closed $(n,n-1)$ forms in the domain in terms of their boundary values. Formulated this way the problem is not well posed, because it may, e.g., well happen that such a form vanishes on $\partial\mathcal{D}$ without being identically zero (just take $h = \bar\partial g$, where g is any $(n,n-2)$ form with compact support in $\mathcal{D}$). Fortunately one can, with no loss of generality put one more condition on the forms, and to formulate this extra condition in a natural way, we first change the formalism a little bit.

Let

$$h = \Sigma\, h_j \widehat{d\bar\zeta}_j$$

be a $(n,n-1)$-form. Here $\widehat{d\bar\zeta}_j$ means the wedge product of all differentials except $d\bar\zeta_j$ with a coefficient chosen to that

$$f \wedge h = \Sigma f_j h_j d\lambda, \; d\lambda = \tfrac{i}{2}\partial\bar\partial|\zeta|^2)^n/n!$$

The Hodge star operator associates to h a $(0,1)$ form γ,

$$\gamma = *\bar{h} = \Sigma \overline{h_j} d\overline{\zeta_j} \; .$$

Hence $f \wedge h = f\cdot\bar\gamma d\lambda$ and $\partial\bar{h} = 0$ means

$$\sum_1^n \frac{\partial\gamma_j}{\partial\zeta_j} = 0 \; .$$

Following Hormander we next introduce a weight factor by

$$e^{\phi}\gamma = \alpha \; .$$

Denoting the scalar product

$$\int_{\mathcal{D}} f\cdot\bar\alpha e^{-\phi} \quad \text{by} \quad < f,\alpha >,$$

we define the formal adjoint of $\bar\partial$, $\bar\partial^*_\phi$, by

$$< \bar\partial u,\alpha > = < u,\bar\partial^*_\phi\alpha > \quad \text{so} \quad \bar\partial^*_\phi\alpha = -e^{\phi}\Sigma\frac{\partial}{\partial\overline{\zeta_j}}e^{-\phi}\alpha_j \; .$$

This gives the list of translations

i) $\int_{\mathcal{D}} f \wedge h = \int f\cdot\bar\alpha e^{-\phi} = < f,\alpha >$

ii) $\bar\partial h = 0 \Leftrightarrow \bar\partial^*_\phi\alpha = 0$

iii) $h^* = \overline{e^{-\phi}\alpha_{\vec{h}}} = \overline{\Sigma\alpha_j\frac{\partial\rho}{\partial\zeta_j}/|\partial\rho|e^{-\phi}}$

where ρ is a defining function for $\mathcal{D}$.

<u>Lemma 4</u>: <u>If</u> $1 < p \le \infty$ $\exists u \in L^p(e^{-\phi}dS)$ <u>such that</u> $\bar\partial_b u = f$ <u>iff</u>

(1) $$\exists C \quad |\int_{\mathcal{D}} f\cdot\bar\alpha\, e^{-\phi}d\lambda| \le C\,\|\alpha_{\vec{h}}\|_{L^q(e^{-\phi}dS)}$$

for all $\alpha \in C^\infty_{(0,1)}(\bar{\mathcal{D}})$ such that $\bar\partial^*_\phi\alpha = 0$, and $\bar\partial\alpha = 0$.

<u>Proof</u>. If (1) holds for all α, such that $\bar\partial^*_\phi\alpha = 0$, we can define an antilinear functional

$$\alpha_{\vec{h}} \longrightarrow L\alpha_{\vec{h}} = \int_{\mathcal{D}} f\cdot\bar\alpha e^{-\phi}d\lambda$$

on the subspace of $L^q(e^{-\phi}dS)$ consisting of all such $\alpha_{\vec{h}}$: s. By the Hahn-Banach theorem it can be extended to all of L^q, and is thus given by a function $u \in L^p(e^{-\phi}dS)$. Thus

$$\int_{\mathcal{D}} f\cdot\bar\alpha e^{-\phi} = \int_{\partial\mathcal{D}} u\bar\alpha_{\vec{h}}e^{-\phi}dS \quad \forall\alpha,\; \bar\partial^*_\phi\alpha = 0 \; .$$

This implies $\bar\partial_b u = f$ by (3) from the introduction. What we need to show is that if (1) holds for all α such that $\bar\partial^*_\phi\alpha = 0$ and $\bar\partial\alpha = 0$, then (1) holds for all α such that $\bar\partial_\phi^*\alpha = 0$ only. Suppose therefore that $\bar\partial_\phi^*\alpha = 0$. Decompose

$$\alpha = \alpha^1 + \alpha^2$$

where $\bar\partial\alpha^1 = 0$ and $\alpha^2 \perp \{\alpha;\ \bar\partial\alpha = 0\}$. It follows from the regularity of the $\bar\partial$-Neumann problem that α^1 and α^2 are smooth. One easily verifies that the fact that α^2 is orthogonal to closed forms means precisely that

$$\bar\partial_\phi^*\alpha^2 = 0 \quad \text{and} \quad \alpha_{\mathfrak{h}}^2 = 0 \ .$$

Hence $\bar\partial_\phi^*\alpha^1 = 0$ and $\alpha_{\mathfrak{h}} = \alpha_{\mathfrak{h}}^1$. Therefore nothing is changed if we replace α by α^1 in (1), which proves Lemma 4.

We can think of forms satisfying $\bar\partial_\phi^*\alpha = 0$ and $\bar\partial\alpha = 0$ as "harmonic forms". If $\phi \equiv 0$ they are precisely the forms that can be written

$$\alpha = \bar\partial g \quad \text{with} \quad g \quad \text{harmonic.}$$

We now state the basic inequality for such forms.

Proposition 5: Let ρ be a defining function for $\mathcal{D}$, $\alpha \in \mathcal{E}^1(\bar{\mathcal{D}})$ and $\phi \in \mathcal{E}^2(\mathcal{D}) \cap \mathcal{E}^1(\bar{\mathcal{D}})$. Then

$$(2) \qquad \int(-\rho)\Sigma\phi_{j\bar k}\alpha_j\bar\alpha_k e^{-\phi} + \int(-\rho)\Sigma\left|\frac{\partial\alpha_j}{\partial\bar\zeta_k}\right|^2 e^{-\phi} + \int\Sigma\rho_{j\bar k}\alpha_j\bar\alpha_k e^{-\phi}$$
$$= \int_{\mathcal{D}}(-\rho)|\bar\partial_\phi^*\alpha|^2 e^{-\phi} + \int_{\mathcal{D}}(-\rho)|\bar\partial\alpha|^2 e^{-\phi}$$
$$+ 2\mathrm{Re}\int_{\mathcal{D}}\bar\partial_\phi^*\alpha\cdot(\overline{\alpha\cdot\partial\rho})e^{-\phi} + \int_{\partial\mathcal{D}}|\alpha_{\mathfrak{h}}|^2 e^{-\phi}|\partial\rho|dS \ .$$

In particular, if $\bar\partial\alpha = 0$ and $\bar\partial_\phi^*\alpha = 0$, then

$$(3) \qquad \int_{\mathcal{D}}(-\rho)\Sigma\phi_{j\bar k}\alpha_j\bar\alpha_k e^{-\phi} + \int_{\mathcal{D}}(-\rho)\Sigma\left|\frac{\partial\alpha_j}{\partial\bar\zeta_k}\right|^2 e^{-\phi} + \int_{\mathcal{D}}\Sigma\rho_{j\bar k}\alpha_j\bar\alpha_k e^{-\phi}$$
$$= \int_{\partial\mathcal{D}}|\alpha_{\mathfrak{h}}|^2 e^{-\phi}|\partial\rho|dS.$$

Proof: This is, of course, a variation on the theme of Hörmander's weighted L^2-estimates. The Hörmander-Kohn-Morrey identity is [4]:

$$\int_{\mathcal{D}}\Sigma\ \psi_{j\bar k}\alpha_j\bar\alpha_k e^{-\psi} + \int_{\mathcal{D}}\Sigma\left|\frac{\partial\alpha_j}{\partial\bar\zeta_k}\right|^2 e^{-\psi} + \int_{\partial\mathcal{D}}\Sigma\rho_{j\bar k}\alpha_j\bar\alpha_k e^{-\psi}\frac{dS}{|\partial\rho|}$$
$$= \int_{\mathcal{D}}|\bar\partial_\psi^*\alpha|^2 e^{-\psi} + \int_{\mathcal{D}}|\bar\partial\alpha|^2 e^{-\psi} \ ,$$

and it holds under the $\bar\partial$-Neumann boundary condition $\alpha_{\mathfrak{h}} = 0$. If this boundary condition does not hold additional boundary terms appear. They are integrals of functions that are

$$O(|\alpha||\nabla\alpha||\nabla\psi|e^{-\psi}) .$$

Now we choose

$$\psi = \phi + \tau \log - (\frac{1}{\rho}) , \tau > 1 .$$

Then $e^{-\psi} = e^{-\phi}(-\rho)^{\tau}$ and

$$\bar{\partial}^*_{\psi}\alpha = \bar{\partial}^*_{\phi}\alpha + \frac{\tau\Sigma\alpha_j\rho_j}{-\rho} .$$

Therefore, all boundary terms will disappear if $\tau > 1$, and we get, for any $\alpha \in \mathcal{E}^1_{(0,1)}(\bar{\mathcal{D}})$

$$\begin{aligned} \int_{\mathcal{D}}(-\rho)^{\tau}\Sigma\phi_{j\bar{k}}\alpha_j\bar{\alpha}_k &+ \tau \int_{\mathcal{D}}(-\rho)^{\tau-1}\Sigma\rho_{j\bar{k}}\alpha_j\bar{\alpha}_k e^{-\phi} \\ &+ \tau\int_{\mathcal{D}}(-\rho)^{\tau-2}|\Sigma\rho_j\alpha_j|^2 e^{-\phi} \\ &+ \int_{\mathcal{D}}(-\rho)^{\tau}\Sigma\left|\frac{\partial\alpha_j}{\partial\bar{\zeta}_k}\right|^2 e^{-\phi} = \int_{\mathcal{D}}(-\rho)^{\tau}|\bar{\partial}^*_{\phi}\alpha|^2 e^{-\phi} \\ &+ \tau^2\int_{\mathcal{D}}(-\rho)^{\tau-2}|\Sigma\rho_j\alpha_j|^2 e^{-\phi} \\ &+ 2\mathrm{Re}\ \tau\int_{\mathcal{D}}(-\rho)^{\tau-1}\bar{\partial}^*_{\phi}\alpha\overline{\Sigma\rho_j\alpha_j}e^{-\phi} \\ &+ \int_{\mathcal{D}}(-\rho)^{\tau}|\bar{\partial}\alpha|^2 e^{-\phi} . \end{aligned}$$

Now we let $\tau \downarrow 1$ after moving the third term on the left hand side to the right. Combined with the second term on the left this gives

$$\lim_{\tau\to 1}(\tau^2-\tau)\int_{\mathcal{D}}(-\rho)^{\tau-2}|\Sigma\rho_j\alpha_j|^2 e^{-\phi}\ dS$$

which equals

$$\int_{\partial\mathcal{D}} |\alpha_h|^2 e^{-\phi}|\partial\rho|dS .$$

This proves (2) and hence (3).

<u>Proposition 6</u>: <u>Let</u> $(g_{i\bar{j}}) = (-\rho)(\phi_{i\bar{j}}) + (\rho_{i\bar{j}})$ <u>and</u> $(g^{i\bar{j}}) = (g_{i\bar{j}})^{-1}$. <u>Suppose</u> ϕ,ρ <u>are plurisubharmonic</u>. <u>Then</u> $\exists u$ <u>such that</u> $\bar{\partial}_b u = f$ <u>and</u>

(4) $$\int_{\partial\mathcal{D}} |u|^2 \frac{e^{-\phi}}{|\partial\rho|} \le \int_{\mathcal{D}} \Sigma g^{i\bar{j}} f_i \bar{f}_j e^{-\phi}$$

<u>provided the right hand side is finite</u>.

<u>Proof</u>: If $\bar{\partial}\alpha = 0$ and $\bar{\partial}_\phi^* \alpha = 0$ we have

$$\left|\int_{\mathcal{D}} f\cdot\bar{\alpha} e^{-\phi}\right|^2 \le \int_{\mathcal{D}} \Sigma g^{i\bar{j}} f_i \bar{f}_j e^{-\phi} \int_{\mathcal{D}} [(-\rho)\Sigma\phi_{i\bar{j}}\alpha_i\bar{\alpha}_j + \Sigma\rho_{j\bar{k}}\alpha_j\bar{\alpha}_k] e^{-\phi}$$

$$\le \int_{\mathcal{D}} \Sigma g^{i\bar{j}} f_i \bar{f}_j e^{-\phi} \int_{\partial\mathcal{D}} |\alpha_n|^2 |\partial\rho| e^{-\phi} dS,$$

by (3). By (the proof of) Proposition 4, this gives (4).

In particular if $(\rho_{i\bar{j}}) \ge \delta I$ we get a solution u such that

(5) $$\int_{\partial\mathcal{D}} |u|^2 \frac{e^{-\phi}}{|\partial\rho|} dS \le \frac{1}{\delta} \int_{\mathcal{D}} |f|^2 e^{-\phi}$$

but using the full strength of $(\phi_{i\bar{j}})$ we can get better estimates, e.g., one can take

$$\phi = -(-\rho)^\eta \quad \text{for} \quad \eta > 0 .$$

Observe also that when $n = 1$ and ϕ is bounded (3) implies Carleson's inequality.

<u>Section 3. Remarks on Estimates in other norms</u>.

Our first observation is that we can relate L^∞-estimates to weighted L^2-estimates.

<u>Proposition 7</u>: <u>There is a</u> u <u>such that</u> $\bar{\partial}_b u = f$ <u>and</u> $\|u\|_\infty \le C$ <u>iff</u> $\forall\phi \in C^\infty(\partial\mathcal{D})$ <u>there is a</u> u <u>(which may depend on</u> ϕ), <u>such that</u> $\bar{\partial}_b u = f$ <u>and</u>

(1) $$\int_{\partial\mathcal{D}} |u|^2 e^{-\phi} dS \le C^2 \int_{\partial\mathcal{D}} e^{-\phi} dS .$$

<u>Remark</u>: Intuitively, the non-trivial direction is an instance of the von Neumann mini-max theorem. To explain this, let

$$F(u,w) = \int_{\partial} |u|^2 w dS$$

be defined on the set of u's such that $\bar{\partial}_b u = f$, and the set of $w > 0$ such that

$$\int_{\partial\mathcal{D}} w dS = 1 .$$

The statement that for each w there is a solution to $\bar{\partial}_b u = f$ with

$$\int_{\partial\mathcal{D}} |u|^2 w \le C^2$$

means that

$$\max_w \min_u F(u,w) \le C^2 .$$

Then the mini-max theorem tells us that

$$\min_u \max_w F(u,w) \le C^2$$

which means that there is <u>one</u> u that works for <u>all</u> w (at least if we replace C by $C+\varepsilon$). But such a u must obviously be bounded. However, in order to be able to apply the mini-max theorem one needs, of course, certain assumptions about compactness, etc., and rather than trying to push such an argument through we give a simple direct proof:

Assume that for each ϕ there is a solution u such that (1) holds. In order to prove that there is a solution with $\|u\|_\infty \le C$ we need to show that

(2) $$\left|\int_{\mathcal{D}} f\cdot\bar{\alpha}\right| \le C\int_{\partial\mathcal{D}} |\alpha_{\hbar}|dS \quad \forall\alpha \quad \text{such that} \quad \bar{\partial}_0^*\alpha = 0 .$$

If u is an arbitrary solution to $\bar{\partial}_b u = f$ we have

$$\int_{\mathcal{D}} f\cdot\bar{\alpha} = \int_{\partial\mathcal{D}} u\bar{\alpha}_{\hbar}\, dS \quad \text{if} \quad \bar{\partial}_0^*\alpha = 0$$

Cauchy's inequality gives, for any ϕ,

$$\left|\int_{\partial\mathcal{D}} u\bar{\alpha}_{\hbar}\, dS\right|^2 \le \int_{\partial\mathcal{D}} |u|^2 e^{-\phi} dS \int_{\partial\mathcal{D}} |\alpha_{\hbar}|^2 e^{-\phi} dS .$$

Now, choose ϕ so that

$$e^{-\phi} = |\alpha_{\hbar}| + \varepsilon$$

and u to be a solution satisfying (1) for this ϕ. Then we get

$$\left|\int_{\partial\mathcal{D}} u\bar{\alpha}_{\hbar}\, dS\right|^2 \le C^2\left(\int_{\partial\mathcal{D}} (|\alpha_{\hbar}| + \varepsilon)dS\right)\left(\int_{\partial\mathcal{D}} \frac{|\alpha_{\hbar}|^2}{|\alpha_{\hbar}|+\varepsilon}\, dS\right)$$

and letting $\varepsilon \longrightarrow 0$ we obtain (2).

Unfortunately it is not at all clear how to prove estimates like (1). Any $\phi \in C^\infty(\partial\mathcal{D})$ can be extended to a plurisubharmonic function in $\mathcal{D}$, but the methods developed in the previous section, estimate the L^2-norm of u on the boundary in terms of an integral involving ϕ over the domain, and not as we would like an integral over the boundary. In one variable, this does not matter since we can choose the extension of ϕ to be harmonic, and apply a Carleson-type inequality for functions to land on the boundary, but in higher dimensions this does not work.

Finally we formulate a conjecture about what an optimal L^∞ result could be. To motivate the condition we have in mind we recall the theorem of Wolff (see [3] p. 322).

Theorem: Suppose f is a $(0,1)$ form in the unit disc and that

$$(1-|\zeta|^2)|f|^2 \in \mathcal{C} \quad \text{and}$$
$$(1-|\zeta|^2)|\partial f| \in \mathcal{C} .$$

Then $\exists u \in L^\infty(\partial\Delta)$ such that $\bar\partial_b u = f$.

Let us now make the same assumption as Wolff but replace the class of Carleson measures $\mathcal{C}$ by $\mathcal{C}^*$. We thus assume

(3) $\quad \exists \phi \in L^\infty(\Delta)$ such that $|f|^2 \le \Delta\phi$ and $|\partial f| \le \Delta\phi$

We note in passing that under this assumption one actually gets a solution u which is bounded in all of Δ, and not just on the boundary. To see this, consider

$$v = u - G[\frac{\partial f}{\partial \zeta}]$$

whereas before G is the Green potential, and u is the solution furnished by Wolff's theorem. Then v is harmonic and bounded on the boundary, hence everywhere. But since $|\partial f| \le \Delta\phi$, $\phi \in L^\infty$, the Green potential is also bounded in Δ. Hence u is bounded.

Now consider a $\bar\partial$-closed $(0,1)$ form f in $\mathcal{D}$. We say that f satisfies condition A if there is a bounded plurisubharmonic function ϕ in $\mathcal{D}$, such that

(i) $\quad \forall a \in \mathbb{C}^n \; |\Sigma f_j \bar a_j|^2 \le \Sigma \phi_{j\bar k} a_j \bar a_k$

(ii) $\quad \forall a,b \in \mathbb{C}^n \; \left|\Sigma \frac{\partial f_j}{\partial \zeta_k} \bar a_j b_k\right|^2 \le \Sigma\phi_{j\bar k} a_j \bar a_k \, \Sigma\phi_{j\bar k} b_j \bar b_k$

uniformly in $\mathcal{D}$. Notice that for $n = 1$ this is just (3). Our conjecture is that if f satisfies A then there is a $u \in L^\infty(\partial\mathcal{D})$ such that $\bar\partial_b u = f$.

Using the same argument as Wolff one can show that if the conjecture holds in a domain $\mathcal{D}$, then the Corona theorem for two generators is true in that domain. Thus it would be of interest to know if the conjecture holds even in the planar case (when $\mathcal{D}$ is multiply connected).

By essentially repeating Wolff's argument for the disc, but

replacing the reference to Carleson measures by the methods indicated in section 2, one can show that if f satisfies A, then for each ϕ on $\partial\mathcal{D}$ which has a plurisubharmonic extension to $\mathcal{D}$ there is a solution u to $\bar\partial_b u = f$, such that (1) holds.

As a final remark we notice that condition A can be reformulated as follows: The function ϕ defines a Kähler metric in $\mathcal{D}$ with fundamental form

$$i\partial\bar\partial\phi$$

(By adding $|\zeta|^2$ if necessary, we can always assume that $i\partial\bar\partial\phi$ is strictly positive.) The condition is then precisely that both forms f and ∂f are uniformly bounded when measured in this metric.

References

[1] Amar, E., Bonami, A., Mesures de Carleson d'ordre α et estima-estimations de solutions du $\bar\partial$. Bull. Soc. Math. France 107 (1979).

[2] Berndtsson, B., Ransford, T., Analytic multifunctions, the $\bar\partial$-equation and the Corona theorem. To appear in Pac. J. Math. (1986).

[3] Garnett, J., Bounded analytic functions, Academic Press, (1981).

[4] Hörmander, L., L^2-estimates and existence theorems for the $\bar\partial$-operator, Acta Math. 113 (1965).

[5] Hörmander, L., L^p-estimates for (pluri)subharmonic functions, Math. Scand. 20 (1967).

[6] Skoda, H., Valeurs au bord pour les solutions de l'operateur d″, et characterisation de zeros des founctions de la classe de Nevanlinna, Bull. Soc. Math. France 104 (1976).

Department of Mathematics
University of Goteborg and CTH
41296 Goteborg
Sweden

BOUNDARY ABSOLUTE CONTINUITY OF FUNCTIONS IN THE BALL ALGEBRA

Joaquim Bruna

1. This is a report of joint work with Prof. P. Ahern (Madison).

A well-known theorem of Privalov states that if f is a holomorphic function in the unit disc Δ and $f' \in H^1$, i.e.

$$\sup_{o<r<1} \int_o^{2\pi} |f'(re^{i\theta})| d\theta < +\infty$$

then f has absolutely continuous boundary values. We can view this result as the limiting case of the following one: if $f' \in H^p$, $p>1$, then f satisfies a Lipschitz condition with exponent $1-1/p$. In the papers [2], [4], the n-dimensional situation is considered. If f is holomorphic in the unit ball B^n of $\mathbb{C}^n$ and if

$$Rf(z) = \sum_1^n z_j \frac{\partial f}{\partial z_j}(z)$$

denotes the radial derivative of f, it is shown that the condition $Rf \in H^p$, i.e.

$$\sup_{o<r<1} \int_S |Rf(s)|^p \, d\sigma(s) < +\infty.$$

(here $S = bB^n$ and $d\sigma$ is Lebesque measure on S), implies, if $p>n$, that f satisfies a Lipschitz condition with exponent $1-n/p$. The limiting case appears thus to be $Rf \in H^n$, but it was shown in [3] that this condition does not even imply $f \in H^\infty$. Recently, F. Beatrous has proved in [1] that the slightly stronger assumption $R^n f \in H^1$ does imply continuity up to the boundary. Moreover, he proves that the restriction of f to any real-analytic transverse curve $\varphi(t)$ on S (i.e., the tangent $\varphi'(t)$ is never complex orthogonal to $\varphi(t)$) is absolutely continuous.

We improve Beatrous results in two directions. First, we prove something stronger than continuity up to the boundary. Namely, if

$$f(z) = \sum_{k=o}^{\infty} f_k(z)$$

is the homogeneous expansion of f, we prove that

$$\sum_k \|f_k\|_\infty < +\infty.$$

(the one dimensional case is of course a consequence of Hardy's inequality).

Secondly, we show that if $\varphi: I \to S$ is <u>any</u> C^1(simple) curve on S, then $f(\varphi(t))$ is absolutely continuous.

In this communication we will sketch the proof of this second result. The details and the study of other regularity properties of these functions will be the object of a forthcoming publication.

2. Let $\varphi: I \to S$ be a C^1 simple curve on S. The idea in studying $f(\varphi(t))$ is classical. Namely we will look at

$$u(r,t) = f(r\varphi(t)) \ , \ u_r(t) = u(r,t)$$

and prove that u_r (which converges uniformly to $f(\varphi(t))$ as $r \to 1$) is "uniformly absolutely continuous", i.e. if

$$H^*(t) = \sup_{o<r<1} \left| \frac{\partial u}{\partial t}(r,t) \right|$$

then $H^* \in L^1(dt)$. The hypothesis is on $F = R^n f$, hence we will look first to $\frac{\partial}{\partial t}(F(r\varphi(t)))$. If Π denotes the surface

$$\Pi = \{ r\varphi(t), \ o \le r \le 1, \ t \in I \}$$

and V the vector field whose value at $r\varphi(t)$ is $\varphi'(t)$, the partial derivative is in fact VF on Π. In the next paragraph we will consider an arbitrary $F \in H^1$, an arbitrary V, and estimate VF in B^n.

3. At this point we need to recall the maximal characterization of H^1. For $\zeta \in S$ and $\alpha > 1$, write

$$D_\alpha(\zeta) = \{ z \in B^n : |1 - z\bar{\zeta}| < \frac{\alpha}{2}(1 - |z|^2) \}.$$

This is the admissible approach regain with vertex ζ. For a function $h: B^n \to \mathbb{C}$, let

$$M_\alpha h(\zeta) = \sup \{ |h(z)| : z \in D_\alpha(\zeta) \}$$

be the corresponding maximal function. Then an holomorphic function in B^n belongs to H^1 if an only if $M_\alpha h \in L^1(S)$ ([5, Thm 5.6.5]).

Suppose now that $F \in H^1$ and V is a vector field on B^n. We want to estimate the function VF(z) in B^n. More concretely, we will make an estimate of $M_\alpha(VF)(\zeta)$. Let $z \in D_\alpha(\zeta)$, $z=r\eta$, $\eta \in S$, let v the value at z of V and assume without loss of generality that $|v|=1$. The classical idea for bounded functions is to estimate the size of the largest complex disc that centered at z in the direction of v is inside B^n, and then use Cauchy inequality. Here the growth condition is in terms of the $M_\beta F$. Hence it is more reasonable here to pick $\beta>\alpha$ and place this disc inside $D_\beta(\zeta)$. Some easy computations show that the size of this disc is

$$\varepsilon(\alpha,\beta)\frac{1-r^2}{(1-r^2)^{1/2}+|v\cdot\bar{\eta}|}.$$

So the size is comparable to $(1-r^2)$ if $v\cdot\bar{\eta}=1$ to $(1-r^2)^{1/2}$ if $r\cdot\bar{\eta}=0$, which are the two extreme well-known cases.

The Cauchy inequality along this disc gives then

$$|VF(z)|\frac{1-r^2}{(1-r^2)^{1/2}+|V(z)\cdot\bar{\eta}|} \le \varepsilon(\alpha.\beta)^{-1}\, M_\beta F(\zeta),\quad z=r\eta\in D_\alpha(\zeta)$$

To put it more compactly, define G=G(F,V) by

$$(1)\quad G(z) = |VF(z)|\,\frac{1-r^2}{(1-r^2)^{1/2}+|V(z)\cdot\bar{\eta}|},\qquad z=r\eta\in B^n.$$

The above says that for $\alpha<\beta$ there is $C(\alpha,\beta)$ such that

$$M_\alpha G \le C(\alpha,\beta)\, M_\beta F,$$

that is, concerning the growth condition, G is as nice as F: $M_\alpha G \in L^1(d\sigma), \forall\alpha$.

4. The next step is to use Carleson's measures. Recall that a positive measure on B^n is called a <u>Carleson</u> measure if for $\zeta\in S$ and $\delta>o$,

$$\mu(B(\zeta,\delta)) \le C\delta^n \quad \text{where } B(\zeta,\delta) = \{z\in B^n : |1-z\bar{\zeta}|<\delta\}.$$

Carleson operate measures on H^1. In fact, as showed by Fefferman and Stein,

(2) $$\int_{B^n} |h|d\mu \le C_\alpha \int_S M_\alpha h \, d\sigma$$

whenever μ is a Carleson measure.

In our case we were interested in the values of VF on Π when V is $\varphi'(t)$ on $r\varphi(t)$. It seems thus quite natural to look for Carleson measures supported on Π. We introduce

$$T(t) = |\varphi'(t)\cdot\overline{\varphi(t)}\,|, \quad t\in I$$

and the measure μ supported on Π defined by

$$\int hd\mu = \int_0^1 \int_I h(r\varphi(t))\,(1-r)^{n-2}\{(1-r)^{1/2} + T(t)\}\,drdt.$$

Then it turns out that $d\mu$ is a Carleson measure.

5. The final step is to use (2) for the measure μ above and the function G defined in (1), with $V=\varphi'(t)$. The result is

$$\int_0^1 \int_I |\tfrac{\partial}{\partial t} F(r\varphi(t)))|(1-r)^{n-1}\,drdt < +\infty.$$

Applying this to $F=R^n f$, and with the notations of paragraph 2, we obtain

$$\int_0^1 \int_I |\frac{\partial^{n+1}}{\partial t\partial r^n}\mu(r,t)|(1-r)^{n-1}\,drdt < +\infty.$$

Finally, it is easy to see that this implies

$$\int_0^1 \int_I |\frac{\partial^2}{\partial t\partial r}\mu(r,t)|\,drdt < +\infty.$$

which in turn implies $H^*\in L^1(dt)$.

References

[1] F. Beatrous, "Boundary continuity of holomorphic functions in the ball", to appear in Proc. Amer. Math. Soc.

[2] I. Graham, "The radial derivative, fractional integrals and the comparative growth of means of holomorphic functions on the unit ball in $\mathbb{C}^n$", Recent developments in Several Complex Variables, Annals of Math. Studies 100(1981), 171-178.

[3] I. Graham, "An H^p-space theorem for the radial derivatives of holomorphic functions on the unit ball in $\mathbb{C}^n$", preprint.

[4] S. Krantz, "Analysis on the Heisenberg group and estimates for functions in Hardy classes of several complex variables", Math. Ann. 244(1979), 243-262.

[5] W. Rudin, "Function theory in the unit ball of $\mathbb{C}^n$", Springer-Verlag, New York 1980.

Universitat Autonoma de Barcelona, Spain
State University of New York at Albany

SUBHARMONIC FUNCTIONS AND MINIMAL SURFACES

Urban Cegrell

1. Introduction.

Subharmonic functions have been used in the study of minimal surfaces already by Rado [5] and Bechenbach and Rado [1]. The purpose of this paper is to study (pluri-)subharmonic functions having graphs with smallest positive area.

Let U be a domain in $\mathbb{R}^n$ or $\mathbb{C}^n$ and let S be a weak*-closed set of subharmonic functions on U. If b is a given continuous function on ∂U, the problem is to find $h \in S$ with $\lim_{\substack{z \in U \\ z \to x}} h(z) = b(x)$, $\forall x \in \partial U$ and such that h minimizes graph area among the functions in S.

2. Existence of solution.

Theorem 1. Let U be a bounded subset of $\mathbb{R}^n$ or $\mathbb{C}^n$. Let b be a continuous function on ∂U and assume there is a convex function φ, which is the envelope of the affine functions it dominates, and a harmonic function h such that $\lim_{\substack{z \in U \\ z \to x}} \varphi(z) = \lim_{\substack{z \in U \\ z \to x}} h(z) = b(x)$, $\forall x \in \partial U$.

Assume that S is a weak*-closed family of subharmonic functions containing all the convex functions and stable under sup. Then there is a function U in S which minimizes graph area among the functions in S.

Proof: Let S_1 denote the functions in S that are minimized by φ and majorized by h. Then S_1 is weak*-compact and we prove first that there is a $v \in S_1$ that minimizes $A(v) = \int \sqrt{1+|\text{grad } v|^2}$. Put $\alpha = \inf_{v \in S_1} A(v)$ and choose $v_j \in S_1$, such that $A(v_j) \searrow \alpha$ as $j \to +\infty$.

Observe that since $v_j \in SH \cap L^\infty_{loc}(U)$ $|\text{grad } U_j|^2 \in L^1_{loc}(U)$ and we can assume that $v_j \to v \in S_1$ (otherwise, pick a subsequence). We claim that $A(v) = \alpha$.

Now $A(v) \geq \alpha$ by the definition of α. Let U' be a relatively compact subset of U. We can then select convex combinations $w_k = \sum_j \theta_j^k v_j$ where $0 \leq \theta_j^k \leq 1$, $\theta_j^k = 0$, $j \leq k$ and $\sum_j \theta_j^k = 1$ such that w_k tends strongly to w in potential theoretic sense (cf. Landkof

[4]). Therefore,

$$\int\sqrt{1+|\text{grad } v|^2}-\int\sqrt{1+|\text{grad } w_k|^2} \le m(U')\left(\int_{U'}\sqrt{1+|\text{grad}(v-wk)|^2}\right)^2 \to 0 ,$$

as $k \longrightarrow \infty$. But since $\sqrt{1+|x|^2}$ is convex we get

$$\int_{U'}\sqrt{1+|\text{grad } v|^2} = \lim_{n\to+\infty}\int_{U'}\sqrt{1+|\text{grad } w_k|^2}$$

$$\le \lim_{k\to+\infty}\sum_j \theta_j^k \int_{U'}\sqrt{1+|\text{grad } v_j|^2} \le \lim_{k\to+\infty}\int_{U'}\sqrt{1+|\text{grad } v_k|^2} \le \alpha.$$

Thus $\int_{U'}\sqrt{1+|\text{grad } v|^2} \le \alpha$ for every relatively compact U' in U which proves the claim.

To prove the theorem, it is enough to prove that if $v \in S$, $\lim_{\substack{z\to x\\ z\in U}} v(z) = b(x)$, $\forall x \in \partial U$ and

$$\int_U\sqrt{1+|\text{grad } v|^2} = \alpha$$

then $v \ge \varphi$.

To get a contradiction, assume that $P_\varepsilon = \{z \in U \text{ , } v \le \varphi-\varepsilon\} \ne \emptyset$ for any $\varepsilon > 0$. Then there is a $z_0 \in P_\varepsilon$ and an affine function g such that $g \le \varphi$ on U with equality at z_0. Therefore $v_\varepsilon = \max(v,g-\frac{\varepsilon}{2}) \in S$, $\lim_{\substack{z\to x\\ z\in U}} v_\varepsilon(z) = b(x)$ and

$$\int_U\sqrt{1+|\text{grad } v_\varepsilon|^2} = \int_{v_\varepsilon=v}\sqrt{1+|\text{grad } v_\varepsilon|^2} + \int_{v_\varepsilon<v}\sqrt{1+|\text{grad } v_\varepsilon|^2} <$$

$$< \int_U\sqrt{1+|\text{grad } v|^2} = \alpha,$$

since

$$\int_{v<v_\varepsilon}\sqrt{1+|\text{grad } v_\varepsilon|^2} < \int_{v<v_\varepsilon}\sqrt{1+|\text{grad } v|^2} ,$$

which is a contradiction.

3. Convex functions.

Theorem 2. Let b be continuous on ∂U, where U is an open, bounded and strictly convex set. Then the function

$$V(z) = \sup\{v(z) : v \in CVX(U) : \overline{\lim_{\substack{z\to x\\ z\in U}}}\, v(x) \le b(x)\ \forall x\in U\}$$

has the following properties:

i) V is continuous on $\bar{U}$, convex on U and equal to b on ∂U; V is the largest convex function with these properties.

ii) The real Monge-Ampère of V vanishes (i.e., $\det\left(\frac{\partial^2 V}{\partial x_i \partial x_j}\right) = 0$ in the generalized sense).

iii) The graph of V has the smallest area within the class of convex functions with boundary values b.

Proof: For i) and ii) see [6] and the references therein. Now iii) follows from i) and Theorem 1.

Corollary 1. Let U be bounded and strictly convex domain and S a weak*-closed set of subharmonic functions on U, stable under sup and containing all convex functions. Then for every $b \in C(\partial U)$ a function $v \in S$ minimizing graph area within the functions in S with boundary values b.

Proof: Since every bounded strictly convex is regular for the Dirichlet problem, the corollary follows from Theorem 1 and Theorem 2.

4. (Pluri-)subharmonic functions.

Let U be a bounded domain, regular for the Dirichlet problem and let S be a convex and weak*-compact family of functions. Assume that $b \in C(\partial U)$ and consider

$$S_b = \{\varphi \in S : \lim_{\substack{z\in U\\ z\to x}} \varphi(z) = b(x)\ ,\ \forall x \in \partial U\}$$

(also assuming S_b contains at least two different functions). Then if $u,v \in S_b \cap C^2$ we have

$$A(tu+(1-t)v) = \int_U \sqrt{1+|\mathrm{grad}(tu+(1-t)v)|^2}$$

and

$$\frac{dA}{dt}(tu(1-t)v) = \int_U \frac{(v-u)M(ta+(1-t)v)}{(1+|\mathrm{grad}(ta-(1-t)v|^2)^{3/2}} \qquad (*).$$

Here, $M(u) = (1+|\text{grad } u|^2)\Delta u - D_iuD_juD_{ij}u$ (the minimal surface operator).

Note that if u is convex then $D_iuD_juD_{ij}u \le |\text{grad } u|^2\Delta u$ so $M(u) \ge 0$ which gives another proof of iii) in Theorem 2. Now, if u is minimizing graph area within S_b, $\left.\frac{dA}{dt}\right|_{t=1} \le 0$ so (*) gives that

$$\int_U \frac{(v-u)M(u)}{(1+|\text{grad } u|^2)^{3/2}} \le 0 \qquad \text{(i)}.$$

Also, note that if $ta+(1-t)v \in S_b$ for t near 1 then $\left.\frac{dA}{dt}\right|_{t=1} = 0$ (ii). With notation as above, and with S = all the subharmonic functions we have the following classification (where u is assumed to be minimizing and C^2):

1) If $M(u) \le 0$ then $M(u) \equiv 0$ so u is the **unique** minimal surface with boundary b. For take $v \in S_b$, $v < u$. Then i) gives $M(u) \equiv 0$.

2) If $M(u)$ has no definite sign, then $\text{supp}\,\Delta u \subset \{M(u) \le 0\}$. For let $z_0 \in \text{supp}\,\Delta u$ be given. If we had $M(u)(z_0 > 0$ on the ball $B(z_0,r)$ then (*) gives

$$\frac{d}{dt}\int_{B(z_0,r)} \sqrt{1+|\text{grad}(tu+(1-t)H)|^2} = \int_{B(z_0,r)} \frac{(H-u)M(u)}{(1+|\text{grad } u|^2)^{3/2}} > 0$$

where H is harmonic on $B(z_0,r)$ with boundary values u. This means that there are points $t_0 < 1$ with

$$\int_{B(z_0,r)} \sqrt{1+|\text{grad}(t_0u+(1-t)t_0H|^2} < \int_{B(z_0,r)} \sqrt{1+|\text{grad } H|^2}.$$

Moreover, since $t_0u+(1-t_0)H \ge u$ on $B(z_0,r)$ with equality on $\partial B(z_0,r)$ the function

$$\Psi = \begin{cases} u & \text{on } U\backslash B(z_0,r) \\ t_0u+(1-t_0)H & \text{on } B(z_0,r) \end{cases}$$

is subharmonic on U, equals b on ∂U and strictly less graph area than u - a contradiction.

Proposition 1. Let H now be the harmonic function with boundary values b. Then if $M(u) \gneq 0$ it follows that $u \equiv H$. Furthermore, if $M(H) \ge 0$, then $u \equiv H$.

Proof: Assume that $M(u) \ge 0$. Then (ii) gives that

$\int_U \frac{(H-u)M(u)}{(1+|\text{grad } u|^2)^{3/2}} = 0$ and since either $H = u$ or $H > u$ on U the assumption $M(u) \gneq 0$ gives $H \equiv u$.

Assume now that $M(H) \geq 0$ and consider $P(t) = A(ta+(1-t)H)$ which is a convex function on $[0,1]$ and

$$P'(0) = \int_U \frac{(H-u)M(H)}{(1+|\text{grad } H|^2)^{3/2}} \geq 0.$$

Therefore $P'(t) \geq 0$, $t \in [0,1]$ so since u is minimizing $P' \equiv 0$. Therefore $H \equiv u$, for if $u < H$ then $M(H) \equiv 0$ so as in 1) we conclude that H is the unique minimal surface.

Definition: A plurisubharmonic function H on U is called **locally minimizing** if for every open set $U' \subset U$, H minimizes graph area among the plurisubharmonic function on U' with boundary values H.

Theorem 3. Let H be a function that is C^2, plurisubharmonic and satisfies $(dd^cH)^n = 0$ on U. Then H is locally minimizing if and only if $M(H) \geq 0$. Furthermore, if $M(H) > 0$ then H is uniquely determined.

Proof: If U' is open in U and if v is plurisubharmonic on U' with boundary values H on $\partial U'$ then $v \leq H$ on U' since $(dd^cH)^k = 0$ (cf. [2]).

Therefore, if H is locally minimizing $M(H) \geq 0$ by (i). On the other hand, if we defined

$$P(t) = \int_{U'} \sqrt{1+|\text{grad}(tv+(1-t)H)|^2}\ , \quad t \in [0,1]$$

then by (*)

$$P'(t) = \int_{U'} \frac{(H-v)M(tv+(1-t)H)}{(1+|\text{grad}(tv+(1-t)H)|^2)^{3/2}}$$

so since P is convex and $P'(0) \geq 0$ it follows that $P(0) \leq P(1)$ so H is minimizing. Furthermore, if v is also minimizing $P(0) = P(1)$ and $P' \equiv 0$ so $M(H) > 0$ only if $u \equiv H$.

5. Examples.

Example 1. Consider the harmonic function $H(x,y) = xy$ in the first quadrant. Then $M(x,y) = -2xy$ so if U is any domain in the first quadrant then it follows from 1) that H cannot minimize graph area among the subharmonic functions on U with boundary values $H|_{\partial U}$. If we also assume that U is strictly convex it is well known that there is a unique C^∞ solution m to $M(u) = 0$; $u = H$ on ∂U; m is the solution to Plateau's problem (cf. [3]). Since $M(H) \le 0$ it follows that $m \le H$ on U (cf. [3]) so there are points z_0 where $\Delta m(z_0) > 0$ because the maximum principle for subharmonic would otherwise give that $m \ge H$. If we take U' to be a strictly convex set contained in $\{z \in U;\ \Delta m > 0\}$ we get an example where $u = m$ is locally minimizing but not harmonic.

Example 2. Let $u(x) = \frac{1}{2}\log \sum_{j=1}^{n} x_j^2$, $x \in \mathbb{R}^n$, $n \ge 2$. Then $u_i = \frac{x_i}{\Sigma x_j^2}$, $u_{ii} = \frac{1}{\Sigma x_j^2} - \frac{2x_i^2}{(\Sigma x_j^2)^2} = \frac{\Sigma x_j^2 - 2x_i^2}{(\Sigma x_j^2)^2}$ and for $i \ne j$ $u_{ij} = \frac{-2x_ix_j}{(\Sigma x_j^2)^2}$ so therefore

$$M(u) = \left(1+\frac{1}{\Sigma x_j^2}\right)\frac{n-2}{\Sigma x_j^2} - \sum_i \frac{(x_j^2-2x_i^2)x_i^2}{(\Sigma x_i^2)^4} + \sum_{i\ne j} \frac{2x_i^2x_j^2}{(\Sigma x_j^2)^4} = \frac{n-2}{\Sigma x_j^2} + \frac{n-3}{(\Sigma x_j^2)^2} +$$

$$\frac{2}{(\Sigma x_i^2)^4}\left[\sum_{i=1}^{n} x_i^4 + \sum_{i\ne j} x_i^2x_j^2\right] = \frac{n-2}{\Sigma x_j^2} + \frac{n-3}{(\Sigma x_j^2)^2} + \frac{2}{(\Sigma x_j^2)^2} = \frac{n-2}{\Sigma x_i^2} + \frac{n-1}{(\Sigma x_i)^2} > 0$$

for $x \ne 0$.

It is well known that in $\mathbb{C}^n$, $m \le n$, $(dd^c\frac{1}{2}\log \sum_{j=1}^{m} |z_j|^2)^n = 0$ outside $\sum_{j=1}^{m} |z_j|^2 = 0$ so by Theorem 3 and the above computation, $\frac{1}{2}\log \sum_{j=1}^{m} |z_i|^2$ is locally minimizing.

Example 3. Let D be the unit disc in $\mathbb{C}$ and take U to be $D\times D\setminus e^{-1}D\times e^{-1}D$. Let u be a graph area minimizing plurisubharmonic function with boundary values 0 on $\partial(D\times D)$ and -1 on $e^{-1}D\times e^{-1}D$. Since both $\max(\log|z_1|,-1)$ and $\max(\log|z_2|,-1)$ are locally minimizing by Example 2, it follows that $u \ge \max(\log|z_1|,\log|z_2|,-1)$. But the right hand side is the largest plurisubharmonic functions with the given boundary values so

$$u = \max(\log|z_1|,\log|z_2|,-1).$$

References

[1] Bechenbach, E.F., and Rado, T., Subharmonic functions and minimal surfaces. Trans. Amer. Math. Soc. vol. 35 (1933), 648-661.

[2] Bedford, E., and Taylor, B.A., The Dirichlet problem for a complex Monge-Ampère equation. Invent. Math. 37, (1976), 1-44.

[3] Gilbarg, D., and Trudinger, N.S., Elliptic Partial Differential Equations of Second Order. Grundlehren der mathematischen Wissenschaften 224, Springer-Verlag Berlin, Heidelberg, New York, Tokyo, 1983.

[4] Landkof, N.S., Foundations of Modern Potential Theory. Grundlehren der mathematischen Wissenschaften 180, Springer-Verlag Berlin, Heidelberg, New York, 1972.

[5] Rado, T., On the problem of Plateau. Verlag von Julius Springer, Berlin 1933. Reprint Springer-Verlag Berlin, Heidelberg, New York, 1971.

[6] Rauch, I., Taylor, B.A., The Dirichlet problem for the multidimensional Monge-Ampère equation. Rocky Mountain Journal of Mathematics 7 no. 2 (1977), 345-364.

Department of Mathematics
University of Umeå
S-90187 Umeå
Sweden

PROPRIETES DE RECOUVREMENT DES SOUS-ENSEMBLES DE LA FRONTIERE D'UN DOMAINE STRICTEMENT PSEUDO-CONVEXE

Anne-Marie Chollet

I - Introduction.

1°) D Désigne un domaine borné de $\mathbb{C}^n$ à frontière ∂D de classe C^∞. On note $\overline{D} = D \cup \partial D$.

E est un sous-ensemble fermé de ∂D de mesure nulle pour la mesure superficielle σ.

On s'intéresse au problème suivant: étant donné un tel ensemble E, trouver des conditions suffisantes sur E pour qu'il soit l'ensemble des zéros d'une fonction F holomorphe dans D possédant des propriétés de régularité dans $\overline{D}$. On considèrera des classes de fonctions comme $A(D)$ la classe des fonctions holomorphes dans D et continues dans $\overline{D}$, comme $A^\infty(D)$ la classe des fonctions holomorphes dans D, continues ainsi que toutes leurs dérivées dans $\overline{D}$ ou comme $G_{1+1/\alpha}(D)$, $0 < \alpha < 1$, la classe des fonctions holomorphes dans D, vérifiant une condition de Gevrey d'indice $1+1/\alpha$ dans $\overline{D}$.

Dans tous les cas, on construit une fonction $\varphi(z)$ holomorphe dans D, de classe C^∞ dans $\overline{D}\backslash E$, de partie réelle strictement positive dans $\overline{D}\backslash E$ et tendant vers $+\infty$ lorsque z tend vers E. Alors, si on pose, pour tout z de $\overline{D}\backslash E$,

$$F(z) = \exp - \varphi(z),$$

il est clair que F peut être étendue en une fonction continue sur $\overline{D}$, nulle sur E. Un bon contrôle de la croissance des dérivées de $\varphi(z)$ lorsque z tend vers E permet éventuellement de montrer que F s'étend en une fonction de classe C^∞ dans $\overline{D}$ dont toutes les dérivées sont nulles sur E.

II - Cas du disque unité du plan complexe.

2 - Rappels P. Fatou en 1906 [8] a montré que tout sous-ensemble fermé de

mesure nulle sur le cercle unité $\mathbb{T}$ est un ensemble de zéros pour $A(D)$ lorsque D est le disque unité du plan complexe en prenant pour φ une fonction définie pour z dans D par

$$\varphi(z) = \frac{1}{2\pi} \int_0^{2\pi} \frac{e^{ix}+z}{e^{ix}-z}\, h(x)\, dx.$$

La fonction $h(x)$ est ici une fonction de $L^1(\mathbb{T})$, positive, de classe C^∞ sur $\mathbb{T}|E$ et tendant vers $+\infty$ lorsque x tend vers E.

Cette méthode qui utilise très fortement le fait que la partie réelle de φ est l'intégrale de Poisson de h a été reprise par de nombreux auteurs, toujours dans le cas $n = 1$. Pour des références bibliographiques plus complètes, on pourra consulter [4].

C'est en 1971 que B. Korenblum [10] a eu l'idée de représenter φ par une série en utilisant un recouvrement de Whitney de $\complement E$ le complémentaire de E dans $\mathbb{T}$. Il considère en effet une suite d'intervalles I_ν de centre ζ_ν et de rayon r_ν tels que l'on ait

$$\complement E = \bigcup_\nu I_\nu$$

et, si $x \in I_\nu$, $d(x,E) \sim r_\nu$.

Alors, si E vérifie l'hypothèse

$$(c_1) \qquad \sum_\nu r_\nu \operatorname{Log} \frac{1}{r_\nu} < \infty \qquad \text{et} \quad \sigma(E) = 0$$

et, si on pose, pour tout z de D,

$$\varphi(z) = \sum_\nu \lambda_\nu\, r_\nu \operatorname{Log} \frac{1}{r_\nu} \quad \frac{r_\nu}{r_\nu+1-\bar{\zeta}_\nu z}$$

où λ_ν est une suite croissante de réels positifs tendant vers $+\infty$ et tels que la série $\sum_\nu \lambda_\nu r_\nu \log \frac{1}{r_\nu} < \infty$, la fonction $F = \exp - \varphi$ appartient à $A^\infty(D)$ et s'annule exactement sur E. De plus toutes les dérivées de F sont nulles sur E.

L'usage d'un recouvrement de Whitney de $\complement E$ permet l'extension de cette définition de φ dans le cas d'un domaine strictement pseudoconvexe de $\mathbb{C}^n$ [2] mais ne fournit aisément le contrôle de la croissance de la partie réelle de $\varphi(z)$ que dans le cas où z tend vers E le long de

∂D. On utilise alors pour conclure la représentation de F comme intégrale de Poisson de ses valeurs au bord. Celle-ci est possible car la positivité dans D de $\operatorname{Re}\varphi$ implique l'appartenance de F à $H^\infty(D)$, la classe des fonctions holomorphes et bornées dans D.

On se propose maintenant de donner une troisième méthode [6] pour construire une fonction φ ayant les propriétés recherchées. On controlera ici très simplement la croissance de la partie réelle de $\varphi(z)$ lorsque z tend vers E dans $\bar{D}\setminus E$. Pour simplifier, on présentera encore cette méthode dans le cas du disque unité du plan complexe.

3 - Définition. Pour tout $\varepsilon > 0$, on désigne par $N_\varepsilon(E)$ le nombre minimal d'intervalles de rayon ε dont la réunion recouvre E. Sans restreindre la généralité, on peut supposer que ces intervalles ont leur centre sur E et que ces centres sont situés à des distances mutuelles supérieures ou égales à ε. On dira de ces intervalles qu'ils forment un ε-recouvrement de E.

4 - Théorème. La condition

$$(c_2) \qquad \int_0^1 N_\varepsilon(E)\, d\varepsilon < \infty$$

est suffisante pour qu'il existe une fonction F de $A^\infty(D)$ nulle seulement sur E. De plus toutes les dérivées de F s'annulent sur E.

Preuve. Si on note $N_{2^{-k}}(E) = N_k$, la condition (c_2) est équivalente à la convergence de la série $\sum_k 2^{-k} N_k$. Il existe donc une suite croissante de réels positifs λ_k tendant vers $+\infty$ tels que l'on ait

$$(4.1) \qquad \sum_k \lambda_k\, 2^{-k} N_k < \infty .$$

Soit φ, la fonction définie, pour tout z dans D, par

$$(4.2) \quad \varphi(z) = \sum_{k=1}^{\infty} \sum_{j=1}^{N_k} \lambda_k \frac{2^{-k}}{2^{-k}+1-\bar{\zeta}_{j,k} z}$$

où, pour chaque entier k, $(\zeta_{j,k})$, $j=1,\ldots,N_k$, désigne la suite des centres des intervalles $I_{j,k}$ d'un 2^{-k}-recouvrement de E.

De la convergene de la série (4.1), on déduit que φ est holomorphe donc de classe C^∞ au voisinage de tout point de $\bar{D}\setminus E$ et que, de plus, pour tout p de $\mathbb{N}$ il existe une constante $C(p)$ telle que, pour tout z de $\bar{D}\setminus E$, on ait

$$(4.3) \qquad |\varphi^{(p)}(z)| \leq \frac{C(p)}{d(z,E)^{p+1}} .$$

Soit z un point de $\bar{D}\setminus E$ vérifiant $d(z,E) \leq 1/4$, alors il existe un entier k_z , $k_z > 1$ tel que l'on ait

$$(4.4) \qquad 2^{-k_z-1} < d(z,E) \leq 2^{-k_z} .$$

On note w_z un point de E réalisant

$$d(z,E) = d(z,w_z) .$$

Alors pour chaque entier k , w_z appartient à au moins un intervalle du 2^{-k}-recouvrement correspondant. Comme, pour tout ζ de ∂D, la partie réelle de $H_\zeta(z) = 1 - \bar{\zeta}z$ est strictement positive dans $\bar{D}\setminus\{\zeta\}$, on peut minorer $\operatorname{Re}\varphi(z)$, en ne gardant, pour chaque k , dans la sommation portant sur j , qu'un seul intervalle du 2^{-k}-recouvrement, un des intervalles qui contient w_z et dont on désigne le centre par $\zeta_{j(z,k),k}$. On a donc

$$\operatorname{Re}\varphi(z) \geq \sum_{k=1}^{\infty} \lambda_k \operatorname{Re} \frac{2^{-k}}{2^{-k}+1-\bar{\zeta}_{j(z,k),k}z} .$$

On vérifie aisément que, pour tout k , $k \leq k_z$, on a

$$\operatorname{Re} \frac{2^{-k}}{2^{-k}+1-\bar{\zeta}_{j(z,k),k}z} \geq 1/4$$

et donc, si on désigne par $[k_z/2]$ la partie entière de $k_z/2$, on a

$$\operatorname{Re}\varphi(z) \geq \frac{1}{4} \sum_{k=[k_z/2]}^{k_z} \lambda_k \geq \frac{1}{4} [k_z/2] \lambda_{[k_z/2]} .$$

Or, d'après (4.4), on a

$$k_z \sim \operatorname{Log} \frac{1}{d(z,E)} \quad \text{et} \quad \lambda_{[k_z/2]} = \omega(d(z,E))$$

où $\omega(x)$ tend vers $+\infty$ lorsque x tend vers 0 d'après la définition de la suite λ_k.

Il existe donc une constante $c > 0$ telle que, pour tout z de $\bar{D}\setminus E$, on ait

$$\text{(4.5)} \qquad \operatorname{Re} \varphi(z) \geqslant c\omega(d(z,E)) \log \frac{1}{d(z,E)} \ .$$

Si, on note, pour tout z de $\bar{D}\setminus E$,

$$F(z) = \exp - \varphi(z)$$

on conclut que F est holomorphe dans D , de classe C^∞ dans $\bar{D}\setminus E$ et que l'on a

$$|F(z)| \leqslant d(z,E)^{c\omega(d(z,E))}$$

c'est-à-dire que $|F(z)|$ tend vers 0 plus vite que n'importe quelle puissance de $d(z,E)$ lorsque z tend vers E. On déduit donc de là, en utilisant (4.3) que l'on peut étendre F en une fonction de $C^\infty(\bar{D})$, nulle seulement sur E et dont toutes les dérivées s'annulent sur E.

5 - Remarque. On sait que, lorsque D est le disque unité du plan complexe, le condition $\sigma(E) = 0$ est nécessaire et suffisante pour que E soit l'ensemble des zéros d'une fonction de $A(D)$. On sait aussi, toujours dans ce cas, que les conditions (c_1) et (c_2) sont équivalentes et caractérisent les ensembles de zéros de $A^\infty(D)$.

III - Cas de certains domaines de $\mathbb{C}^n$, $n > 1$.

6 - Dans tout ce qui suit, D désigne un domaine borné de $\mathbb{C}^n$, à frontière ∂D de classe C^∞. D est donc défini par la donnée d'une fonction r de classe C^∞ dans $\mathbb{C}^n$ telle que l'on ait

(1) $\quad D = \{z \ ; \ r(z) < 0\}$,

(2) $\quad \operatorname{grad} r \neq 0$ sur ∂D.

Si β est un réel strictement positif, on note

$$D_\beta = \{z \ ; \ r(z) < \beta\}.$$

La méthode décrite précédemment s'étend à tout domaine D de $\mathbb{C}^n$ à condition qu'il vérifie les propriétés (H_1), (H_2) et (H_3) suivantes :

(H_1) : Il existe sur $\bar{D} \times \bar{D}$ une fonction $\rho(z,w)$ positive telle que l'on ait

(a) $\rho(z,w) = 0$ si et seulement si $(z,w) \in \partial D \times \partial D$ et $z = w$,

(b) $\rho(z,w) = \rho(w,z)$ pour z et w dans $\bar{D}$,

(c) il existe une constante $K > 0$ telle que l'on ait

$\rho(z,t) \leqslant K[\rho(z,w) + \rho(w,t)]$ pour z,t et w dans $\bar{D}$.

On dit alors que ρ définit une pseudo-distance sur ∂D.

(H_2) : Il existe $\beta > 0$ et il existe sur $\partial D \times D_\beta$ une fonction $H(\zeta,z)$ telle que

(a) H est de classe C^∞ dans $\partial D \times D_\beta$ et, pour chaque ζ de ∂D, la fonction $H_\zeta : z \to H_\zeta(z) = H(\zeta,z)$ est holomorphe dans D_β.

(b) Pour tout ζ de ∂D, $H_\zeta(z)$ ne s'annule dans $\bar{D}$ que pour $z = \zeta$ et $\mathrm{Re}\, H_\zeta$ est strictement positive dans $\bar{D} \setminus \{\zeta\}$.

On dit que, pour tout ζ de ∂D, H_ζ est une fonction-support au point ζ.

(H_3) : Il existe des constantes A et B, strictement positives, et δ, $\delta > 1$, telles que, pour tout (ζ,z) de $\partial D \times \bar{D}$, on ait

$$A\rho(\zeta,z)^\delta \leqslant |H(\zeta,z)| \leqslant B\,\rho(\zeta,z).$$

On sait que les conditions (H_1), (H_2) et (H_3) sont réalisées lorsque D est un domaine strictement pseudo convexe de $\mathbb{C}^n$ c'est-à-dire lorsque la fonction r qui le définit est strictement plurisousharmonique. Dans le cas de la boule unité de $\mathbb{C}^n$, par exemple, si on note $\langle\, ,\, \rangle$ le produit hermitien dans $\mathbb{C}^n$, on peut prendre

$$H(\zeta,z) = 1 - \langle z,\zeta \rangle \quad \text{et} \quad \rho(\zeta,z) = |H(\zeta,z)|.$$

7 - <u>Théorème</u> [5][7]. Soit D un domaine vérifiant les propriétés (H_1), (H_2) et (H_3). Soit E un sous-ensemble fermé de ∂D et $N_\varepsilon(E)$, le nombre minimal de boules pour la pseudo distance ρ sur ∂D, de rayon ε dont la réunion recouvre E, alors

1°) si on a $\lim_{\varepsilon\to 0} \varepsilon N_\varepsilon(E) = 0$, alors E est un ensemble de zéros pour $A(D)$,

2°) si on a $\int_0^1 N_\varepsilon(E)\, d\varepsilon < \infty$, alors E est un ensemble de zéros pour $A^\infty(D)$.

<u>Preuve</u>. 1) si on a $\lim_{\varepsilon\to 0} \varepsilon N_\varepsilon(E) = 0$, on a alors $\lim_{k\to\infty} 2^{-k} N_k = 0$ et il existe donc une suite d'entiers $k(i)$ tendant vers $+\infty$ en croissant telle que l'on ait

$$\sum_{i=1}^{\infty} 2^{-k(i)} N_{k(i)} < \infty .$$

On vérifie alors que, si on désigne, pour tout entier k , par $(\zeta_{j,k})$ $j=1,\dots,N_k$ la suite des centres des boules pour la pseudo distance ρ sur ∂D d'un 2^{-k}-recouvrement de E , la fonction

$$F(z) = \exp - \sum_{i=1}^{\infty} \sum_{j=1}^{N_{k(i)}} \frac{2^{-k(i)}}{2^{-k(i)}+H(\zeta_{j,k(i)},z)}$$

appartient à $A(D)$ et s'annule sur E.

2) Si on a $\int_0^1 N_\varepsilon(E)d\varepsilon < \infty$ et donc une suite de réels positifs λ_k tendant en croissant vers $+\infty$ tels que $\sum_k \lambda_k 2^{-k} N_k$ converge alors c'est la fonction

$$F(z) = \exp - \sum_{k=1}^{\infty} \lambda_k \sum_{j=1}^{N_k} \frac{2^{-k}}{2^{-k}+H(\zeta_{j,k},z)}$$

qui appartient à $A^\infty(D)$, s'annule exactement sur E et a toutes ses dérivées nulles sur E.

8 - <u>Théorème</u> [1]. Soit D un domaine borné strictement pseudoconvexe de $\mathbb{C}^n$ et E un sous-ensemble fermé de ∂D. On note $N_\varepsilon(E)$ le nombre minimal de boules pour la pseudo-distance usuelle sur ∂D, de rayon ε, dont la réunion recouvre E. Soit α , $0<\alpha<1$. S'il exsite une fonction δ positive et croissante sur $[0,1]$ telle que l'on ait

a) $\delta(x) - \delta(y) \leqslant \delta(x-y)$ pour tout (x,y) de $[0,1]^2$, $x > y$

b) $\int_0^1 \frac{\delta(t)}{t}\, dt < \infty$

c) $\int_0^1 \frac{N_\varepsilon(E)}{\varepsilon^{\alpha}\delta(\varepsilon)^{\alpha}} < \infty$

alors E est un ensemble de zéros pour $G_{1+1/\alpha}(D)$. Ce résultat généralise à $\mathbb{C}^n$ un théorème de S.V. Hruscev [9].

9 - Dans une perspective un peu différente en s'inspirant des idées de S.V. Hruscev on peut démontrer la proposition suivante

Proposition. Soit E un sous-ensemble fermé de la frontière d'un domaine D strictement pseudo-convexe de $\mathbb{C}^n$ et $N_\varepsilon(E)$ le nombre minimal de boules pour la pseudo distance usuelle sur ∂D dont la réunion recouvre E. On suppose que E vérifie la condition :

$$(K_\alpha) \qquad \int_0^1 (N_\varepsilon(E)\ \varepsilon^{1-\alpha})^{1/1+\alpha} \frac{d\varepsilon}{\varepsilon} < \infty$$

On suppose, de plus, qu'il existe des constantes β et C, strictement positives, tels que

(a) $\varepsilon \to N_\varepsilon(E).\ \varepsilon^{\beta}$ soit croissante

(b) pour tout r, $0 < r \leq 1$ et toute boule B_r, on ait

$$\int_{B_r} \rho(r,E)^{\frac{2-\beta}{1+\alpha} - n + 1}\, d\sigma(t) \leq C r^{1+\frac{2-\beta}{1+\alpha}}.$$

Alors E est un ensemble de zéros pour $G_{1+1/\alpha}(D)$.

Remarques. Si E est situé sur une courbe dont la tangente en chaque point est transverse au plan tangent complexe à ∂D, les conditions (a) et (b) sont trivialement vérifiées si l'on choisit $\beta = 2$. On retrouve ainsi dans le cas $n = 1$ le théorème 8.1 de S.V. Hruscev [9] et on le généralise à $\mathbb{C}^n$, $n > 1$.

Si on est dans $\mathbb{C}^2$ et si on pose $\beta = 1 - \alpha$, la condition (b) est clairement vérifiée. On obtient alors un résultat voisin de celui du théorème 8. On peut aussi montrer, toujours dans le cas $n = 2$, que le théorème 8 reste vrai sous les seules hypothèses 8 b) et 8 c).

On peut vérifier [9] que 8 b) et 8 c) impliquent (K_α) on ne sait pas, dans le cas général, si la condition (K_α) suffit pour assurer que E soit un ensemble de zéros pour $G_{1+1/\alpha}$.

On rappelle enfin [3] qu'un sous-ensemble fermé E de ∂D vérifiant

$$(C_\alpha) \quad \int_0^1 N_\varepsilon(E)\, \varepsilon^{-\alpha}\, d\varepsilon < \infty$$

et

(c) il existe une constante $C > 0$ telle que, pour tout r, $0 < r < 1$, et toute boule B_r on ait

$$\int_{B_r} \rho(r, E)^{-\alpha - n + 1}\, d\sigma(r) \leq C r^{1-\alpha}$$

est un ensemble de zéros pour $G_{1+1/\alpha}(D)$.
La condition (C_α) est moins restrictive que (K_α) alors que la condition (c) est plus forte que (b).

IV - Une extension de la méthode de construction.

10 - Dans le paragraphe 6, pour chaque ζ de ∂D, la fonction H_ζ est une fonction support au point ζ. On peut adapter la construction précédente dès lors que l'on est assuré de l'existence de fonctions supports le long de compacts du bord de ∂D.

11 - Lemme. Soit D un domaine borné strictement pseudo-convexe de $\mathbb{C}^n$. Soit p un point de ∂D. Il existe ω un voisinage de p dans ∂D et il existe une famille de sous-variétés de classe C^∞, $(N_x)_{x \in [0,1]^n}$ de dimension réelle $n-1$, dépendant régulièrement de x tels que, $\bigcup_{x \in [0,1]^n} N_x$ soit un feuilletage de ω et que, pour tout $x \in [0,1]^n$ et pour

tout p de N_x, l'espace tangent en p à N_x soit contenu dans l'espace tangent complexe à ∂D.

12 - Lemme [6] les notations sont celles du lemme précédent. Pour tout compact K de ω, il existe une fonction $H(x,z)$ définie sur $[0,1]^n \times D_\beta$ vérifiant les propriétés suivantes.

a) H est de classe C^∞ dans $[0,1]^n \times D_\beta$ et, pour chaque x de $[0,1]^n$, la fonction $H_x : z \to H_x(z) = H(x,z)$ est holomorphe dans D_β.

b) Pour tout x de $[0,1]^n$, H_x ne s'annule dans $\bar{D}$ que sur $K_x = K \cap N_x$ et $\operatorname{Re} H_x$ est strictement positive dans $\bar{D} \setminus K_x$.

13 - Théorème [6] les notations et les hypothèses sont celles des lemmes 11 et 12. Soit $E = \{x \in [0,1]^n \,;\, K_x \neq \emptyset\}$.

On désigne par $N_\varepsilon(E)$ le nombre minimal de boules euclidiennes de rayon ε dont la réunion recouvre E. Alors

1°) si on a $\lim\limits_{\varepsilon \to 0} \varepsilon N_\varepsilon(E) = 0$ alors K est un ensemble de zéros pour $A(D)$,

2°) si on a $\int_0^1 N_\varepsilon(E)\, d\varepsilon < \infty$, alors K est un ensemble de zéros pour $A^\infty(D)$.

Preuve. Si on désigne, pour tout entier k, par $(x_{j,k})$ $j=1\ldots N_k$ la suite des centres des boules euclidiennes d'un 2^{-k}-recouvrement de E, la fonction de $A(D)$ qui s'annule sur K s'il vérifie 1°) n'est autre que

$$F(z) = \exp - \sum_{i=1}^{\infty} \sum_{j=1}^{N_{k(i)}} \frac{2^{-k(i)}}{2^{-k(i)} + H(x_{j,k(i)}, z)}$$

où $(k(i))_{i \in \mathbb{N}}$ est une suite croissante d'entiers vérifiant $\sum_{i=1}^{\infty} 2^{-k(i)} N_{k(i)} < \infty$. Si K vérifie l'hypothèse 1°) et si $(\lambda_k)_{k \in \mathbb{N}}$ est une suite de réels tendant en croissant vers $+\infty$ tels que $\sum_{k=1}^{\infty} \lambda_k 2^{-k} N_k$ converge alors

$$F(z) = \exp - \sum_{k=1}^{\infty} \lambda_k \sum_{j=1}^{N_k} \frac{2^{-k}}{2^{-k}+H(x_{j,k},z)}$$

appartient à $A^{\infty}(D)$, s'annule exactement sur K et a toutes ses dérivées nulles sur K.

Bibliographie

[1] Ababou-Boumaaz R. Ensembles de zéros et ensembles pics pour des classes de fonctions holomorphes dans des domaines strictement pseudo-convexes C.R. Acad. Sc. Paris 302 (1986) p.507-510.

[2] Chollet A.M. Ensembles de zéros à la frontière de fonctions analytiques dans des domaines strictement pseudo-convexes Ann. Inst. Fourier 26 (1976) p.51-80.

[3] Chollet A.M. Zéros à la frontière de fonctions analytiques d'une ou plusieurs variables complexes. Thèse Orsay 1976.

[4] Chollet A.M. Carleson sets in $\mathbb{C}^n$, $n \geq 1$. Aspects of contemporary Complex Analysis. Proceedings London Math. Soc. Durham 1980.

[5] Chaumat J. et Chollet A.M. Ensemble de zéros et d'interpolation à la frontière de domaines strictement pseudo-convexes Ark.for Mat. 1986 (à paraître).

[6] Chaumat J. et Chollet A.M. Dimension de Hausdorff des ensembles de zéros et d'interpolation pour $A^{\infty}(D)$. Trans. Math. Soc. (à paraître).

[7] Davie A.M. et Øksendal B.K. Peak interpolation sets for some algebras of analytic functions Pacific J. math. 41 (1972) p.81-87.

[8] Fatou P. Séries trigonométriques et séries de Taylor Acta. Math. 30 (1906) p.335-400.

[9] Hruscev S.V. Sets of uniqueness for the Gevrey classes Ark. Mat. 15 (1977) p.253-304.

[10] Korenbulm B.I. Holomorphic functions in a disk and smooth in its closure Soviet Math. Dokl. 12 (1971) p.1312-1315.

Université de Paris-Sud
Unité Associée 757
Analyse Harmonique
Mathématique (Bat. 425)
91405 Orsay Cedex
FRANCE

SOME PROPERTIES OF THE CANONICAL MAPPING OF A COMPLEX SPACE INTO ITS SPECTRUM

S.Coen (*)

SUMMARY. *Conditions on the cohomology and on the singular locus of a complex space* X *are given for the canonical mapping of* X *into its spectrum being surjective or a homeomorphism. Especially, the case of the unbounded dimension is studied.*

Let $(X, \mathcal{O})$ be a complex space and ${}_X\xi : X \to \mathrm{Sp}\, X$ be the canonical mapping of X in its spectrum Sp X. If $\dim X < \infty$, it is known that ξ is a homeomorphism iff X is Stein and that X is Stein iff X is holomorphically separable and $H^q(X, \mathcal{O})$ has countable dimension as a vector space on $\mathbb{C}$ for every $q \geq 1$ (see |F|, |J|). When the dimension of X is unbounded, some results of Markoe |M| and Ephraim |E| are known about the Michael's conjecture on X; but, to our knowledge, little more is known (e.g. see |H| pag.50).

In the first part of this paper some properties of the mapping ξ are given for Stein spaces of unbounded dimension. The following result (see (1.6)) is proved together with various of its improvements. *Let* X *be a reduced holomorphically spreadable complex space. Assume that the singular locus* S(X) *of* X *is a Stein subspace which verifies Michael's conjecture* (e.g. S(X) is a finite dimensional

(*) This work was partly supported by M.P.I. (40%, 60%) and by C.N.R..

Stein space). *Then* X *is a Stein space iff* $\xi : X \to \mathrm{Sp}\, X$ *is a homeomorphism or iff* $\dim_{\mathbb{C}} H^q(X, \mathcal{O})$ *is countable for every* $q \geq 1$. Later we shall see that *if* X *is a Stein space, then* $\xi : X \to \mathrm{Sp}\, X$ *is a homeomorphism* (equivalently: X verifies Michael's conjecture) *iff* ${}_{S(X)}\xi : S(X) \to \mathrm{Sp}\, S(X)$ *is a homeomorphism* (equiv: S(X) verifies Michael's conjecture).

In |H| some interesting characterizations of the surjectivity of ξ onto the continous spectrum are given in the finite dimensional case. Here, from another point of view we are interested in studying how cohomological conditions regarding the structure sheaf imply the surjectivity of ξ and whether it is possible to use rather elementary methods.

Effectively in the last paragraphs we find some results improving |M| , |E| without explicit Steinness assumptions. Although some techniques of |B| , |E| and |J| are used, we do not employ the results of |J| or of |FN| in the main result which is the following one. *Let* $(X, \mathcal{O})$ *be a holomorphically spreadable complex space; let* $\dim_{\mathbb{C}} H^q(X, \mathcal{O})$ *be countable for every* $q \geq 1$. *Let* L *be a closed subspace of* X *such that the connected components of* $X \setminus L$ *are finite dimensional open subspaces of* X . *Let* $\phi \in \mathrm{Sp}\, X$. If $h \in \mathcal{O}(X)$ *exists s.t.* $\phi(h) \neq 0$ *and* $h(y) = 0$ *for every* $y \in L$, *then* ϕ *is a point character*, i.e. $x \in X \setminus L$ *exixts s.t.* $\phi(f) = f(x)$ *for every* $f \in \mathcal{O}(X)$. Some consequences about the surjectivity of ξ are deduced from this result.

Some of the results here obtained were the subject of the exposé |C| .

§ 0. PRELIMINARIES

Complex spaces $(X, \mathcal{O})$ that we are going to consider, have countable Hausdorff topology; they may be nonreduced. A *character* ϕ of $(X, \mathcal{O})$ is a homomorphism $\phi : \mathcal{O}(X) \to \mathbb{C}$ of $\mathbb{C}$-algebras, $\phi(1) = 1$; we do not assume ϕ

continous. The *spectrum* of $(X, \mathcal{O})$ is the space of all the characters of $(X, \mathcal{O})$; it will be denoted by $\operatorname{Sp} X$. The *continous spectrum* $\operatorname{Sp}_c X$ of X is defined by $\operatorname{Sp}_c X = \{ \phi \in \operatorname{Sp} X \mid \phi$ is continuous $\}$. We say that $(X, \mathcal{O})$ *verifies Michael's conjecture* if $\operatorname{Sp} X = \operatorname{Sp}_c X$ (i.e. if every character of X is continuous). For every $\phi \in \operatorname{Sp} X$, $\varepsilon > 0$ and $f_1, \dots, f_s \in \mathcal{O}(X)$ we set

$$W(\phi;(f_1,\dots,f_s), \varepsilon) := \{\sigma \in \operatorname{Sp} X \mid |\sigma(f_j) - \phi(f_j)| < \varepsilon \quad j = 1,\dots,s\}$$

The topology of $\operatorname{Sp} X$ is defined such that for every $\phi \in \operatorname{Sp} X$ the family of the sets $W(\phi;(f_1,\dots,f_s);\varepsilon)$, when $f_1,\dots,f_s \in \mathcal{O}(X)$ and $\varepsilon \in \mathbb{R}^+$, is a fundamental system of neighborhoods of ϕ in $\operatorname{Sp} X$. The *canonical mapping* ${}_X\xi = \xi : X \to \operatorname{Sp} X$ associates each $x \in X$ with the *point character* ${}_X\xi_x$ where for every $f \in \mathcal{O}(X)$ we have ${}_X\xi_x(f) = f(x)$ (value of f in x); ξ is continuous.

For every $f \in \mathcal{O}(X)$ let us set

$$V(f;X) := \{x \in X \mid f(x) = 0\} \quad .$$

Let us recall some known results that will be used later.

(0.1) THEOREM (|F| , |GR|). *Let* $(X, \mathcal{O})$ *be a complex space of dimension* n. *The canonical mapping* $\xi : X \to \operatorname{Sp} X$ *is a homeomorphism iff* X *is a Stein space.*

By the methods employed by Siu |S| in the regular case, the following theorem can be deduced (see |B| , |C| , |J|):

(0.2) THEOREM (|S|). *Suppose* F *is a coherent analytic sheaf on a complex space* $(X, \mathcal{O})$. *Then for every non-negative integer* p , $H^p(X, F)$ *as a vector space on* $\mathbb{C}$ *cannot be countably infinite dimensional.*

We need also the following theorem proved by Siu-Trautmann (see Theor. (1.11), (1.18) of |ST| and |BS|).

(0.3) THEOREM (|ST|) *Suppose that* $(X, \mathcal{O})$ *is a complex space and let* F *be a coherent analytic sheaf on* X. *For every non-negative integer* k *the "singularity subvariety"*

$$S_k(F) := \{x \in X \mid \text{prof } F_x \leq k\}$$

is a subvariety of X *and* $\dim S_k(F) \leq k$.

If $f \in \mathcal{O}(X)$, f_x *is not a zero-divisor of* F_x *for every* $x \in X$ *iff*

$$\dim(V(f;\, X) \cap S_{k+1}(F)) \leq k \qquad \textit{for all } k.$$

(for the meaning of prof F_x see e.g. |BS|).

The following is Theor. 2.3 of | E | .

(0.4) THEOREM (|E|). *Let* X *be a Stein space. Suppose that one can find a sequence of analytic subsets of* X, $X = Y_o \supset Y_1 \supset \ldots \supset Y_n = \emptyset$ *such that for any* i, $0 \leq i < n$, *all the connected components of* $Y_i \setminus Y_{i+1}$ *are finite dimensional. Then every multiplicative linear functional* $\alpha : \mathcal{O}(X) \to \mathbb{C}$ *is continuous*

Finally we state the theorem of Jennane, quoted in the introduction.

(0.5) THEOREM (|J|). *Let* X, Ω *be analytic spaces and let* $\pi : X \to \Omega$ *be an analytic morphism with Stein fibres. Assume that* Ω *is holomorphically separable, that* $\dim X < \infty$ *and that* $\dim_{\mathbb{C}} H^p(X, \mathcal{O})$ *is countable for* $1 \leq p \leq n-1$. *Then* X *is a Stein space.*

Let us recall that a complex space $(X, \mathcal{O})$ is said to be *holomorphically spreadable* if for every $x \in X$ an open neighborhood A_x of x in X and $f_1, \ldots, f_{s(x)} \in \mathcal{O}(X)$ exist such that

$$A_x \cap \{ y \in X \mid f_1(y) = \ldots = f_{s(x)}(y) = 0\} = \{x\}$$

It is known (Grauert's Theorem, see |W|) that if $(X, \mathcal{O})$ is a holomorphically spreadable complex space of finite dimension then a holomorphic mapping $g: X \to \mathbb{C}^n$ is given such that all fibres of g are discrete subsets of X. In particular theorem (0.5) holds (without any assumption on Ω) if X is a holomorphically spreadable complex space of finite dimension.

§ 1 THE CASE OF STEIN SPACES

(1.1) THEOREM. *Let* $(X, \mathcal{O})$ *be a reduced holomorphically spreadable complex space. Assume the singular locus* $S(X)$ *of* X *be a Stein subspace of* X . *The following conditions are equivalent*

a) X *is a Stein space;*

b) $H^q(X, \mathcal{O}) = 0$ *for every* $q \geq 1$

c) $\dim_{\mathbb{C}} H^q(X, \mathcal{O})$ *is countable for every* $q \geq 1$.

Proof

By Cartan-Serre's Theorem B we only have to prove c) ==> a).

Let $\{A_i\}_{i \in I}$ be the family of the irreducible components of X; for every $i \in I$ we set

$$B_i := \bigcup_{j \neq i, j \in I} A_j$$

For every $i \in I$ we denote by $\mathcal{A}_i$ (resp: $\mathcal{B}_i$) the nullstellen ideal sheaf of A_i (resp: B_i). Moreover we set

$$\mathcal{O}_i' := \mathcal{O}/\mathcal{A}_i \quad , \quad \mathcal{O}_i'' := \mathcal{O}/\mathcal{B}_i \quad , \quad \mathcal{O}_i''' := \mathcal{O}/\mathcal{A}_i + \mathcal{B}_i$$

for every $i \in I$ and we denote by $\pi_i: \mathcal{O} \to \mathcal{O}_i'$, $\sigma_i: \mathcal{O} \to \mathcal{O}_i''$, $\tau_i: \mathcal{O} \to \mathcal{O}_i'''$ the corresponding residue epimorphisms.

For every $i \in I$ let us consider the following sequence

$$0 \to \mathcal{O} \xrightarrow{\alpha_i} \mathcal{O}_i' \oplus \mathcal{O}_i'' \xrightarrow{\beta_i} \mathcal{O}_i''' \to 0$$

where $\alpha_i(f) = (\pi_i(f), \sigma_i(f))$, $\beta_i(\pi_i(h), \sigma_i(g)) = \tau_i(h) - \tau_i(g)$. We have

$\mathscr{A}_i \cap \mathscr{B}_i = (0)$ for every $i \in I$; then it is a straightforward verification that these Mayer-Vietoris sequences are exact.

For every $i \in I$ we have

$$A_i \cap B_i = \text{zero-set of } (\mathscr{A}_i + \mathscr{B}_i)$$

Thus, if $\mathscr{C}_i$ is the nullstellen ideal sheaf of $A_i \cap B_i$ in X, by Hilbert's Nullstellensatz we have $\mathscr{C}_i = \mathrm{rad}(\mathscr{A}_i + \mathscr{B}_i)$. If $\mathscr{N}_i$ is the nilradical of $\mathscr{O}_i'''$ we have

$$\mathscr{N}_i = \mathscr{C}_i / \mathscr{A}_i + \mathscr{B}_i$$

Hence, the reduction of the space $(A_i \cap B_i, \mathscr{O}_i'''|_{A_i \cap B_i})$ is the space $(A_i \cap B_i, (\mathscr{O} / \mathscr{C}_i)|_{A_i \cap B_i})$.

Now, we have $A_i \cap B_i \subset S(X)$ for every $i \in I$, then $(A_i \cap B_i, (\mathscr{O} / \mathscr{C}_i)|_{A_i \cap B_i})$ as a closed subspace of a Stein space is itself a Stein space. By a well-known (see |GR|) theorem of Grauert $(A_i \cap B_i, \mathscr{O}_i'''|_{A_i \cap B_i})$ is a Stein space, too.

By Cartan-Serre's Theorem B we have

$$H^q(X, \mathscr{O}_i''') = H^q(A_i \cap B_i, \mathscr{O}_i'''|_{A_i \cap B_i}) = 0 \quad q \geq 1,\ i \in I.$$

The exact long cohomology sequences associated with the short ones considered above are the following ones

$$H^1(X, \mathscr{O}) \to H^1(X, \mathscr{O}_i' \oplus \mathscr{O}_i'') \simeq H^1(A_i, \mathscr{O}_i'|_{A_i}) \oplus H^1(B_1, \mathscr{O}_i''|_{B_i}) \to 0 \quad \ldots\ldots \to$$
$$\to \ldots\ldots\ldots \to$$
$$\to H^q(X, \mathscr{O}) \to H^q(A_i, \mathscr{O}_i'|_{A_i}) \oplus H^q(B_i, \mathscr{O}''_i|_{B_i}) \to 0 \to \ldots\ldots$$

The exactness of these sequences implies that $H^q(A_i, \mathscr{O}_i'|_{A_i})$ as a vector space on $\mathbb{C}$ has countable dimension for $q \geq 1$, $i \in I$. Hence, by Theorem (0.5), $(A_i, \mathscr{O}_i'|_{A_i})$ is a Stein space for every $i \in I$.

Now a Narasimhan's Theorem (see also |AN|) assures that X is a Stein space.

q.e.d.

The following remark will be improved later in (3.7).

(1.2) REMARK. *Let* $(X, \mathcal{O})$ *be a Stein space. Then the following conditions ase equivalent*

a) $(X, \mathcal{O})$ *verifies Michael's conjecture;*

b) *the canonical mapping* $\xi: X \to \mathrm{Sp}\, X$ *is surjective;*

c) *the canonical mapping* $\xi: X \to \mathrm{Sp}\, X$ *is a homeomorphism.*

Proof.

a) ==> b). Let $\phi \in \mathrm{Sp}\, X$ be a character of $(X, \mathcal{O})$. Let us consider the zero-set of ker ϕ

$$V(\ker \phi) = \{x \in X \mid f(x) = 0 \text{ for every } f \in \ker \phi\}$$

Of course, ker ϕ is a proper ideal of $\mathcal{O}(X)$ because we assume $\phi \neq 0$. By a), ker ϕ is a closed proper ideal of $\mathcal{O}(X)$. Then, because $(X, \mathcal{O})$is a Stein space,by a well known Cartan's Theorem we have $V(\ker \phi) \neq \emptyset$. Let $x \in V(\ker \phi)$. For every $f \in \mathcal{O}(X)$ we have $f - \phi(f) \in \ker \phi$; hence $f(x) = \phi(f)$ for every $f \in \mathcal{O}(X)$ and $\phi = \xi_x$.

b) ==> c). It is enough to prove that ξ is open. For the sake of completeness we are going to repeat more or less the proof given in |GR| .

Let $(X, \mathcal{O}^R)$ be the reduced space associated with $(X, \mathcal{O})$. We may see directly that the reduction mapping Red: $\mathcal{O}(X) \to \mathcal{O}^R(X)$ induces a natural homeomorphism $\mathrm{Red}^* : \mathrm{Sp}(X, \mathcal{O}^R) \to \mathrm{Sp}(X, \mathcal{O})$.

We have $(\mathrm{Red}^*) \circ {}_{(X, \mathcal{O}^R)}\xi = {}_{(X, \mathcal{O})}\xi$. Then it sufficies to prove that ${}_{(X, \mathcal{O}^R)}\xi$ is an open mapping and we may assume without loss of generality that $(X, \mathcal{O})$ is a reduced space.

We are going to prove that for every $p \in X$ a fundamental system of open neighborhoods $\{U_j\}_{j \geq 1}$ of p in X is given such that for every $j \geq 1$ $\xi(U_j)$ is an open neighborhood of ξ_p in Sp X.

Let $p \in X$. Let X' be the closed analytic subset of X of all the irreducible components of X containing p; X' is a finite dimensional Stein space. We denote by X'' the closed analytic subset of X constituted by all the other irreducible components of X.

The imbedding Theorem for Stein spaces implies the existence of $f_1,\dots,f_m \in \mathcal{O}(X')$ such that $(f_1,\dots,f_m): X' \to \mathbb{C}^m$ is a homeomorphism onto a closed subspace of $\mathbb{C}^m$ and $f_1(p) = \dots = f_m(p) = 0$. Because X is a Stein space, every f_i may be extended to a function $F_i \in \mathcal{O}(X)$ and a function $F_{m+1} \in \mathcal{O}(X)$ is given with $F_{m+1}(p) = 0$, $F_{m+1}(x) = 1$ for every $x \in X''$.

For every $j \in \mathbb{N}$, $j > 1$, let us set

$$U_j := \{x \in X \mid |F_i(x)| < \frac{1}{j}\ ,\ i = 1,\dots,m+1\}$$

$$V_j := f^{-1}(P(\frac{1}{j}) \cap f(X'))$$

where $P(\frac{1}{j})$ is the polydisk $P(\frac{1}{j}) := \{z \in \mathbb{C}^m \mid |z_i| < \frac{1}{j}\ \ i = 1,\dots,m\}$.

The family $\{V_j\}_{j \geq 1}$ is a fundamental system of neighborhoods of p in X. We may observe that for every $j > 1$ we have

$$U_j = V_j \cap \{x \in X \mid F_{m+1}(x) < \frac{1}{j}\}$$

Then $\{U_j\}_{j > 1}$ is a fundamental system of neighborhoods of p in X.

Now it sufficies to remark that for every $j > 1$.

$$\xi(U_j) = W(\xi_p;(F_1,\dots,F_{m-1}); \frac{1}{j})$$

(1.3) PROPOSITION. *Let* $(X,\mathcal{O})$ *be a complex space. Let a sequence* $(X_o,\mathcal{O}_o),\dots,(X_n,\mathcal{O}_n) = (X,\mathcal{O})$ *of complex spaces be given such that for every* $i = 0,\dots,n-1$, $(X_1,\mathcal{O}_i)$ *is a closed complex subspace of* $(X_{i+1},\mathcal{O}_{i+1})$.

Let us assume that $\dim X_o < \infty$ *and that the connected components of* $X_i \setminus X_{i-1}$

are finite dimensional open subspaces of $(X_i, \mathcal{O}_i)$ *for every* $i = 1,\dots,n$.

The following conditions are equivalent

a) X *is a Stein space;*

b) *the canonical mapping* $\xi : X \to \operatorname{Sp} X$ *is a homeomorphism.*

Proof.

The implication b) ==> a) is known (see e.g. | GR |). Thus, by the previous remark, it is enough to prove that $(X, \mathcal{O})$ verifies Michael's conjecture. If follows from Ephraim's Theorem (0.4).

q.e.d.

Later in (3.4), we shall see that the condition $\dim X_o < \infty$ may be substituted by the requirement that the mapping ${}_{X_o}\xi : X_o \to \operatorname{Sp} X_o$ is surjective.

(1.4) REMARK. Let $(S_1, \mathcal{O}_1)$ be the singular locus of the reduction of the space $(X, \mathcal{O})$; for $j > 1$ let $(S_j, \mathcal{O}_j)$ be the singular locus of $(S_{j-1}, \mathcal{O}_{j-1})$. If an integer $n \geq 1$ exists such that $\dim S_n < \infty$, then the conditions of (1.3) are satisfied $(X_o = S_n, \dots, X_{n-1} = S_1)$.

Now, we are able to summarize some of the previous results by the following corollary.

(1.5) COROLLARY. *Let* $(X, \mathcal{O})$ *be a reduced holomorphically spreadable complex space. Let* $(S_1, \mathcal{O}_1)$ *be the singular locus of* $(X, \mathcal{O})$; *for* $j > 1$ *let* $(S_j, \mathcal{O}_j)$ *be the singular locus of* $(S_{j-1}, \mathcal{O}_{j-1})$.

Let us consider the following conditions;

a) $\dim_{\mathbb{C}} H^q(X, \mathcal{O})$ *is countable for every* $q \geq 1$;

b) *an integer* $n \geq 1$ *exists such that* $\dim_{\mathbb{C}} H^q(S_j, \mathcal{O}_j)$ *is countable for every* $q \geq 1$, $1 \leq j \leq n-1$ *and* $(S_n, \mathcal{O}_n)$ *is a Stein space;*

c) *an integer* $n \geq 1$ *exists such that* $\dim_{\mathbb{C}} H^q(S_j, \mathcal{O}_j)$ *is countable for every* $q \geq 1$, $1 \leq j \leq n$ *and* $\dim S_n < \infty$.

If conditions a),b) *hold,* $(X, \mathcal{O})$ *is a Stein space. If conditions* a), c) *hold,* $(X, \mathcal{O})$ *is a Stein space and the canonical mapping* $\xi: X \to \mathrm{Sp}\, X$ *is a homeomorphism.*

Proof.

Let a), b) hold. It is appropriate to set $(S_o, \mathcal{O}_o) := (X, \mathcal{O})$. By Theorem (1.1) the space $(S_{n-1}, \mathcal{O}_{n-1})$ is Stein. The iterated use of (1.1) implies the Steinness of $(X, \mathcal{O})$.

Let a), c) hold. By Jennane's Theorem (0.5) $(S_n, \mathcal{O}_n)$ is a Stein space, b) also holds and $(X, \mathcal{O})$ is a Stein space. By Theorem (0.1) and Proposition (1.3) ξ is a homeomorphism.

q.e.d.

(1.6) REMARK. For the sake of completeness we give a result which follows from our third paragraph and which improves slightly (1.5). Let us consider the following condition

d) an integer $n \geq 1$ exists such that $\dim_{\mathbb{C}} H^q(S_j, \mathcal{O}_j)$ is countable for every $q \geq 1$, $1 \leq j \leq n-1$ and $(S_n, \mathcal{O}_n)$ is a Stein space which verifies Michael's conjecture.

Let X *be a reduced holomorphically spreadable complex space. If conditions* a) *of* (1.5) *and* d) *are satisfied, then* X *is a Stein space and the canonical mapping* $\xi: X \to \mathrm{Sp}\, X$ *is a homeomorphism.*

Indeed, by (1.5) X is a Stein space; by (3.5) or (3.7) ξ is a homeomorphism.

§ 2 FIRST RESULTS AND LEMMATA IN THE GENERAL CASE.

(2.1) DEFINITION. *Let* $(X, \mathcal{O})$ *be a complex space. Let* x *be a point of* X *and* A_x *be an open neighborhood of* x *in* X. *The point* x *is said to be (holomorphically) defined by* $\mathcal{O}(X)$ *in* A_x *if finitely many sections* $f_1,\dots,f_s \in \mathcal{O}(X)$ *exist such that*

$$\{y \in A_x \mid f_1(y) = \dots = f_s(y) = 0\} = \{x\}.$$

Observe that $(X, \mathcal{O})$ is holomorphically spreadable iff every point $x \in X$ is defined by $\mathcal{O}(X)$ in some open neighborhood A_x of x in X.

(2.2) LEMMA. *Let* $(X, \mathcal{O})$ *be a holomorphically speadable complex space. Let* D *be a discrete subset of* X. *Let* $\{T_i\}_{i \in I}$ *be a countable family of irreducible closed analytic positive-dimensional subsets of* X.

For every $i \in I$ *let us pick a point* $x_i \in T_i$ *and an open neighborhood* U_i *of* x_i *in* X *in which* x_i *is defined by* $\mathcal{O}(X)$. *For every* $i \in I$ *let us pick a point* $y_i \in U_i \cap T_i$, $y_i \neq x_i$ *(it is always possible).*

Then a section $f \in \mathcal{O}(X)$ *exists such that*

a) $f(x_i) \neq f(y_i)$ *for every* $i \in I$

b) $f(x) \neq 0$ *for every* $x \in D$

c) $\dim(V(f;X) \cap T_i) < \dim T_i$ *for every* $i \in I$.

Proof.

For any $x,y \in X$ with $x \neq y$ let us set

$$F_{x,y} := \{f \in \mathcal{O}(X) \mid f(x) \neq f(y)\} = (\xi_x - \xi_y)^{-1}(\mathbb{C}^*)$$

$$F_x := \{f \in \mathcal{O}(X) \mid f(x) \neq 0\} = \xi_x^{-1}(\mathbb{C}^*).$$

Point characters are continous, then F_x and $F_{x,y}$ are open subsets of $\mathcal{O}(X)$ for any $x,y \in X$ with $x \neq y$. Furthermore we have $F_x \neq \emptyset$ for every $x \in X$.

Let us note that $F_{x,y}$ is a dense subset of $\mathcal{O}(X)$ if $F_{x,y} \neq \emptyset$. Indeed, the $\mathbb{C}$-linear mapping $\xi_x - \xi_y : \mathcal{O}(X) \to \mathbb{C}$ is surjective because if $g \in F_{x,y}$ then for any $c \in \mathbb{C}$ we have

$$c = (\xi_x - \xi_y)\left(\frac{c}{g(x) - g(y)}g\right).$$

Therefore, by the Banach Open Mapping Theorem, $\xi_x - \xi_y$ is an open mapping. Let A be an open subset of $\mathcal{O}(X)$; because of the property $(\xi_x - \xi_y)(A) \cap \mathbb{C}^* \neq \emptyset$, a section $h \in A$ exists s.t. $h(x) \neq h(y)$ i.e. we have $A \cap F_{x,y} \neq \emptyset$ as we wanted. In a similar manner it may be proved that F_x is dense in $\mathcal{O}(X)$.

Let us fix an index $j \in I$ and a point $x_j \in T_j$ Because X is spreadable it is possible to choose a neighborhood U_j as required in the statement Because $\dim (T_j)_{x_j} > 0$, x_j is not an isolated point in T_j, then we can pick a suitable point $y_j \in U_j \cap T_j$, $y_j \neq x_j$. The choice of U_j implies

$$F_{x_j, y_j} \neq \emptyset ;$$

then F_{x_j,y_j} is an open and dense subset of $\mathcal{O}(X)$.

By Baire's Theorem it follows that the set

$$F := \Big(\bigcap_{i \in I} F_{x_i, y_i} \Big) \bigcap \Big(\bigcap_{x \in D} F_x \Big)$$

is not empty . Any $f \in F$ satisfies conditions a), b).

From a) it follows that $V(f;X) \cap T_i$ is a proper analytic subset of T_i for every $i \in I$; hence also c) holds.

q.e.d.

By the previous lemma we can deduce a vanishing theorem for cohomology, that we shall use later. The theorem was proved in $|A_1|$ in the reduced case; one can see also |B| and |J| . In the special case in which X is an open subset of a finite dimensional perfect Stein space and $\mathcal{F}$ is locally free the proposition was proved in |AB| .

(2.3) PROPOSITION. *Let* $(X, \mathcal{O})$ *be a holomorphically spreadable complex space and* $\mathcal{F}$ *be a coherent analytic sheaf on* X; $\dim X = n < \infty$. *Assume that for a suitable integer* $q \geq 1$, $H^q(X,\mathcal{F})$ *is finite dimensional as a vector space on* $\mathbb{C}$ *and that* $H^{q+s}(X,\mathcal{F}) = 0$ *for* $s \geq 1$.

Then $H^q(X,F) = 0$.

Proof.

For any $k \in \mathbb{N}$ let us set

$$S_k := \{x \in X \mid \operatorname{prof} F_x \leq k\} .$$

We may suppose that X is a connected space.

The proof is by induction on the dimension n of $(X, \mathcal{O})$.

Let $n = 0$; our theorem holds because we have $H^j(X,F) = 0$ for every $j \geq 1$.

Let us suppose the theorem holds for spreadable spaces $(Y, {}_Y\mathcal{O})$ with $\dim Y \leq n-1$.

Let $\{Y_i\}_{i \in I}$ be the family of positive dimensional irreducible components of every S_k, $k > 0$; I is countable. For every $i \in I$, let us choose points x_i, y_i as in Lemma (2.2) and let $f \in \mathcal{O}(X)$ be a section chosen as in Lemma (2.2).

For every holomorphic "section" $g \in \mathcal{O}(\mathbb{C})$ we shall denote by $g \circ f \in \mathcal{O}(X)$ the section on X which is canonically associated with the morphism $(X, \mathcal{O}) \to \mathbb{C}$ obtained by composition of the morphisms $(X, \mathcal{O}) \to \mathbb{C}$, $\mathbb{C} \to \mathbb{C}$ respectively canonically associated with f and with g.

Now, we are going to remark that for every choice of $g \in \mathcal{O}(\mathbb{C})$, $g \neq 0$, the germ $(g \circ f)_x$ is not a zero-divisor for $x \in X \setminus S_o$. Indeed, if g is constant, but $g \neq 0$, then $(g \circ f)$ is invertible in $\mathcal{O}_x$ for every $x \in X$. Otherwise if g is not a constant function, g is open. Let as pick $k \geq 0$ and let T be an irreducible component of S_{k+1} of positive dimension. By the choice of f we know that f is nonconstant on T. Then f is open on T and $g \circ f$, as a mapping, is open on T. In particular, $g \circ f$ is nonconstant on T and we have

$$\dim(V(g \circ f;\ X) \cap T) < \dim T \leq k+1.$$

From this inequality, it follows for every $k \in \mathbb{N}$

$$\dim(V(g\circ f;\, X)\cap S_{k+1}) \leq k;$$

then, by Theorem (0.3), $(g\circ f)_x$ is not a zero-divisor of $\mathcal{F}_x$ for every $x \in X \setminus S_o$.

Let us consider $g \in \mathcal{O}(\mathbb{C})$, $g \neq 0$, and the morphism $\pi_g = \pi$

$$\pi : \mathcal{F} \to (g\circ f)\cdot \mathcal{F}$$

defined, for every $x \in X$, by $\pi_x(h_x) = (g\circ f)_x h_x$. From what we have just seen it follows that π_x is injective for $x \notin S_o$; then the support of $\ker \pi$ is discrete. For every $j > 1$ the homomorphism

$$\pi^j : H^j(X,\mathcal{F}) \to H^j(X,(g\circ f)\cdot \mathcal{F})$$

$\pi^j = \pi^j_g$, induced by π, is an isomorphism. In particular we have

$$H^{q+s}(X,(g\circ f)\cdot\mathcal{F}) = 0 \qquad \text{for } s \geq 1.$$

Let us set $Y := \{x\in X \mid g\circ f(x) = 0\}$, ${}_Y\mathcal{O} := (\mathcal{O}/(g\circ f)\,\mathcal{O})_{|Y}$; $(Y, {}_Y\mathcal{O})$ is a holomorphically spreadable complex space. As $g\circ f$ is nonconstant on any irreducible component of X of positive dimension, we have $\dim Y \leq n-1$.
Let us write $\mathcal{G} := (\mathcal{F}/(g\circ f)\cdot\mathcal{F})_{|Y}$ and remark that $\mathcal{G}$ is a ${}_Y\mathcal{O}$-coherent analytic sheaf on Y.

Now, let us consider the short exact sequence

$$0 \to (g\circ f)\cdot\mathcal{F} \xrightarrow{\sigma} \mathcal{F} \to \mathcal{F}/(g\circ f)\cdot\mathcal{F} \to 0$$

where $\sigma = \sigma_g$ is the canonical injection. By our hypotheses and by the cohomological properties already verified for the sheaf $(g\circ f)\cdot\mathcal{F}$, we have

$$H^{q+s}(Y,\mathcal{G}) \simeq H^{q+s}(X, \mathcal{F}/(g\circ f)\cdot\mathcal{F}) = 0 \qquad \text{for } s \geq 1,$$

$$\dim H^q(Y,\mathcal{G}) = \dim H^q(X, \mathcal{F}/(g\circ f)\cdot\mathcal{F}) \leq \dim H^q(X,\mathcal{F}) < \infty \quad .$$

Then, by the inductive hypothesis we have $H^q(X, \mathcal{F}/(g\circ f)\cdot\mathcal{F}) = H^q(Y,\mathcal{G}) = 0$; the homomorphism

$$\sigma^q : H^q(X,(g\circ f)\cdot\mathcal{F}) \to H^q(X,\mathcal{F})$$

is surjective. Now, if $\tau = \tau_g$ is the morphism

$$\tau : \mathcal{F} \to \mathcal{F}$$

defined by $\tau(h_x) = (g\circ f)_x \cdot h_x$, we have $\tau = \sigma\circ\pi$, $\tau^j = \sigma^j\circ\pi^j$ for $j \geq 1$. Then for every $g \in \mathcal{O}(\mathbb{C})$, $g \neq 0$, the homomorphism

$$\tau_g^q : H^q(X,\mathcal{F}) \to H^q(X,\mathcal{F})$$

is a surjective endomorphism, therefore it is a $\mathbb{C}$-linear automorphism.

Let us suppose that there is a cohomology class $\gamma \in H^q(X,\mathcal{F})$, $\gamma \neq 0$. Let us consider the mapping $A: \mathcal{O}(\mathbb{C}) \to H^q(X,\mathcal{F})$ defined by $A(0) = 0$ and by $A(g) = \tau^q_{g\circ f}(\gamma)$ for $g \neq 0$. This mapping is $\mathbb{C}$-linear. For every $g \neq 0$ we have $A(g) \neq 0$, then A is injective. This is a contradiction because $H^q(X,\mathcal{F})$ is finite dimensional as a vector space on $\mathbb{C}$.

q.e.d.

(2.4) REMARK. A holomorphically spreadable complex space $(X,\mathcal{O})$ has no compact irreducible component of positive dimension. Then, by a theorem of Siu $|S_2|$, if $\mathcal{F}$ is a coherent analytic sheaf on $(X,\mathcal{O})$, $\dim X = n > 0$, one has $H^j(X,\mathcal{F}) = 0$ for $j \geq n$. Hence in the previous statement the hypothesis "$H^{q+s}(X,\mathcal{F}) = 0$ for $s \geq 1$" may be substituted by the hypothesis "$\dim_{\mathbb{C}} H^{q+s}(X,\mathcal{F}) < \infty$ for $1 \leq s \leq n-q-1$". In particular *if* $(X,\mathcal{O})$ *is a finite dimensional holomorphically spreadable complex space and if* $\dim_{\mathbb{C}} H^j(X,\mathcal{F}) < \infty$ *for* $j \geq 1$, *then one has* $H^j(X,\mathcal{F}) = 0$ *for* $j \geq 1$.

(2.5) LEMMA. *Let* $(X,\mathcal{O})$ *be a complex space and* $f \in \mathcal{O}(X)$. *For every* $t \in \mathbb{C}$ *let us set*

$$Y_t := V(f-t;\ X) \qquad {}_{Y_t}\mathcal{O} := (\mathcal{O}/(f-t)\cdot\mathcal{O})_{|Y_t}$$

$D_t := \{x \in X \mid (f-t)_x$ is not a zero-divisor in $\mathcal{O}_x\}$.

Let $t \in \mathbb{C}$ *be a number such that* D_t *is a discrete subspace of* X. *Then one has*

i) $((f-t)\cdot\mathcal{O})(X) = (f-t)\cdot\mathcal{O}(X)$;

ii) *for every* $q \geq 1$ *the homomorphism* $H^q(X,\mathcal{O}) \to H^q(X(f-t)\cdot\mathcal{O})$ *induced by the "product by* $f-t$*" is an isomorphism.*

Now, let us assume $H^1(X,\mathcal{O}) = 0$. *Then one has*

iii) *if* D_t *is a discrete subspace of* X, *then the natural projection* $r_t : \mathcal{O}(X) \to {}_{Y_t}\mathcal{O}(Y_t)$ *is surjective;*

iv) *if* D_t *is a discrete subspace of* X *for every* $t \in \mathbb{C}$, *then the natural mapping* $\xi : X \to \mathrm{Sp}\,X$ *is surjective (*resp: *injective, bijective) iff for every* $t \in \mathbb{C}$ *the mapping* ${}_{Y_t}\xi : Y_t \to \mathrm{Sp}\,Y_t$ *is surjective (* resp. *injective, bijective).*

Proof.

For every $t \in \mathbb{C}$ let us denote by

$$\pi_t : \mathcal{O} \to (f-t)\cdot\mathcal{O}$$

the product by $f-t$; i.e. for every $g_x \in \mathcal{O}_x$ we have $\pi_t(g_x) = (f-t)_x g_x$. The support of the sheaf $\ker\pi_t$ is contained in D_t, then if D_t is discrete this support is a discrete subspace of X too. In this case we have

$$H^s(X, \ker\pi_t) = 0 \qquad \text{for } s \geq 1.$$

Now, let us prove every step of our statement.

i) If D_t is discrete, it follows from the property already proved, that the homomorphism $\pi_t^o : \mathcal{O}(X) \to ((f-t)\cdot\mathcal{O})(X)$, induced by π_t, is surjective. Therefore, we have $((f-t)\cdot\mathcal{O})(X) \subset (f-t)\cdot\mathcal{O}(X)$. Of course, the other inclusion holds, too.

ii) It sufficies to consider the cohomology sequence induced by the short exact sequence

$$0 \to \ker \pi_t \longrightarrow \mathcal{O} \overset{\pi_t}{\longrightarrow} (f-t)\mathcal{O}\cdot \longrightarrow 0$$

and the equalities $H^s(X, \ker \pi_t) = 0$ for every $s \geq 1$.

iii) Let us consider the exact sequence

$$0 \to (f-t)\cdot\mathcal{O} \overset{\sigma_t}{\longrightarrow} \mathcal{O} \twoheadrightarrow \mathcal{O}/(f-t)\cdot\mathcal{O} \to 0$$

where σ_t is the natural injection and $t \in \mathbb{C}$. By our hypotheses and by ii) we have $H^1(X, (f-t)\cdot\mathcal{O}) = 0$. Then, from the cohomology sequence associated with the previous sequence, the surjectivity of the projection

$$r_t : \mathcal{O}(X) \longrightarrow \mathcal{O}/(f-t)\cdot\mathcal{O}(X) \simeq {}_{Y_t}\mathcal{O}(Y_t)$$

follows.

iv) Let us make a preliminary remark. Let $\phi \in \mathrm{Sp}\,X$ and $t := \phi(f)$. Then one and only one character $\phi_t \in \mathrm{Sp}\,Y_t$ exists s.t. $\phi = \phi_t \circ r_t$. Indeed we have $\phi(f-t) = 0$; then, by i), $((f-t)\cdot\mathcal{O})(X) \subset \ker \phi$. Hence we can define ϕ_t as the quotient homomorphism of ϕ in

$$\begin{array}{ccc} \mathcal{O}(X) & \overset{r_t}{\longrightarrow} & {}_{Y_t}\mathcal{O}(Y_t) \simeq \mathcal{O}(X)/((f-t)\cdot\mathcal{O})(X) \\ & \overset{\phi}{\searrow} & \downarrow \phi_t \\ & & \mathbb{C} \end{array}$$

Now, let us assume ${}_{Y_t}\xi$ surjective for every $t \in \mathbb{C}$. Let $\phi \in \mathrm{Sp}\,X$, $t := \phi(f)$. A point $x \in Y_t$ exists with $\phi_t = {}_{Y_t}\xi_x$. Then for every $g \in \mathcal{O}(X)$ we have $\phi(g) = \phi_t \circ r_t(g) = r_t(g)(x) = g(x)$, $\phi = \xi_x$. Viceversa, let ξ be surjective. Let $t \in \mathbb{C}$ and $\lambda \in \mathrm{Sp}\,Y_t$. We have $\lambda \circ r_t(f-t) = \lambda(0) = 0$, $\lambda \circ r_t(f) = t$. If $\lambda \circ r_t = \xi_x$, we have $f(x) = t$, $x \in Y_t$. By iii) we deduce that for every $h \in {}_{Y_t}\mathcal{O}(Y_t)$ a section $\tilde{h} \in \mathcal{O}(X)$ is given with $r_t(\tilde{h}) = h$, $\lambda(h) = \lambda \circ r_t(\tilde{h}) = \tilde{h}(x) = h(x)$.

Hence $\lambda = {}_{Y_t}\xi_x$.

Now, let us assume that ${}_{Y_t}\xi$ is injective for every $t \in \mathbb{C}$. Let $x,y \in X$, $x \neq y$. If $f(x) = f(y)$ then $x,y \in Y_t$. If $t := f(x)$, a $g \in {}_{Y_t}\mathcal{O}(Y_t)$ is given s.t. $g(x) \neq g(y)$ By iii) $\tilde{g} \in \mathcal{O}(X)$ exists s.t. $\tilde{g}(x) \neq \tilde{g}(y)$.

q.e.d.

(2.6) PROPOSITION. *Let* $(X, \mathcal{O})$ *be a complex space. Let* $\{X_i\}_{i\in I}$ *be the family of the connected components of* X.

The canonical mapping ${}_X\xi : X \to \operatorname{Sp} X$ *is surjective* (resp.: *injective*) *iff for every* $i \in I$ *the canonical mapping* ${}_{X_i}\xi : X_i \to \operatorname{Sp} X_i$ *is surjective* (resp.:*injective*)

Proof.

For every $i \in I$, let $r_i : \mathcal{O}(X) \to \mathcal{O}(X_i)$ be the restriction; r_i is surjective. Let us assume ${}_X\xi$ surjective. Let $i \in I$ and $\nu \in \operatorname{Sp} X_i$. A point $x \in X$ is given s.t. $\nu \circ r_i = {}_X\xi_x$. Let $h \in \mathcal{O}(X)$ be a section s.t. $r_j(h) = 0$ if $j \neq i$ and s.t. $r_i(h) = 1$. We have $\nu(r_i(h)) = 1$; then $x \in X_i$. Every section $g_i \in \mathcal{O}(X_i)$ may be extended to a suitable $g \in \mathcal{O}(X)$. Then $\nu(g_i) = \nu \circ r_i(g) = g_i(x) = {}_{X_i}\xi_x(g_i)$.

Viceversa, let us assume that ${}_{X_i}\xi$ is surjective for every $x \in X_i$. Let $\phi \in \operatorname{Sp} X$. If sufficies to prove that $j \in I$, $\phi_j \in \operatorname{Sp} X_j$ exist s.t.

$$\phi = \phi_j \circ r_j .$$

Indeed, in such a case we have $\phi_j = {}_{X_j}\xi_x$ for a suitable point $x \in X_j$ and $\phi = {}_X\xi_x$. Then it is sufficient to prove that an index $j \in I$ exists s.t. $\ker r_j \subset \ker \phi$. Effectively, in this case, the character ϕ_j will be obtained as a quotient homomorphism of ϕ.

We may assume that I is a finite or infinite segment of $\mathbb{N}^*$, $I = (1,2,\ldots)$.

It may be useful to identify $\mathcal{O}(X)$ with the product

$$\Gamma := \prod_{i\in I} \mathcal{O}(X_i)$$

as $\mathbb{C}$-algebras, via the natural mapping

$$f \to (r_1(f),\ r_2(f),\dots)$$

Now, let us remark that for every $\alpha = (\alpha_1, \alpha_2,\dots,) \in \Gamma$ with $\alpha_i \in \mathbb{C}$ for each $i \in I$, we have

$$\phi(\alpha) \in \{ \alpha_1, \alpha_2, \dots \} .$$

Indeed, let us set $g: = (\alpha_1 - \phi(\alpha),\ \alpha_2 - \phi(\alpha),\dots)$. We have $\phi(g) = \phi(\alpha - -\phi(\alpha)(1,1,\dots)) = 0$. Then g is not an invertible element of Γ . But if $\phi(\alpha) \notin \notin \{ \alpha_1, \alpha_2,\dots\}$, we have $g \cdot ((\alpha_1 - \phi(\alpha))^{-1},\ (\alpha_2 - \phi(\alpha))^{-1},\dots) = (1,1,\dots)$. This is a contradiction.

Let j be

$$j: = \phi(1,2,\dots) \in I.$$

Let us set $e_j: = (0,\dots,0,\underset{j}{1},0,\dots)$; then $\phi^2(e_j) = \phi(e_j, e_j) = \phi(e_j)$ and we have $\phi(e_j) = 0$ or $\phi(e_j) = 1$. In the first case let us denote by β the element $\beta : = (1,2,\dots,j-1,\ 0, j+1,\dots)$. By what we have just seen, we have $\phi(\beta) \in I \setminus \{j\}$; then $\phi(\beta) \neq j$. But

$$j = \phi(\beta + e_j) = \phi(\beta) + \phi(e_j);$$

This is a contradiction; hence $\phi(e_j) = 1$.

For every $f \in \mathcal{O}(X)$ let us note $f^{(j)}: = (r_1(f),\dots,r_{j-1}(f),0,r_{j+1}(f),\dots)$. Then

$$\phi(f^j) = \phi(f^{(j)} \cdot e_j) = \phi(0) = 0$$

and

$$\phi(f) = \phi(f^{(j)} + r_j(f) \cdot e_j) = \phi(0,\dots,0,r_j(f),\ 0,\dots)$$

Then $\ker r_j \subset \ker \phi$.

q.e.d.

(2.7) The previous Proposition holds when we substitute $\mathrm{Sp}\,X$, $\mathrm{Sp}\,X_i$ by the continuous spectra $\mathrm{Sp}_c X$, $\mathrm{Sp}_c X_i$. Indeed, it is enough to remark that the character ϕ_j

(defined in the proof) is continous iff ϕ is continous (by Banach's Open Mapping Theorem).

Next lemma follows from (0.5) and (0.1). Here we are going to give a direct and elementary proof of this lemma (but other techniques may be used too).

(2.8) LEMMA. *Let* $(X, \mathcal{O})$ *be a holomorphically spreadable complex space,* $\dim X = n$. *Let the dimension of* $H^q(X, \mathcal{O})$ *as a vector space on* $\mathbb{C}$ *be countable for* $1 \leq q \leq n-1$.

Then the canonical mapping ${}_X\xi : X \to \operatorname{Sp} X$ *is bijective.*

Proof.

By Remark (2.4) we have $H^q(X, \mathcal{O}) = 0$ for $q \geq 1$. Then $H^q(X_i, \mathcal{O}) = 0$ for every connected component X_i of X and every $q \geq 1$. By Proposition (2.6) one deduces that X may be assumed connected.

The proof is by induction on the dimension n of X.

Case $n = 0$, X is constituted by only one point $X = \{x\}$. If $(X, \mathcal{O})$ has the reduced structure, then $\mathcal{O}(X) \simeq \mathbb{C}$. Only one character $\mathbb{C} \to \mathbb{C}$ exists and it is the identity. Then $\operatorname{Sp} X = \{\xi_x\}$. If $(X, \mathcal{O})$ is nonreduced, let $\{x\}$ be its associated reduced space and let $\mathcal{N}$ be the nilpotent sheaf of $(X, \mathcal{O})$. Let $\phi \in \operatorname{Sp} X$; one has $\phi(g) = 0$ for every $g \in \mathcal{N} \simeq \mathcal{N}(X)$; then ϕ induces a character on $\{x\}$. This last character is necessarily ${}_{\{x\}}\xi_x$; then $\phi = {}_X\xi_x$.

Let $n > 0$; in this case all the irreducible components of X have positive dimension. Let us assume the thesis holds for holomorphically spreadable complex spaces of dimension $< n$.

For every $k \in \mathbb{N}$, let us note

$$S_k := \{x \in X \mid \operatorname{prof} \mathcal{O}_x \leq k\} .$$

Let $\{T_i\}_{i \in I}$ be the family of all the irreducible components of X and all the positive dimensional irreducible components of all the S_k.

For every $i \in I$, let x_i be a point in T_i and let y_i be another point of T_i $y_i \neq x_i$, chosen as it is indicated in the statement of (2.2).

It follows from Lemma (2.2) that $f \in \mathcal{O}(X)$ is given with $f(x_i) \neq f(y_i)$ for every $i \in I$. For every $t \in \mathbb{C}$ and every $i \in I$, we have $(f-t)(x_i) \neq (f-t)(y_i)$; then

$$\dim(V(f-t;\, X) \cap S_{k+1}) \leq k \qquad k \geq 0.$$

By Theorem (0.3) we deduce that f-t does not induce germs which are zero-divisors out of S_o, for every $t \in \mathbb{C}$. Then by Lemma (2.5) it is enough to prove that, for every $t \in \mathbb{C}$ the mapping ${}_{Y_t}\xi$ is surjective.

By our cohomological hypotheses, Lemma (2.5) ii), Theorem (0.2) and Remark (2.4), it follows that for every $t \in \mathbb{C}$ and every $q \geq 1$ one has $H^q(X,(f-t)\cdot\mathcal{O}) = 0$. Let us consider the cohomology sequence induced by the short exact sequence

$$0 \to (f-t)\cdot\mathcal{O} \xrightarrow{j_t} \mathcal{O} \longrightarrow \mathcal{O}/\,(f-t)\cdot\mathcal{O} \to 0$$

where j_t is the natural morphism. One has

$$H^q(Y_t,\, {}_{Y_t}\mathcal{O}) \simeq H^q(X,\, \mathcal{O}/\,(f-t)\cdot\mathcal{O}) = 0$$

for every $t \in \mathbb{C}$, $q \geq 1$.

Now, we only have to prove that for every $t \in \mathbb{C}$ one has $\dim Y_t \leq n-1$ and to use the inductive hypothesis.

Directly by the definition of f, one has that the reduction Red f of f is nonconstant on any irreducible component of X. Then for every $t \in \mathbb{C}$, Red f-t does not vanish on any irreducible component of X and one has $\dim Y_t \leq n-1$.

q.e.d.

Proposition (2.6) and Lemma (2.8) imply the following result.

(2.9) REMARK. *Let* $(X, \mathcal{O})$ *be a holomorphically spreadable complex space: let* $\{X_i\}_{i \in I}$ *be the family of the connected components of* X.

Assume that for every $i \in I$, X_i *is a finite dimensional open subspace of* X *and that for every* $q \geq 1$ $\dim_{\mathbb{C}} H^q(X, \mathcal{O})$ *is countable.*

Then the canonical mapping ${}_X\xi : X \to \mathrm{Sp}\, X$ *is bjective.*

Indeed, it sufficies to observe that for every $q \geq 1$ one has

$$H^q(X, \mathcal{O}) \simeq \prod_{i \in I} H^q(X_i, \mathcal{O}_{|X_i})$$

§ 3. SOME RESULTS IN THE GENERAL CASE

(3.1) THEOREM. *Let* $(X, \mathcal{O})$ *be a holomorphically spreadable complex space; let* L *be a closed subset of* X.

Assume that $H^q(X, \mathcal{O})$ *as a vector space on* $\mathbb{C}$ *has countable dimension for every* $q \geq 1$ *and that the connected components of* $X \setminus L$ *are finite dimensional open subspaces of* $(X, \mathcal{O})$.

Let $\phi \in \mathrm{Sp}\, X$ *be a character s.t.* $h \in \mathcal{O}(X)$ *exists with* $\phi(h) \neq 0$, $h(y) = 0$ *for every* $y \in L$. *Then a point* $x \in X \setminus L$ *is given with*

$$\phi = {}_X\xi_x .$$

<u>Proof.</u>

First step. We want to prove that we may choose a section $f \in \mathcal{O}(X)$ such that $\phi(f) = \phi(h)$, s.t. $f(y) = 0$ for every $y \in L$ and s.t. the germ $f_x - t$ is not a zero-divisor in $\mathcal{O}_x$ for every $t \in \mathbb{C}^*$ and every $x \in X$ with prof $\mathcal{O}_x > 0$.

Once again, for every $k \in \mathbb{N}$ we set

$$S_k := \{x \in X \mid \text{prof}\, \mathcal{O}_x \leq k\} .$$

Let $\{T_i\}_{i\in I}$ be the family of the irreducible components of all the analytic subsets S_k under the condition that for every $i \in I$ one has $\dim T_i > 0$ and that h does not vanish identically on T_i.

Of course, I is countable. For every $i \in I$ let us pick a point $x_i \in T_i$ with $h(x_i) \neq 0$. For every $i \in I$ let us fix an open neighborhood U_i of x_i on which x_i is defined by $\mathcal{O}(X)$ and let us pick a point $y_i \in U_i$ with $x_i \neq y_i$.

By Lemma (2.2) a section $g \in \mathcal{O}(X)$ exists s.t.

$$g(x_i) \neq g(y_i) \qquad \text{for every } i \in I.$$

Now, we may choose $c \in \mathbb{C}$ such that $\phi(g) + c \neq 0$ and such that for every index $i \in I$ with $h(x_i) \neq h(y_i)$, we have

$$\text{(A)} \qquad \frac{h(x_i)g(x_i) - h(y_i)g(y_i)}{h(y_i) - h(x_i)} \neq c\,.$$

Let us define $\ell,\ f \in \mathcal{O}(X)$ by

$$\ell := \frac{1}{\phi(g) + c}\,(g + c)$$

$$f := h \cdot \ell$$

We have $\phi(\ell) = 1$, $\phi(f) = \phi(h)$; moreover we have $f(y) = 0$ for every $y \in L$. Let us fix $k \in \mathbb{N}$ and let $\tilde{T}$ be an irreducible component of S_{k+1}, with $\dim \tilde{T} > 0$. Let $t \in \mathbb{C}^*$. We have

$$\dim(V(f - t;\ X) \cap \tilde{T}) < \dim \tilde{T} \leq k + 1.$$

Indeed, if h vanishes identically on $\tilde{T}$, then $V(f-t;\ X) \cap \tilde{T} = \emptyset$. Otherwise, we have $\tilde{T} = T_i$ for a suitable index $i \in I$. It sufficies to prove that $f(x_i) \neq f(y_i)$. If $h(x_i) = h(y_i)$, one has $h(x_i) \neq 0$ and $\ell(x_i) \neq \ell(y_i)$ and hence the desired inequality. If $h(x_i) \neq h(y_i)$, our inequality $f(x_i) \neq f(y_i)$ follows from (A).

Then we have

$$\dim(V(f-t;\ X) \cap S_{k+1}) < \dim S_{k+1} \leq k+1 \qquad k \in \mathbb{N},\ t \in \mathbb{C}^*;$$

by Theorem (0.3) we deduce that $f_x - t$ is not a zero-divisor in $\mathcal{O}_x$ for every $x \in X \setminus S_o$, $t \in \mathbb{C}^*$.

Second step. The thesis.

Let $a := \phi(h) \in \mathbb{C}^*$. By the first step we have the following inclusion

$$D_a := \{x \in X \mid f_x - a \text{ is a zero-divisor in } \mathcal{O}_x\} \subset S_o,$$

we have $\dim S_o \leq 0$; then D_a is a discrete subset of X.

By Theorem (0.2) we have $\dim_{\mathbb{C}} H^q(X, \mathcal{O}) < \infty$ for every $q \geq 1$. By Lemma (2.5) ii) we have $H^q(X, \mathcal{O}) \simeq H^q(X, (f-a)\cdot \mathcal{O})$ for every $q \geq 1$. Then by the short exact sequence

$$0 \to (f-a)\cdot \mathcal{O} \xrightarrow{j_a} \mathcal{O} \xrightarrow{r_a} \mathcal{O} / (f-a)\cdot \mathcal{O} \to 0.$$

(where j_a, r_a are the natural morphisms) we deduce

$$\dim_{\mathbb{C}} H^q(X, \mathcal{O} / (f-a)\cdot \mathcal{O}) = \dim H^q(Y_a, {}_{Y_a}\mathcal{O}) < \infty$$

for $q \geq 1$; as usually we set $Y_a := V(f-a;\ X)$, ${}_{Y_a}\mathcal{O} := (\mathcal{O} / (f-a)\cdot \mathcal{O})_{|Y_a}$.

Because of the inequality $a \neq 0$, we have $Y_a \subset X \setminus L$. Then the connected components of Y_a are finite dimensional and by Lemma (2.9) we deduce that the canonical mapping ${}_{Y_a}\xi : Y_a \to \mathrm{Sp}\, Y_a$ is bijective.

Now, we have only to prove that a character $\phi_a \in \mathrm{Sp}\, Y_a$ is given such that $\phi = \phi_a \circ r_a^o$ (where $r_a^o : \mathcal{O}(X) \to {}_{Y_a}\mathcal{O}(Y_a)$ is the mapping induced by r_a). Indeed, since we have $\phi_a = {}_{Y_a}\xi_x$ for a suitable point $x \in Y_a$, the existence of ϕ_a implies $\phi = {}_X\xi_x$.

Let us observe that r_a^o is surjective (as it is seen by Lemma (2.5)) and that $\ker r_a^o = \mathrm{Im}\, j_a^o = ((f-a)\cdot \mathcal{O})(X)$ where j_a^o is induced by j_a. Then, by

Lemma (2.5) we have $\ker r_a^o = (f-a) \cdot \mathcal{O}(X)$. By this last remark we deduce the inclusion $\ker r_a^o \subset \ker \phi$. Hence the desired existence of ϕ_a.

q.e.d.

(3.2.) COROLLARY. *Let* $(X, \mathcal{O})$ *be a holomorphically spreadable complex space. Let a sequence* $(X_o, \mathcal{O}_o),\dots,(X_n, \mathcal{O}_n) = (X, \mathcal{O})$ *of complex spaces be given such that, for every* $i = 0,\dots,n-1$, $(X_i, \mathcal{O}_i)$ *is a closed complex subspace of* $(X_{i+1}, \mathcal{O}_{i+1})$.

Assume that

a) $\dim_{\mathbb{C}} H^q(X_i, \mathcal{O}_i)$ *is countable for every* $q \geq 1$, $i = 1,\dots,n$;

b) *the natural mapping* $r_i : \mathcal{O}(X) \to \mathcal{O}(X_i)$ *is surjective for every* $i = 0,\dots,n-1$;

c) *the connected components of* $X_i \setminus X_{i-1}$ *are finite dimensional subspaces of* $(X_i, \mathcal{O}_i)$ *for every* $i = 1,\dots,n$;

d) *the canonical mapping* ${}_{X_o}\xi : X_o \to \mathrm{Sp}\, X_o$ *is surjective.*

Then the canonical mapping $\xi : X \to \mathrm{Sp}\, X$ *is surjective.*

<u>Proof.</u>

Let us denote by $r_n : \mathcal{O}(X) \to \mathcal{O}(X)$ the identical mapping. For every $i = 0,\dots,n-1$ let $\rho_i : \mathcal{O}_{i+1}(X_{i+1}) \to \mathcal{O}_i(X_i)$ be the natural restriction, we have $r_i = \rho_i \circ \dots \circ \rho_{n-1}$.

Let $\phi \in \mathrm{Sp}\, X$.

Then either $\ker \phi \supset \ker r_o$ or an index $j \in \{1,\dots,n\}$ is given such that $\ker \phi \supset \ker r_j$, $\ker \phi \not\supset \ker r_{j-1}$.

In any case, property b) implies the existence of a character $\phi_j \in \mathrm{Sp}\, X_j$ with $\phi = \phi_j \circ r_j$.

If $j = 0$, it follows from d) that a suitable point $x \in X_o$ exists with $\phi_o = {}_{X_o}\xi_x$; then $\phi = {}_X\xi_x$. Let us assume $j > 0$. Let $h \in \ker r_{j-1}$, $h \notin \ker \phi$; we have $\rho_{j-1}(r_j(h)) = \theta$, $\phi_j(r_j)(h)) \neq 0$. By properties a) and c) we may apply Theorem (3.1) to the space $(X_j, \mathcal{O}_j)$. Then a point $w \in X_j \setminus X_{j-1}$ is given such that $\phi_j = {}_{X_j}\xi_x$, $\phi = {}_X\xi_x$.

q.e.d.

Condition d) holds if $(X_o, \mathcal{O}_o)$ is a finite dimensional Stein space.

(3.3) Conditions employed in (3.2) are rather ponderous; especially fastidicus are the extension properties employed. Ronghly speaking it may be because the techniques used are rather elementary. Indeed, in the Stein case the use of standard Stein techniques may assure a simpler proof of (3.1).

In any case, if we consider Stein spaces, previous conditions are generally satisfied and we may improve some results of the first paragraph.

(3.4) PROPOSITION. *Let* $(X, \mathcal{O})$ *be a complex space. Let a sequence* $(X_o, \mathcal{O}_o), \ldots, (X_n, \mathcal{O}_n) = (X, \mathcal{O})$ *of complex spaces be given such that for every* $i = 0, \ldots, n-1$. $(X_i, \mathcal{O}_i)$ *is a closed complex subspace of* $(X_{i+1}, \mathcal{O}_{i+1})$.

Assume that $(X_o, \mathcal{O}_o)$ *verifies Michael's conjecture and that the connected components of* $X_i \setminus X_{i-1}$ *are finite dimensional open subspaces of* $(X_i, \mathcal{O}_i)$ *for every* $i = 1, \ldots, n$.

The following conditions are equivalent

a) X *is a Stein space;*

b) *the canonical mapping* $\xi : X \to \mathrm{Sp}\, X$ *is a homeomorphism.*

Proof.

We have already remarked that implication b) ==> a) is known. Viceversa let $(X, \mathcal{O})$ be Stein, then $(X_o, \mathcal{O}_o)$ is Stein. By (1.2) ${}_{X_o}\xi$ is surjective. By (3.2) ${}_X\xi : X \to \mathrm{Sp}\, X$ is surjective; by (1.2) this is enough to assume that b) holds.

q.e.d.

(3.5) COROLLARY. *Let* $(X, \mathcal{O})$ *be a complex space. Let* $(S_1, \mathcal{O}_1)$ *be the singular locus of the reduction of* $(X, \mathcal{O})$. *For every* $j > 1$ *let* $(S_j, \mathcal{O}_j)$ *be the singular locus of* $(S_{j-1}, \mathcal{O}_{j-1})$. *Assume that an integer* $n \geq 1$ *is given such that* $(S_n, \mathcal{O}_n)$

verifies Michael's conjecture.

Then $(X, \mathcal{O})$ *is a Stein space iff the canonical mapping* ${}_X\xi : X \to \mathrm{Sp}\, X$ *is a homeomorphism.*

(3.6) PROPOSITION. Let $(X, \mathcal{O})$ be a Stein space. Let a sequence $(X_o, \mathcal{O}_o), \ldots, (X_n, \mathcal{O}_n) = (X, \mathcal{O})$ be given such that, for every $i = 0, \ldots, n-1$, $(X_i, \mathcal{O}_i)$ is a closed complex subspace of $(X_{i+1}, \mathcal{O}_{i+1})$ and such that the connected components of $X_i \setminus X_{i-1}$ are finite dimensional open subspaces of $(X_i, \mathcal{O}_i)$ for every $i = 1, \ldots, n$.

The following conditions are equivalent

a) the canonical mapping ${}_{X_o}\xi : X_o \to \mathrm{Sp}\, X_o$ is a homeomorphism;

b) an integer m exists, $0 \le m \le n$, such that the canonical mapping ${}_{X_m}\xi : X_m \to \mathrm{Sp}\, X_m$ is a homeomorphism;

c) the canonical mapping $\xi : X \to \mathrm{Sp}\, X$ is a homeomorphism.

Proof.

b) ==> c) $(X_m, \mathcal{O}_m)$ verifies Michael's conjecture; then we may apply Proposition (3.5).

c) ==> a) By (1.2) it is enough to prove that $(X_o, \mathcal{O}_o)$ verifies Michael's conjecture. Let $r: \mathcal{O}(X) \to \mathcal{O}_o(X_o)$ be the natural mapping. Let $\phi_o \in \mathrm{Sp}\, X_o$. Because $(X, \mathcal{O})$ verifies Michael's conjecture, $\phi_o \circ r$ is continous, $\phi_o \circ r \in \mathrm{Sp}_c\, X$. By Banach Open Mapping Theorem, r is open and ϕ_o is also continuous.

(3.7) THEOREM. *Let* $(X, \mathcal{O})$ *be a Stein space. Let* $(S_1, \mathcal{O}_1)$ *be the singular locus of the reduction of* $(X, \mathcal{O})$. *For every* $j > 1$, *let* $(\mathcal{O}_j, S_j)$ *be the singular locus of* $(S_{j-1}, \mathcal{O}_{j-1})$. *The following conditions are equivalent.*

a) *the canonical mapping* ${}_{S_1}\xi : S_1 \to \mathrm{Sp}\, S_1$ *is a homeomorphism;*

b) *an integer* m *exists such that the mapping* ${}_{S_m}\xi : S_m \to \mathrm{Sp}\, S_m$ *is a homeomorphism;*

c) *the mapping* $\xi: X \to \mathrm{Sp}\, X$ *is a homeomorphism.*

Of course, by (1.2) each of the previous conditions a), b), c) of (3.6) are equivalent to each of the following conditions: d) $(X_o, \mathcal{O}_o)$ verifies Michael's conjecture, e) ${}_{X_o}\xi$ is surjective, f) $(X_m, \mathcal{O}_m)$ verifies Michael's conjecture, g) ${}_{X_m}\xi$ is surjective, h) $(X, \mathcal{O})$ verifies Michael's conjecture, i) ξ_X is surjective. Analogous remarks hold for conditions a), b), c) of (3.7).

Moreover if $\dim X_m < \infty$ (resp.: $\dim S_m < \infty$) then conditions a), b), c) of (3.6) (resp.: (3.7)) hold.

REFERENCES

|AB| A.Andreotti - C.Bănică *"Relative Duality on Complex Spaces I"*, Rev.Roum.Math., Tome XX, N° 9, 981-1041, 1975.

|AN| A.Andreotti - R.Narasimhan *"Oka's Heftungslemma and the Levi Problem for Complex Spaces"*, Trans.Am.Math.Soc., 111, 345-366, 1964.

|Al| L.Alessandrini *"On the Cohomology of a Holomorphically Separable Complex Analytic Space"*, Boll. U.M.I. (6), 1-A, 261-268, 1982.

|B| E.Ballico *"Finitezza e annullamento di gruppi di coomologia su uno spazio complesso"*, Boll. U.M.I.(6), 1-B, 131-142, 1982.

|BS| C.Bănică - O.Stănăşilă *"Méthodes algébriques dans la théorie globale des espaces complexes, Vol.1"* Gauthier-Villars, 1977.

|C| S.Coen *"Sull'uso di un metodo induttivo in variabile complessa"* Seminarî di Geometria 1984, Dip.Mat.Univ.Bologna, 49-91, 1985.

|E| R.Ephraim *Multiplicative Linear Functionals of Stein Algebras"* Pacif J. of Math., 78, 89-93, 1978.

|F |O.Forster *"Uniqueness of Topology in Stein Algebras"* in "Function Algebras" F.Bartel ed., 157-167, Scott-Foresmann, 1966.

|FN|J.E.Fornaess - R.Narasimhan *"The Levi Problem on Complex Spaces with Singularities"* Math.Ann. 248, 47-72, 1980.

|GR| H.Grauert - R.Remmert *"Theory of Stein Spaces"* Springer, 1979.

|J | B.Jennane *"Problème de Levi et espaces holomorphiquement séparés"* Math. Ann. 268, 305-316, 1984.

|M | A.Markoe *"Maximal Ideals of Stein Algebras"* in "Conference on Complex Analysis" W.Zame ed., Dept.Math.St.Univ.N.Y.Buffalo, 25-39, 1973

|S | Y.T.Siu *"Non Countable Dimensions of Cohomology Groups of Analytic Sheaves and Domains of Holomorphy"* Math.Z.102, 17-29, 1967.

|S_2| Y.T.Siu *"Analytic Sheaf Cohomology Groups of Dimension* n *of* n*-dimensional Complex Spaces"* Trans.A.M.S., 143, 77-94, 1963.

|ST| Y.T.Siu - G.Trautmann *"Gap-Sheaves and Extension of Coherent Analytic Subsheaves"*, Springer, Lect.Notes in Math., 1971.

|W | K.W.Wiegmann *"Einbettungen komplexer Räume in Zahlenräume"*, Invent.Math., 1, 229-242, 1966.

S.Coen
Dipartimento di Matematica
Università di Bologna
Piazza di Porta S.Donato, 5
I 40127 Bologna - Italia.

SCALAR BOUNDARY INVARIANTS AND THE BERGMAN KERNEL

C. Robin Graham

This paper continues the program of Fefferman [11], the goal of which is to invariantly express the asymptotic expansion of the Bergman kernel of a strictly pseudoconvex domain Ω in $\mathbb{C}^n$, $n\geq 2$, in terms of the local biholomorphic geometry of the boundary. If ρ is a defining function for Ω with $\Omega=\{\rho>o\}$, then by [9], [3], the Bergman kernel on the diagonal may be written in the form $K=\frac{\varphi}{\rho^{n+1}} + \psi \log \rho$, where $\varphi,\psi\in C^{\infty}(\bar{\Omega})$. Also K mod $C^{\infty}(\bar{\Omega})$ is determined locally by $b\Omega$ so that the full Taylor expansion of ψ and that of φ modρ^{n+1} at a point $z\in b\Omega$ are determined by the Taylor expansion of ρ at z. The goal is to write down the expansions of φmod ρ^{n+1} and ψ in an invariant manner. One conceives of expanding in powers of ρ with coefficients which can be identified as invariants of the biholomorphic geometry of $b\Omega$. Thus a first portion of this program must consist of constructing local scalar invariants of strictly pseudoconvex boundaries, the coefficients-to-be in the above expansion. The coefficients of different powers of ρ must transform differently under a biholomorphic change, so one speaks of an invariant of weight w, which will transform as a coefficient of ρ^w in the expansion of φ. A preliminary definition is: An invariant P of weight w is an assignment to each C^{∞} strictly pseudoconvex hypersurface $M\subset\mathbb{C}^n$ of a function $P_M\in C^{\infty}(M)$ so that P_M is locally determined in the sense that $P_M(z)$ depends only on derivatives at $z\in M$ of a defining function ρ for M, and such that if Φ is a biholomorphism of a neighborhood of M to a neighborhood of $\tilde{M}$, then $P_{\tilde{M}}\circ\Phi=|\det\Phi'|^{\frac{-2w}{n+1}} P_M$. A more precise definition will be given in §1, and in it $P_M(z)$ will also be required to depend polynomially on derivatives of the defining function. The definition of weight used here differs by a factor of 2 from that of [11].

This research was partially supported by NSF grant #DMS-8501754.

Fefferman's "geometric problem" is to construct all invariants of weight $w \geq 0$. For $w \leq n$ he succeeded in devising a construction of invariants of weight w which one might expect gives all such invariants. This is exactly the range of weights required to express $\varphi \bmod \rho^{n+1}$. He also showed that if $w \leq n-20$ then in fact all invariants of weight w arise from his construction and $\varphi \bmod \rho^{n-19}$ can be expressed in terms of these invariants. However Fefferman's construction terminates when w=n. Although invariants of higher weight may be obtained, for example, by multiplying Fefferman's low weight invariants, these invariants cannot be expected to generate all higher weight invariants as they involve no derivatives of the defining function of order greater than 2n+2. In §1 of this paper a construction of invariants of arbitrarily high weight, involving arbitrarily high order derivatives of the defining function, is described. These new invariants arise from the higher, log term, asymptotics of solutions to the complex Monge-Ampère equation $J(u)=1$, $u|_{b\Omega}=0$, where $J(u)=(-1)^n \det\begin{pmatrix} u & u_{\bar{j}} \\ u_i & u_{i\bar{j}} \end{pmatrix}$, somewhat analogous to Fefferman's invariants which are constructed from smooth approximate solutions to this equation. However we still do not have all the invariants: an example is presented of an invariant not arising from this construction. Some of the results concerning the higher asymptotics of the Monge-Ampère solution were obtained independently by J. Bland and J. Lee. In particular, Theorem 1.6a) is contained in [2] and is implicit in the calculations of [12], [13], and the fact that the coefficient of the first log term, $b\eta_1$ in our notation, is an invariant of weight n+1 is in [12], as are Proposition 1.8, Proposition 2.2, and a result similar to Proposition 1.10.

§2 presents some invariant-theoretic arguments which result in the explicit identification of all invariants of weight ≤ 2 in dimensions $n \geq 3$, and of all invariants of weight ≤ 5 in $\mathbb{C}^2$. Some of this material is well-known to workers in the field; other of it is new. §3 shows how the invariant theory of §2 can be applied to obtain concrete information about the asymptotic expansion of the Bergman kernel, thus providing a different point of view on some results of Christoffers and Diederich. (A discussion of these results and other topics overlapping with the subject matter of this paper may be found in [1].) More recent work of D. Burns on the log term of the Bergman kernel in $\mathbb{C}^2$ is in a similar spirit; this too is described in §3.

We also present some new results on the first log term coefficients of the Bergman kernel and of the Monge-Ampère solution. One knows that these invariants are distinguished analytically by their respective occurrences as log term coefficients, and it is shown that they possess an invariant-theoretic property which distinguishes them on purely algebraic grounds as well. A consequence of this is the fact that, as polynomials in the Moser normal form coefficients, these two invariants have linear parts which agree up to a multiple. When n=2 the invariants themselves are linear, so are multiples of one another. It would be interesting to determine whether or not the full invariants so agree when $n \geq 3$.

The influence of the work of Charles Fefferman on this paper is evident. The conceptual framework for and formulation of the problems discussed here all come from his paper [11]. I am deeply indebted to him for valuable advice and encouragement and for explaining his work to me. Also I am grateful to Dan Burns for discussions of the material in §§2,3 and for permission to describe his unpublished work.

1. Construction of Scalar Invariants

We begin by reformulating the definition of invariants given above, following [11]. The pseudogroup of all biholomorphic changes of coordinates is too large and unwieldy to work with, so we use Moser's normal form to reduce to the study of the action of a finite dimensional group. Recall ([7]) that a real analytic strictly pseudoconvex hypersurface containing $0 \in \mathbb{C}^n$ is in normal form if it is defined by an equation of the form

$$2u = |z|^2 + \sum_{\substack{|\alpha|,|\beta| \geq 2 \\ \ell \geq 0}} A^{\ell}_{\alpha\bar{\beta}} z^{\alpha} \bar{z}^{\beta} v^{\ell}, \tag{1.1}$$

where $z \in \mathbb{C}^{n-1}$, $w = u+iv \in \mathbb{C}$ so that $(z,w) \in \mathbb{C}^n$, α, β are lists of indices between 1 and n-1, and the coefficients $A^{\ell}_{\alpha\bar{\beta}} \in \mathbb{C}$ satisfy:

i) $A^{\ell}_{\alpha\bar{\beta}}$ is symmetric with respect to permutations of the indices in α and β, respectively.

ii) $\overline{A^{\ell}_{\alpha\bar{\beta}}} = A^{\ell}_{\beta\bar{\alpha}}$

iii) $\operatorname{tr} A^{\ell}_{2\bar{2}} = 0$, $\operatorname{tr}^2 A^{\ell}_{2\bar{3}} = 0$, $\operatorname{tr}^3 A^{\ell}_{3\bar{3}} = 0$.

Here, for $p,q \geq 2$, $A^{\ell}_{p\bar{q}}$ is the symmetric tensor $\left(A^{\ell}_{\alpha\bar{\beta}} \right)_{\substack{|\alpha|=p \\ |\beta|=q}}$ on $\mathbb{C}^{n-1}$, and the traces are the usual tensorial traces with respect to $\delta_{i\bar{j}}$. If M is any real analytic strictly pseudoconvex hypersurface and $p \in M$, then there is a biholomorphic mapping Φ defined near p which maps p to 0 and M to a hypersurface in normal form. However the mapping Φ and resulting normal form are not uniquely determined. In order to describe the nonuniqueness let $D = \{2u > |z|^2\} \subset \mathbb{C}^n$ be the Siegel domain and $G = \mathrm{Aut}(D)$ be the group of biholomorphic self mappings of D. Also let $H \subset G$ be the isotropy group of $0 \in bD$ in G; i.e. H consists of all those automorphisms h of D which extend holomorphically to a neighborhood of 0 and which satisfy $h(0)=0$. Then there is an action of H on the space of normal forms whose orbits are exactly the equivalence classes of normal forms under biholomorphic equivalence preserving 0. Otherwise stated, two normal forms are biholomorphically equivalent iff one is mapped to the other by some element of H. Thus the study of invariants of strictly pseudoconvex domains under biholomorphic equivalence is reduced to the study of this action of H on the space of normal forms.

The structure of the group H and its action on normal forms will now be described. First H contains as a subgroup a copy of $\mathbb{R}_+$, the multiplicative group of positive real numbers, embedded in $G = \mathrm{Aut}(D)$ via $\delta(z,w) = (\delta z, \delta^2 w)$, $\delta \in \mathbb{R}_+$. Since $\det \delta' = \delta^{n+1}$, it follows that H can be written $H = \mathbb{R}_+ H_o$, where the subgroup H_o of H is defined by $H_o = \{h \in H: |\det h'(0)| = 1\}$. And H_o itself can be written as the product of two subgroups $H_o = U(n-1) \cdot \mathbb{H}^{n-1}$. $U(n-1)$ denotes the unitary group of $|z|^2$ in $\mathbb{C}^{n-1}$, which clearly acts on D preserving 0. $\mathbb{H}^{n-1}$ is the Heisenberg group, which we identify with bD with multiplication $(\xi,\eta)\cdot(\xi',\eta') = (\xi+\xi', \eta+\eta' + \xi\cdot\bar{\xi}')$. (Here $\xi \in \mathbb{C}^{n-1}$, $\eta \in \mathbb{C}$, and $2\,\mathrm{Re}\,\eta = |\xi|^2$ so that $(\xi,\eta) \in bD$. Similarly for (ξ',η'). Also $\xi\cdot\bar{\xi}' =$

$\sum_{j=1}^{n-1} \xi_j \bar{\xi}'_j)$. $\mathbb{H}^{n-1}$ acts on D by fractional linear transformations: for $h=(\xi,\eta)\in\mathbb{H}^{n-1}$, define $h(z,w) = \frac{(z+w\xi,w)}{1+z\cdot\bar{\xi}+w\bar{\eta}}$. One checks easily that $h:D\to D$, $h(0)=0$, and $|\det h'(0)|=1$, so $h\in H_o$.

Now let N be a normal form, i.e. a hypersurface of the form (1.1), which we identify with the coefficients $\left(A^{\ell}_{\alpha\bar{\beta}}\right)_{\substack{|\alpha|,|\beta|\geq 2\\ \ell\geq 0}}$. If $h\in H$ then $h^*N=\{(z,w): h(z,w)\in N\}$ is a hypersurface which can be written in the form $h^*N = \{2u=|z|^2+f(z,\bar{z},v)\}$, where f is a real analytic function whose Taylor expansion at 0 involves only terms of strength ≥ 4, where strength (z_j)=strength $(\bar{z}_j)=1$, strength (w)=strength $(\bar{w})$= strength $(v)=2$. But $h^*(N)$ is not necessarily in normal form. However, it follows from [7] that there is a unique holomorphic change of coordinates $T(z,w) = (\tilde{z},\tilde{w})$ satisfying

$$(1.2) \qquad T(0)=0,\ T'(0)=\mathrm{Id.},\ \mathrm{Im}\ \frac{\partial^2\tilde{w}}{\partial w^2}(0)=0,$$

and $T^*(h^*(N))$ is in normal form. Set $Nh=T^*(h^*(N))=(h\circ T)^*N$; then clearly Nh is a normal form biholomorphically equivalent to N, and by [7] any normal form equivalent to N is of the form Nh for some $h\in H$. The mapping $(N,h)\to Nh$ can be shown to be a right action of H on the space of normal forms.

If $\delta\in\mathbb{R}_+\subset H$ as above, then δ^*N is already in normal form, so $\left(A^{\ell}_{\alpha\bar{\beta}}\right)\delta = \left(\delta^{2w}A^{\ell}_{\alpha\bar{\beta}}\right)$, where $w=1/2(|\alpha|+|\beta|)+\ell-1$ is the weight of the coefficient $A^{\ell}_{\alpha\bar{\beta}}$. And if $U\in U(n-1)\subset H$ then again U^*N is in normal form so $\left(A^{\ell}_{\alpha\bar{\beta}}\right)U= \left(\tilde{A}^{\ell}_{\alpha\bar{\beta}}\right)$ where $\tilde{A}^{\ell}_{\alpha\bar{\beta}}$ is obtained by transforming $A^{\ell}_{\alpha\bar{\beta}}$ as a tensor in $\alpha,\bar{\beta}$, i.e. for each $p,q\geq 2$, $\ell\geq 0$, $\sum_{\substack{|\alpha|=p\\|\beta|=q}} \tilde{A}^{\ell}_{\alpha\bar{\beta}}\, z^{\alpha}\bar{z}^{\beta} = \sum_{\substack{|\alpha|=p\\|\beta|=q}} A^{\ell}_{\alpha\bar{\beta}}\, (Uz)^{\alpha}\, (\overline{Uz})^{\beta}$. Unfortunately the action of $\mathbb{H}^{n-1}$ cannot be written down so explicitly. In §2 it will be calculated in $\mathbb{C}^2$ on the first few terms in the normal form.

Finally we are in a position to give the precise definition of scalar invariants.

<u>Definition 1.3</u>. An invariant of weight $w\geq 0$ is a polynomial P in the normal form coefficients $\left(A^{\ell}_{\alpha\bar{\beta}}\right)$ so that

$$P(Nh) = |\det h'(0)|^{\frac{2w}{n+1}} P(N), \quad h \in H.$$

By the above, this is equivalent to the two conditions:

$$\text{i) } P(N\delta) = \delta^{2w} P(N) \quad , \delta \in \mathbb{R}_+$$

$$\text{ii) } P(Nh) = P(N) \quad , h \in H_o.$$

An invariant P(N) in the sense of Definition 1.3 gives rise to an invariant in the sense of the introduction as follows. Given an analytic strictly pseudoconvex hypersurface M and $p \in M$, choose a biholomorphism $\Phi: M \to N$, $\Phi(p)=0$, where N is a hypersurface in normal form, and set $P_M(p) = |\det \Phi'(0)|^{\frac{2w}{n+1}} P(N)$. Then Definition 1.3 insures that this value is independent of the choice of Φ,N. Since P_M depends only on finite order information about M at p, this procedure defines P_M for general C^∞ M too. Conversely, an invariant in the sense of the introduction gives rise to an invariant in the sense of Definition 1.3 by evaluation at 0 on a normal form, except that in 1.3 the resulting function of $\left(A^{\ell}_{\alpha\bar{\beta}}\right)$ is additionally required to be a polynomial. Specific examples of invariants will be given in §2.

Next we recall Fefferman's "Weyl invariants", which are constructed from a solution of the complex Monge-Ampère equation. Let $M \subset \mathbb{C}^n$ be a strictly pseudoconvex hypersurface, V a neighborhood of M in $\mathbb{C}^n$, and ρ a defining function for M chosen so that the side of M given by $V_+ = \{\rho>0\}$ is strictly pseudoconvex. In [10], Fefferman showed that there is a function $r \in C^\infty(V)$ uniquely determined modulo functions vanishing to order n+2 on M by the conditions $J(r)=1+O(\rho^{n+1})$, $r|_M=0$, $r>0$ in V_+. r is biholomorphically invariant in the sense that if $\Phi: V \to \tilde{V}$ is a biholomorphism with $\Phi(M)=\tilde{M}$, then $\tilde{r}\circ\Phi = |\det\Phi'|^{\frac{2}{n+1}} r$ modulo ρ^{n+2}. And r is locally determined by M; in fact there is an algorithm to construct r from ρ.

From r Fefferman constructs the Kähler-Lorentz metric g on $\mathbb{C}^* \times V$ with Kähler form $i\partial\bar{\partial}(|z_o|^{\frac{2}{n+1}} r(z))$, where $z_o \in \mathbb{C}^* = \mathbb{C}-\{0\}$, $z \in V$. By the above transformation law for r under biholomorphisms it follows that if z_o is transformed as a fiber variable in the canonical bundle of $\mathbb{C}^n$ (i.e. for a biholomorphism $\Phi: V \to \tilde{V}$, set $\tilde{z}_o = z_o(\det \Phi')^{-1}$), then

$|z_o|^{\frac{2}{n+1}} r(z)$ is an invariantly defined function on $\mathbb{C}^* \times V$, so the Kähler metric g is a biholomorphic invariant of M.

Now associated to any Kähler metric are its scalar invariants, which are complete contractions of tensor products of the curvature tensor and its covariant derivatives. Thus let $R = R_{i\bar{j}k\bar{\ell}}$ be the curvature tensor of g and $\nabla^m R$ be one of its covariant derivatives of order m. So $\nabla^m R$ is a covariant tensor of rank m+4 on $\mathbb{C}^* \times V$, holomorphic in some indices and antiholomorphic in others. A function of the form $W = \mathrm{contr}(\nabla^{m_1} R \otimes \ldots \otimes \nabla^{m_s} R) \in C^\infty(\mathbb{C}^* \times V)$ is called a Weyl invariant of g, where the contraction is taken with respect to $g_{i\bar{j}}$ for some pairing of holomorphic with antiholomorphic indices and all indices are assumed to be contracted so that one obtains a scalar. The weight of W is by definition $w = s + (1/2) \sum_{i=1}^{s} m_i$. It follows easily from the form of the metric g that W is of the form $W(z_o, z) = |z_o|^{\frac{-2w}{n+1}} f(z)$, where $f \in C^\infty(V)$ is some function depending only on z. In order to obtain local invariants of M one would like to restrict f to M. However if W involves too many differentiations then the restriction of f to M will not be independent of the ambiguity of r at order n+2, so cannot give rise to a boundary invariant. So not all Weyl invariants W can be used to construct boundary invariants this way. Fefferman showed that if $w \le n$ then $W|_{\substack{z_o=1 \\ z \in b\Omega}}$ depends only on the Taylor expansion of r through order n+1 along M and defines an invariant of the hypersurface M of weight w. One might hope that every invariant of weight $\le n$ is a linear combination of such "Weyl invariants", and Fefferman showed that this is true for invariants of weight $\le n-20$.

We focus now on the problem of constructing invariants of weight $>n$. It is clear that some such invariants are defined by the above procedure, for if $m_i \le n-3$, $1 \le i \le s$, then a simple derivative count shows that each of the tensors $\nabla^{m_i} R$ is well-defined on $\mathbb{C}^* \times M$ independent of the ambiguity in r, so $W|_{\substack{z_o=1 \\ z \in b\Omega}}$ defines an invariant whose weight $w = s + (1/2) \sum_{i=1}^{s} m_i$ can be arbitrarily large. Also the arguments of the Ambiguity Lemma of [11] can be extended to show that if n=2 and $w \le 2s$ or if $n \ge 3$ and $w \le n+s-1$, then $W|_{\substack{z_o=1 \\ z \in b\Omega}}$ is well defined independent of the ambiguity in r. All of these restrictions, however, have as a consequence that the resulting invariants involve only the coefficients $A^{\ell}_{\alpha\bar{\beta}}$ with $1/2(|\alpha|+|\beta|)+\ell \le n+3/2$; invariants which would involve higher coefficients cannot be expected to be independent of the ambiguity in r.

In order to construct invariants involving the remaining $(A^{\ell}_{\alpha\bar{\beta}})$ we study the higher asymptotics of solutions of the Monge-Ampère equation. Cheng-Yau [6] and Lee-Melrose [13] investigated the global existence and regularity questions for this equation. Taken together, their results can be stated as follows.

Theorem 1.4. If $\Omega\subset\mathbb{C}^n$ is a smooth bounded strictly pseudoconvex domain with smooth defining function ρ, then there is a unique solution $0\le u\in C^{\infty}(\Omega)\cap C^{n+2-\varepsilon}(\bar{\Omega})\forall\varepsilon>0$ to the problem $J(u)=1$, $u|_{b\Omega}=0$. Near $b\Omega$ u has the asymptotic expansion

$$(1.5) \qquad u\sim\rho\sum_{k=0}^{\infty} \eta_k(\rho^{n+1} \log |\rho|)^k, \text{ where } \eta_k\in C^{\infty}(\bar{\Omega}).$$

Some basic remarks concerning expansions of the form (1.5) are in order. First, (1.5) means that for $N\ge0$, $u-\rho\sum_{k=0}^{N} \eta_k(\rho^{n+1}\log |\rho|)^k\in C^{(n+1)(N+1)}(\bar{\Omega})$ and vanishes along with all of its derivatives of order $\le(n+1)(N+1)$ at $b\Omega$. u vanishes to infinite order at $b\Omega$ iff all η_k vanish to infinite order for $k\ge0$, and $u\in C^{\infty}(\bar{\Omega})$ iff the η_k so vanish for $k\ge1$. It is easily seen that the class of functions having an expansion of the form (1.5) is independent of the choice of smooth defining function ρ. And given $\eta_k\in C^{\infty}(\bar{\Omega})$, $k\ge0$, there is a function u, uniquely determined modulo smooth functions vanishing to infinite order at $b\Omega$, having the prescribed expansion.

The smooth part of u is $\rho\eta_0\in C^{\infty}(\bar{\Omega})$, and from the fact that $J(u)=1$ it follows that $J(\rho\eta_0)=1+O(\rho^{n+1})$. Consequently $\rho\eta_0$ is one of Fefferman's approximate solutions r; in particular $\rho\eta_0$ is locally determined mod $O(\rho^{n+2})$ by $M=b\Omega$. The full exact solution u, however, is the unique solution of a global problem on Ω so that the rest of the asymptotic expansion of u cannot be expected to be locally determined by $b\Omega$. Nonetheless a part of it is and can be used to construct further invariants, as is described next.

We return to the local setup of a strictly pseudoconvex hypersurface $M \subset V \subset \mathbb{C}^n$, and ask which formal expansions of the form (1.5) can solve $J(u)=1$ to infinite order along M. It turns out that the expansion is uniquely determined once one chooses $T^{n+1}\eta_o$ on M, where T is a vector field transverse to M. Thus $T^{n+1}\eta_o$ can be thought of as the second piece of local Cauchy data for the equation $J(u)=1$. In order to precisely state the result, we rewrite (1.5) in terms of one of Fefferman's special defining functions r. So choose some $r \in C^\infty(V)$ with $J(r)=1+O(\rho^{n+1})$; if $u \sim r \sum_{k=o}^{\infty} \eta_k (r^{n+1} \log |r|)^k$ is to solve $J(u)=1$ to infinite order then we know by Fefferman that $\eta_o = 1+O(r^{n+1})$.

<u>Theorem 1.6:</u> a) Let $a \in C^\infty(M)$. Then there is a unique expansion $u \sim r \sum_{k=o}^{\infty} \eta_k (r^{n+1} \log |r|)^k$ satisfying $J(r)=1$ to infinite order along M, for which $\eta_o = 1+ar^{n+1}+O(r^{n+2})$.

b) For $k \geq 1$, each η_k mod $O(r^{n+1})$ is independent of the choice of a.

c) For $k \geq 1$, each η_k mod $O(r^{n+1})$ is also independent of the choice of approximate solution r to $J(r)=1+O(r^{n+1})$.

The proof of Theorem 1.6 will appear elsewhere. It shows that the η_k can be constructed inductively by solving the equation $J(u)=1$ to higher and higher order along M.

Because of Theorem 1.6, we now have the locally determined functions η_k mod ρ^{n+1}, $k \geq 1$, in addition to the approximate solution r mod ρ^{n+2} with which to construct invariants. From the biholomorphic invariance of the Monge-Ampère equation one knows that if $\Phi: V \to \tilde{V}$ is a biholomorphism with $\Phi(M)=\tilde{M}$ and u is a solution to $J(u)=1$ to infinite order along M, then $\tilde{u}$ defined by $\tilde{u} \circ \Phi = |\det \Phi'|^{\frac{2}{n+1}} u$ is a solution of $J(\tilde{u})=1$ to infinite order along $\tilde{M}$. As $\tilde{r} \circ \Phi = |\det \Phi'|^{\frac{2}{n+1}} r$ mod ρ^{n+2}, it follows upon transforming the expansion (1.5) that

$$\tilde{\eta}_k \circ \Phi = |\det \Phi'|^{-2k} \eta_k \bmod \rho^{n+1}, \quad k \geq 1. \tag{1.7}$$

In particular, $b\eta_k = \eta_k|_M$ is an invariant of weight $k(n+1)$ for $k \geq 1$. These are the simplest examples of new invariants coming from the higher asymptotics of the Monge-Ampère solution. More complicated examples involving derivatives of the η_k at M may be constructed as follows.

First observe that the transformation law (1.7) for η_k implies that $|z_0|^{-2k}\eta_k(z)$ is invariantly defined on the canonical bundle $\mathbb{C}^*\times V$. Thus the tensors $\nabla^{\ell}(|z_0|^{-2k}\eta_k(z))$ are also invariantly defined, where again ∇^{ℓ} is a covariant derivative of order ℓ with respect to the metric g, holomorphic in some indices and antiholomorphic in others. Hence an expression of the form

$$\text{contr}\,(\nabla^{m_1}R\otimes\ldots\otimes\nabla^{m_s}R\otimes\nabla^{\ell_1}(|z_0|^{-2k_1}\eta_{k_1})\otimes\ldots\otimes\nabla^{\ell_t}(|z_0|^{-2k_t}\eta_{k_t})),$$

where $\nabla^{m_i}R$ are as before the covariant derivatives of curvature of g and contr indicates complete contraction with respect to $g_{i\bar{j}}$ for some pairing of the indices, will, when restricted to $z_0=1$, $z\in M$, give rise to an invariant of weight

$$w=s+(1/2)\sum_{i=1}^{s} m_i + \sum_{j=1}^{t}\left[(n+1)k_j+(1/2)\ell_j\right],\ \text{so long as it is}$$

independent of the ambiguity of r and of the η_k. As with Fefferman's invariants this can be insured in a variety of ways; for example if $w\le n+\sum_{j=1}^{t}(n+1)k_j$, or if $m_i\le n-3$, $1\le i\le s$, and $\ell_j\le n$, $1\le j\le t$, then this procedure produces a well defined invariant. Unfortunately, as will be shown in §2, not all invariants arise this way.

As previously mentioned, the simplest examples of these new invariants are $b\eta_k=\eta_k|_M$. The first of these, $b\eta_1$, which is the coefficient of the first log term on M, is of particular importance. As the next result shows, it governs whether or not any log terms occur at all in the Monge-Ampère solution u.

Proposition 1.8: Let M be a strictly pseudoconvex hypersurface, $p\in M$, and suppose that $b\eta_1=0$ near p. If u is any asymptotic expansion of the form (1.5) solving $J(u)=1$ to infinite order on M, then $\eta_k=0$ to infinite order near p, $k\ge 1$. In particular $u\in C^{\infty}$ near p.

In light of Proposition 1.8 one would like to know how restrictive a condition on M it is that $b\eta_1=0$ in an open set. Of course any invariant must vanish on a hypersurface which is locally biholomorphically equivalent to the sphere, since all $A^{\ell}_{\alpha\bar{\beta}}=0$ for the sphere, so this provides one example. However there are other such hypersurfaces.

Proposition 1.9: There are real analytic strictly pseudoconvex hypersurfaces M which are not locally biholomorphically equivalent to the sphere, for which $b\eta_1=0$.

Computationally it is a difficult task to identify explicitly $b\eta_1$ as a polynomial in the Moser normal form coefficients. It is possible to give an explicit formula for its linear part though. As a polynomial in the variables $A^{\ell}_{\alpha\bar{\beta}}$, $b\eta_1$ may be decomposed into homogeneous pieces so that its linear part is well defined. Since the weight of $b\eta_1$ is n+1, by condition i) in Definition 1.3, the linear part must be a linear combination of $A^{\ell}_{\alpha\bar{\beta}}$ with $|\alpha|+|\beta|+2\ell=2n+4$. From the collection of all such $A^{\ell}_{\alpha\bar{\beta}}$ one may form the polynomial $\sum_{|\alpha|+|\beta|+2\ell=2n+4} A^{\ell}_{\alpha\bar{\beta}} z^{\alpha}\bar{z}^{\beta}v^{\ell}$, which is one piece of the Taylor expansion of a hypersurface in normal form. The formula for the linear part of $b\eta_1$ will be given in the form of a differential operator applied to this polynomial. For $\gamma\in\mathbb{C}$ define a second order differential operator L_γ on $\mathbb{C}^n = \{(z,w)\in\mathbb{C}^{n-1}\times\mathbb{C}\}$ by

$$L_\gamma f= \sum_{j=1}^{n-1} (f_{z_j\bar{z}_j} + z_j f_{z_j\bar{w}} + \bar{z}_j f_{\bar{z}_j w}) + 2uf_{w\bar{w}} + (\gamma-1)f_u.$$

Recalling the notion of the strength of a monomial, one sees easily that L_γ maps polynomials of strength S to polynomials of strength S-2.

Proposition 1.10. The linear part of $b\eta_1$ is

$$[(n+2)!(n+1)!]^{-1} L_{n+1}L_n\cdots L_1L_o \Big(\sum_{|\alpha|+|\beta|+2\ell=2n+4} A^{\ell}_{\alpha\bar{\beta}}\, z^{\alpha}\bar{z}^{\beta}v^{\ell}\Big).$$

Since the strength of $z^{\alpha}\bar{z}^{\beta}v^{\ell}$ is $|\alpha|+|\beta|+2\ell=2n+4$, the expression in Proposition 1.10 is a polynomial of strength 0, so is just a linear combination of the $A^{\ell}_{\alpha\bar{\beta}}$. It will be made more explicit in Corollary 2.6. Proposition 1.10 is proved by linearizing the algorithm used to solve J(u)=1.

2. Explicit Identification of Invariants

In this section all invariants of sufficiently low weight will be identified. Some information can already be obtained simply by considering the homogeneity with respect to $\mathbb{R}_+$ and the invariance under U(n-1). These results are then extended by an analysis of the action of $\mathbb{H}^{n-1}$.

We begin by recalling that classical invariant theory identifies all of the polynomials in the variables $(A^{\ell}_{\alpha\bar\beta})_{|\alpha|,|\beta|\geq 2,\ell\geq 0}$ which are invariant under U(n-1). In fact since U(n-1) acts on $(A^{\ell}_{\alpha\bar\beta})$ as tensors in $\alpha,\bar\beta$, it follows that any polynomial in $(A^{\ell}_{\alpha\bar\beta})$ which is U(n-1)-invariant must be a linear combination of the polynomials $\operatorname{contr}(A^{\ell_1}_{p_1\bar q_1}\otimes\ldots\otimes A^{\ell_s}_{p_s\bar q_s})$, where $p_i,q_i\geq 2$, $\ell_i\geq 0$ and the contraction is taken with respect to $\delta_{i\bar j}$ for some pairing of the lower indices.

In counting weights, recall that the weight of $A^{\ell}_{\alpha\bar\beta}$ is $1/2(|\alpha|+|\beta|)+\ell-1$, the weight of a monomial $A^{\ell_1}_{\alpha_1\bar\beta_1}\ldots A^{\ell_s}_{\alpha_s\bar\beta_s}$ is $w=\sum_{i=1}^{s}\left[(1/2)(|\alpha_i|+|\beta_i|)+\ell_i-1\right]$, and that a polynomial $P(A^{\ell}_{\alpha\bar\beta})$ has weight w if it is a sum of monomials each of which has weight w. Clearly $0\leq w\in(1/2)\mathbb{Z}$ and since each $A^{\ell}_{\alpha\bar\beta}$ has weight ≥ 1 the only polynomials of weight 0 are constants. Also note that if $P(A^{\ell}_{\alpha\bar\beta})$ has weight w and is U(n-1)-invariant, then necessarily $w\in\mathbb{Z}$. In fact the weight of $\operatorname{contr}(A^{\ell_1}_{p_1\bar q_1}\otimes\ldots\otimes A^{\ell_s}_{p_s\bar q_s})$ is integral since one must have $\sum_1^s p_i=\sum_1^s q_i$ in order that a complete pairing of the indices be possible, so $1/2\sum_1^s(p_i+q_i)\in\mathbb{Z}$ follows.

<u>Theorem 2.1:</u> a) $n\geq 3$. There are no nonzero invariants of weight 1. The space of weight 2 invariants is 1-dimensional and is spanned by $\|A^{0}_{2\bar 2}\|^2=\sum_{\substack{|\alpha|=2\\|\beta|=2}}|A^{0}_{\alpha\bar\beta}|^2$.

b) $n=2$. There are no nonzero invariants of weights 1 or 2. For $w=3,4,5$ the space of weight w invariants is 1-dimensional and is spanned by $A^{0}_{4\bar 4}$, $|A^{0}_{2\bar 4}|^2$, and $5|A^{0}_{2\bar 5}|^2+9|A^{0}_{3\bar 4}|^2-2\operatorname{Im}A^{0}_{4\bar 2}A^{1}_{2\bar 4}$, resp. (When $n=2$ each tensor $A^{\ell}_{p\bar q}$ is actually a scalar, so we abuse the notation and identify the tensor with its one component.)

<u>Proof</u>: a) The only monomials of weight 1 are $A^0_{\alpha\bar\beta}$ with $|\alpha|=|\beta|=2$. So any polynomial of weight 1 is of the form $\sum_{|\alpha|=|\beta|=2} c_{\alpha\bar\beta} A^0_{\alpha\bar\beta}$ for some coefficients $c_{\alpha\bar\beta}$. If such a polynomial is to be U(n-1)-invariant it must be a linear combination of terms of the form contr $(A^0_{2\bar2})$. But by the symmetry in $\alpha,\bar\beta$ there is effectively only one way to make the pairing so the polynomial must be a multiple of $\mathrm{tr}^2 A^0_{2\bar2}$. However by Moser's normalization $\mathrm{tr}\, A^0_{2\bar2}=0$ so also $\mathrm{tr}^2 A^0_{2\bar2}=0$, and the invariant vanishes.

The only monomials of weight 2 are of the form $A^0_{\alpha\bar\beta}$, $|\alpha|+|\beta|=6$; $A^1_{\alpha\bar\beta}$, $|\alpha|=|\beta|=2$; or $A^0_{\alpha\bar\beta}\, A^0_{\gamma\bar\delta}$, $|\alpha|=|\beta|=|\gamma|=|\delta|=2$. Thus as above all weight 2 polynomials which are U(n-1)-invariant are linear combinations of polynomials of the form contr $A^0_{3\bar3}$, contr $A^1_{2\bar2}$, contr $(A^0_{2\bar2} \otimes A^0_{2\bar2})$. Now contr $A^0_{3\bar3} = \mathrm{tr}^3 A^0_{3\bar3}=0$ and contr $A^1_{2\bar2} = \mathrm{tr}^2 A^1_{2\bar2} = 0$ by Moser's normalizations. In pairing the indices in $A^0_{2\bar2} \otimes A^0_{2\bar2}$, if any two indices in either $A^0_{2\bar2}$ term are paired with one another then the corresponding contraction leaves one with $\mathrm{tr}\, A^0_{2\bar2} = 0$. Thus the only nonzero contraction arises from pairing both holomorphic indices of the first $A^0_{2\bar2}$ term with both antiholomorphic indices of the second $A^0_{2\bar2}$ and vice versa, giving contr $(A^0_{2\bar2} \otimes A^0_{2\bar2}) = \sum_{|\alpha|=|\beta|=2} A^0_{\alpha\bar\beta}\, A^0_{\beta\bar\alpha} = \|A^0_{2\bar2}\|^2$. Hence $\|A^0_{2\bar2}\|^2$ is the only weight 2 polynomial which is U(n-1)-invariant. It remains to show that it is invariant also under $\mathbb{H}^{n-1}$. If $h=(\xi,\eta)\in bD=\mathbb{H}^{n-1}$, recall that the corresponding fractional linear transformation is $h(z,w)= \frac{(z+w\xi,w)}{1+z\cdot\bar\xi+w\bar\eta} = (z^*,w^*)$. Since $2u^*-|z^*|^2 = \frac{2u-|z|^2}{|1+z\cdot\bar\xi+w\bar\eta|^2}$, if N is the normal form $2u-|z|^2 = \sum A^\ell_{\alpha\bar\beta}\, z^\alpha \bar z^\beta v^\ell = f(z,\bar z,v)$, then h^*N is the hypersurface $2u-|z|^2=|1+z\cdot\bar\xi+w\bar\eta|^2 f(z^*,\bar z^*,v^*)$. But $z^*=z+(\text{strength} \ge 2)$, $w^*=w+(\text{strength} \ge 3)$, so it follows that h^*N is of the form $2u-|z|^2= \sum_{|\alpha|=|\beta|=2} A^0_{\alpha\bar\beta}\, z^\alpha \bar z^\beta$ + (terms of strength ≥ 5); in particular is in normal form through terms of strength 4. Hence if $T(z,w)=(\tilde z,\tilde w)$ is the unique mapping satisfying (1.2) so that $T^*(h^*N)$ is in normal form, then $\tilde z=z+(\text{strength} \ge 4)$, $\tilde w=w+(\text{strength} \ge 5)$, so Nh has exactly the same $A^0_{2\bar2}$ term as N. Thus the polynomial $\|A^0_{2\bar2}\|^2$ is $\mathbb{H}^{n-1}$-invariant.

b) The arguments given above hold just as well in case n=2. However when n=2 the trace conditions force $A^{\ell}_{2\bar{2}} = A^{\ell}_{2\bar{3}} = A^{\ell}_{3\bar{3}} = 0$, $\ell \geq 0$. Hence $\|A^{0}_{2\bar{2}}\|^{2}=0$ and there are no nontrivial invariants of weights 1 or 2. The lowest weight coefficient in $\mathbb{C}^2$ is $A^{0}_{2\bar{4}} = \overline{A^{0}_{4\bar{2}}}$, of weight 2. Thus any quadratic monomial must have weight at least 4, so the only weight 3 polynomials are linear and are thus linear combinations of the coefficients $A^{\ell}_{p\bar{q}}$ of weight 3. A quick check shows that the only such possibilities which are U(1)-invariant are $A^{0}_{4\bar{4}}$, $A^{1}_{3\bar{3}}$, $A^{2}_{2\bar{2}}$, but the last two vanish by the trace conditions, so $A^{0}_{4\bar{4}}$ is the only U(1)-invariant polynomial of weight 3. We need to show that $A^{0}_{4\bar{4}}$ is also $\mathbb{H}^1$-invariant. This will be established by direct calculation in Lemma 2.8. However we can see now that $A^{0}_{4\bar{4}}$ must be fully invariant, as follows. In $\mathbb{C}^2$, by the analysis of §1, the coefficient of the first log term in the solution of the Monge-Ampère equation, $b\eta_1$, is an invariant of weight 3. From Proposition 1.10 it can be seen that that $b\eta_1 \neq 0$. Hence there is a nonzero invariant of weight 3. We have shown that $A^{0}_{4\bar{4}}$ is the only possibility; it follows that it must be invariant and must be a multiple of $b\eta_1$. Actually, the fact that the linear part of $b\eta_1$ is a multiple of $A^{0}_{4\bar{4}}$ can be directly calculated from Proposition 1.10: setting n=2 and computing shows that the linear part of $b\eta_1$ is $4A^{0}_{4\bar{4}}$. Of course, by the above we know that $b\eta_1$ must equal its linear part, so we have

<u>Proposition 2.2</u>: When n=2, $b\eta_1 = 4A^{0}_{4\bar{4}}$.

Now return to Theorem 2.1, where we are up to the consideration of weight 4 invariants in $\mathbb{C}^2$. The only possible quadratic terms involve $A^{0}_{2\bar{4}}$ and $A^{0}_{4\bar{2}}$, and the only one of these which is U(1)-invariant is $|A^{0}_{2\bar{4}}|^2$. Since $2\mathrm{Re}A^{0}_{2\bar{4}}z^2\bar{z}^4$ is the lowest strength term occurring in a normal form, the same argument that was used in a) to show that $\mathbb{H}^{n-1}$ acts trivially on the $A^{0}_{2\bar{2}}$ coefficients in $\mathbb{C}^n$ when $n \geq 3$ applies here to show that $\mathbb{H}^1$ acts trivially on $A^{0}_{2\bar{4}}$. Thus $|A^{0}_{2\bar{4}}|^2$ is an invariant of weight 4. Any other polynomial of weight 4 must be a linear combination of coefficients of weight 4; taking into account the trace conditions the only possibilities which are unitarily invariant are $A^{0}_{5\bar{5}}$, $A^{1}_{4\bar{4}}$. That no linear combination of these is fully invariant is a consequence of part a) of the next theorem.

Theorem 2.3. Let $P(A^{\ell}_{\alpha\bar{\beta}})$ be an invariant polynomial of weight w in $\mathbb{C}^n$, $n\geq 2$.

a) If $w\neq n+1$, then the linear part of P vanishes.

b) If $w=n+1$, then the linear part of P is a multiple of

$$\mathcal{L} = \sum_{j=o}^{\left[\frac{n-2}{2}\right]} 2^{-2j} \frac{\binom{n}{j}}{\binom{n+2}{2j}} \operatorname{tr}^p (A^{2j}_{p\bar{p}}), \text{ where we have set } p=n+2-2j.$$

We prove Theorem 2.3 now, returning later to the consideration of weight 5 invariants in Theorem 2.1.

Proof: The linear part of P is a linear combination of coefficients of weight w. And since each $A^{\ell}_{\alpha\bar{\beta}}$ has weight at least 1, the quadratic and higher terms of P can only involve $A^{\ell}_{\alpha\bar{\beta}}$ of weight at most w-1. Consequently P must be independent of coefficients $A^{\ell}_{\alpha\bar{\beta}}$ of weight w-1/2. So P vanishes on any normal form N of the form $2u-|z|^2 = \sum_{1/2(|\alpha|+|\beta|)+\ell=w+1/2} A^{\ell}_{\alpha\bar{\beta}} z^{\alpha}\bar{z}^{\beta}v^{\ell}$. We will compute Nh for $h\in\mathbb{H}^{n-1}$, and will find that Nh has no nonzero coefficients of weight $\leq w-1$, but does have nontrivial coefficients of weight w. Since P is invariant, its linear part must vanish on the weight w coefficients of Nh. This gives a number of conditions on the linear part of P, from which we will be able to deduce the conclusion.

We proceed with the calculation of Nh. For our purposes it will suffice to take only certain special N of the type described above, and only certain $h\in\mathbb{H}^{n-1}$. For $z\in\mathbb{C}^{n-1}$, write $z=(z_1,z')$ where $z_1\in\mathbb{C}$, $z'\in\mathbb{C}^{n-2}$; similarly $\xi=(\xi_1,\xi')$. The $h=(\xi,\eta)$ that we will use have $\xi'=0$, so we take $\xi=(\zeta,0)$, $\zeta\in\mathbb{C}$.

Lemma 2.4: Let N be the normal form $2u-|z|^2=2\mathrm{Re}\, z_1^k\bar{z}_1^{k+1}v^{\ell}$, where $k\geq 3$, $\ell\geq 0$, and let $h=(\xi,\eta)\in\mathbb{H}^{n-1}$, where $\xi=(\zeta,0)$.

a) If $k\geq 4$ then Nh is given by

$$2u-|z|^2=2\mathrm{Re}\, z_1^k\bar{z}_1^{k+1}v^{\ell}$$

$$+ 2\mathrm{Re}\left\{\frac{\ell i}{4}\left(\bar\zeta|z_1|^{2k+2}|z|^2 - \zeta z_1^{k}\bar z_1^{k+2}|z|^2\right)v^{\ell-1}\right.$$

$$+ 1/2\left(\bar\zeta|z_1|^{2k}((k+1)|z|^2-(2k+\ell-2)|z_1|^2)\right.$$

$$\left.+\zeta z_1^{k-1}\bar z_1^{k+1}(k|z|^2-(2k+\ell)|z_1|^2\right)v^{\ell}$$

$$\left.+ i\left(-(k+1)\bar\zeta|z_1|^{2k}+k\zeta z_1^{k-1}\bar z_1^{k+1}\right)v^{\ell+1}\right\} + (\text{terms of strength} \geq 2k+2\ell+3)$$

b) if $k=3$, $\ell=0$, then Nh is given by

$$2u-|z|^2=2\ \mathrm{Re}\ z_1^{3}\bar z_1^{4}$$

$$+ 2\ \mathrm{Re}\left\{1/2\left[4\bar\zeta|z_1|^6(|z|^2-|z_1|^2) + 3\zeta z_1^{2}\bar z_1^{4}\ (|z|^2-2|z_1|^2)\right]\right.$$

$$\left.+ i\left(4\bar\zeta\left(\frac{6}{(n+1)n(n-1)}|z|^6-|z_1|^6\right) + 3\zeta z_1^{2}\bar z_1^{4}\right)v\right\}$$

$$+ (\text{terms of strength} \geq 9)$$

<u>Proof</u>: The fractional linear transformation corresponding to h is $h(z,w)=(1+z_1\bar\zeta+w\bar\eta)^{-1}(z_1+w\zeta,\ z',w)$. It is straightforward to compute h^*N in the form $2u-|z|^2=g(z,\bar z,w,\bar w)$ modulo terms of strength $\geq 2k+2\ell+3$. One checks easily that h^*N may be rewritten in the form $2u-|z|^2=f(z,\bar z,v)$ + (strength $\geq 2k+2\ell+3$) by simply substituting $u=1/2|z|^2$ in g. The resulting formula for f is given in a) above. If $k\geq 4$ then this expression is already in normal form, so $Nh=h^*N$ modulo strength $\geq 2k+2\ell+3$ and a) is proved. But if $k=3$ then the expression in a) is not a normal form because of the $|z_1|^6v^{\ell+1}$ term which violates the trace condition $\mathrm{tr}^3A^{\ell+1}_{3\bar 3}=0$. Thus if $k=3$ one knows only that h^*N is given by a). However it is found that if $k=3$, $\ell=0$, and one sets $T(z,w)=(\tilde z,\tilde w)$ for $\tilde z=z(1-2i\lambda w^3)$, $\tilde w=w(1-i\lambda w^3)$, and $\lambda=\frac{24}{(n+1)n(n-1)}\mathrm{Im}\zeta$, then T satisfies (1.2) and $T^*(h^*N)$ is given by the expression in b) above. As that expression is a normal form it follows that $T^*(h^*N)=Nh$ mod strength ≥ 9, and b) is proved. (A method of computing traces which can be used, for example, to check that the expression in b) above satisfies $\mathrm{tr}^3\ A^1_{3\bar 3}=0$, is described below).

We continue the proof of Theorem 2.3. The linear part of P is a $U(n-1)$-invariant linear combination of coefficients of weight w, so it must be a sum of complete contractions; as a complete contraction which is linear must be a full trace it follows that the linear part of P is of the form $\sum_{p=4}^{w+1} c_p \operatorname{tr}^p(A_{p\bar{p}}^{w+1-p})$ for some coefficients c_p. Moser's normalizations have again been used to eliminate the terms corresponding to $p=2,3$. The traces which occur in this sum can be calculated for specific examples by using the observation that $(p!)^2\operatorname{tr}^p A_{p\bar{p}}^{\ell} = \Delta^p \left(\sum_{|\alpha|=|\beta|=p} A_{\alpha\bar{\beta}}^{\ell} z^{\alpha}\bar{z}^{\beta}\right)$, where $\Delta = \sum_{j=1}^{n-1} \frac{\partial^2}{\partial z_j \partial \bar{z}_j}$. In the following we abuse notation and identify the tensor $A_{p\bar{p}}^{\ell}$ with the polynomial $\sum_{|\alpha|=|\beta|=p} A_{\alpha\bar{\beta}}^{\ell} z^{\alpha}\bar{z}^{\beta}$.

One finds without difficulty that $\operatorname{tr}^p |z_1|^{2p}=1$,

$$\operatorname{tr}^{p+1}|z_1|^{2p}|z|^2 = \frac{n+p-1}{p+1} \text{ and } \operatorname{tr}^p |z|^{2p} = \binom{n+p-2}{p}.$$

In proving Theorem 2.3 we may assume that $w \geq 3$ since in Theorem 2.1 all invariants of weights 1 and 2 have been identified.

Begin with $w=3$. So the linear part of P is $c_4 \operatorname{tr}^4(A_{4\bar{4}}^0)$. As previously discussed, P vanishes on the normal form $N: 2u-|z|^2 = 2\operatorname{Re} z_1^3\bar{z}_1^4$. So if $h \in \mathbb{H}^{n-1}$ then $P(Nh)=0$. If h is of the form $h=(\xi,\eta)$ with $\xi=(\zeta,0)$, then Nh is given by Lemma 2.4b). On Nh, P equals its linear part, so $0=P(Nh)=4c_4\operatorname{Re}\zeta \operatorname{tr}^4\left[|z_1|^6(|z|^2-|z_1|^2)\right] = (n-2)c_4\operatorname{Re}\zeta$. As $\operatorname{Re}\zeta$ is arbitrary, $c_4=0$ unless $n=2$. When $n=2$ we already know that $A_{4\bar{4}}^0$ is the only invariant of weight 3; this agrees with Theorem 2.3b). Thus Theorem 2.3 is proved in case $w=3$.

Now let $w \geq 4$. First consider the normal form N: $2u-|z|^2=2\operatorname{Re} z_1^w\bar{z}_1^{w+1}$. Transforming N as in Lemma 2.4a) and arguing as in the case $w=3$, one deduces that

$$c_{w+1}\operatorname{Re}\zeta \operatorname{tr}^{w+1}\left[|z_1|^{2w}\left((w+1)|z|^2-(2w-2)|z_1|^2\right)\right]$$

$- 2(w+1) c_w \operatorname{Im}\zeta \operatorname{tr}^w|z_1|^{2w}=0$, or $(n+1-w)c_{w+1}\operatorname{Re}\zeta-2(w+1)c_w\operatorname{Im}\zeta=0$. Thus $c_w=0$ always and $c_{w+1}=0$ unless $w=n+1$. Next fix k,ℓ with $k+\ell=w$, $k\geq 4$, $\ell \geq 1$, and apply Lemma 2.4a) with ζ chosen so that $\operatorname{Re}\zeta=0$. Then $P(Nh)=0$

gives

$1/2\ell\ \mathrm{Im}\zeta\ c_{k+2}\ \mathrm{tr}^{k+2}(|z_1|^{2k+2}|z|^2)-2(k+1)\ \mathrm{Im}\zeta\ c_k\ \mathrm{tr}^k|z_1|^{2k}=0$, so

$$\frac{\ell(n+k)}{2(k+2)}\, c_{k+2} - 2(k+1)c_k=0, \text{ and}$$

$$c_k = \frac{(w-k)(n+k)}{4(k+2)(k+1)}\, c_{k+2}\ ,\ 4\le k\le w-1. \tag{2.5}$$

If $w\ne n+1$, then starting from $c_{w+1}=c_w=0$, (2.5) inductively gives $c_p=0$, $p\ge 4$. Hence the linear part of P vanishes and Theorem 2.3a) is proved.

For $w=n+1$, from $c_w=0$ and (2.5) one obtains $c_p=0$ if $p\ge 4$ and $w-p=n+1-p$ is even. If $n+1-p$ is odd we may write $n+2-p=2j$ where $0\le j\le \left[\frac{n-2}{2}\right]$; the recursion (2.5) can then be written in terms of j and solved to conclude the proof of Theorem 2.3b).

<u>Corollary 2.6</u>: The linear part of the first Monge-Ampère log term coefficient $b\eta_1$ is $(n+2)\mathcal{L}$, where $\mathcal{L}$ is as in Theorem 2.3b). In particular there exist invariants whose linear term is a nonzero multiple of $\mathcal{L}$.

<u>Proof</u>: $b\eta_1$ is an invariant of weight n+1, so by Theorem 2.3b) its linear part is a multiple of $\mathcal{L}$. A formula for the linear part is given in Proposition 1.10; evaluating that formula on a normal form $\sum_{|\alpha|=|\beta|=n+2} A^0_{\alpha\bar\beta}\, z^\alpha\bar z^\beta$ gives $\left[(n+2)!(n+1)!\right]^{-1}\Delta^{n+2}\left[\sum_{|\alpha|=|\beta|=n+2} A^0_{\alpha\bar\beta}\, z^\alpha\bar z^\beta\right]$ $=(n+2)\ \mathrm{tr}^{n+2}\ A^0_{(n+2)(\overline{n+2})}$. Comparison with $\mathcal{L}$ shows that the multiple is (n+2).

We return now to the identification of the weight 5 invariants in $\mathbb{C}^2$ in Theorem 2.1b). The argument relies on a formula for the action of $\mathbb{H}^1$ on the low weight terms of a normal form. A normal form N in $\mathbb{C}^2$ may be written

$$\begin{aligned} 2u-|z|^2 &= 2\mathrm{Re}A^0_{2\bar4}\, z^2\bar z^4 + 2\mathrm{Re}(A^0_{2\bar5}\, z^2\bar z^5 + A^0_{3\bar4}\, z^3\bar z^4\,) \\ &\quad + 2\mathrm{Re}(A^0_{2\bar6}\, z^2\bar z^6 + A^0_{3\bar5}\, z^3\bar z^5 + A^1_{2\bar4}\, z^2\bar z^4 v) + A^0_{4\bar4}|z|^8 \\ &\quad + (\text{terms of strength } \ge 9). \end{aligned} \tag{2.7}$$

Lemma 2.8: Let N be given by (2.7) and $h=(\xi,\eta)\in\mathbb{H}^1$, where $\eta=1/2|\xi|^2+i\mu, \mu\in\mathbb{R}$. Nh is a normal form so may be written in the form (2.7) with coefficients $B^{\ell}_{p\bar{q}}$ replacing $A^{\ell}_{p\bar{q}}$. The B's are given by:

$$B^o_{2\bar{4}} = A^o_{2\bar{4}}$$

$$B^o_{2\bar{5}} = A^o_{2\bar{5}} - \xi A^o_{2\bar{4}}$$

$$B^o_{3\bar{4}} = A^o_{3\bar{4}} + 1/3\bar{\xi} A^o_{2\bar{4}}$$

$$B^o_{2\bar{6}} = A^o_{2\bar{6}} - 2\xi A^o_{2\bar{5}} + \xi^2 A^o_{2\bar{4}}$$

$$B^o_{3\bar{5}} = A^o_{3\bar{5}} + 3/2\bar{\xi} A^o_{2\bar{5}} - 3/2\xi A^o_{3\bar{4}} - (|\xi|^2+i\mu)A^o_{2\bar{4}}$$

$$B^1_{2\bar{4}} = A^1_{2\bar{4}} - 5i\bar{\xi} A^o_{2\bar{5}} + 3i\xi A^o_{3\bar{4}} + i(3|\xi|^2+4i\mu)A^o_{2\bar{4}}$$

$$B^o_{4\bar{4}} = A^o_{4\bar{4}} .$$

Proof: The computations are lengthy; we outline the steps and leave the verification of the details to the reader. As in the last part of the proof of Theorem 2.1a), if we let $f(z,\bar{z},v)$ denote the right hand side of (2.7) and $h(z,w)=(z^*,w^*)$ be the fractional linear transformation associated to h, then h^*N is the hypersurface $2u-|z|^2 = |1+z\bar{\xi}+w\bar{\eta}|^2 f(z^*,\bar{z}^*,v^*)$. One substitutes for (z^*,w^*), Taylor expands everything, and collects terms according to strength, obtaining h^*N in the form $2u-|z|^2=g(z,\bar{z},w,\bar{w})$. It is easily determined that modulo terms of strength ≥ 9, h^*N may be written in the form $2u-|z|^2=\tilde{f}(z,\bar{z},v)$ by simply substituting $u=1/2|z|^2$ in g. The result is that h^*N is given through terms of strength ≤ 8 by:

$$2u-|z|^2 = 2\mathrm{Re}\, A^o_{2\bar{4}}\, z^2\bar{z}^4$$

$$+ 2\mathrm{Re}\Big[(A^o_{2\bar{5}} - 2\xi A^o_{2\bar{4}})\, z^2\bar{z}^5 + (A^o_{3\bar{4}} + \bar{\xi}A^o_{2\bar{4}})z^3\bar{z}^4 + 2i\xi A^o_{2\bar{4}} z\bar{z}^4 v - 4i\bar{\xi}A^o_{2\bar{4}}\, z^2\bar{z}^3 v\Big]$$

$$+ 2\mathrm{Re}\Big\{\Big[A^o_{2\bar{6}} - 3\xi A^o_{2\bar{5}} + \frac{13}{4}\,\xi^2 A^o_{2\bar{4}}\Big] z^2\bar{z}^6$$

$$+ \Big[A^o_{3\bar{5}} + \frac{3}{2}\bar{\xi}A^o_{2\bar{5}} - \frac{3}{2}\,\xi A^o_{3\bar{4}} - (3|\xi|^2+i\mu)A^o_{2\bar{4}}\Big] z^3\bar{z}^5$$

$$+ \left[A^1_{2\bar4} - 5i\bar\xi A^0_{2\bar5}+3i\xi A^0_{3\bar4}+i(11|\xi|^2+4i\mu)A^0_{2\bar4}\right]z^2\bar z^4 v\Bigg\} + \left[A^0_{4\bar4} + \mathrm{Re}(A_{2\bar4}^0\bar\xi^2)\right]|z|^8$$

$$+ 2\mathrm{Re}\Bigg\{\left[2i\xi A^0_{2\bar5} - 5i\xi^2 A^0_{2\bar4}\right]z\bar z^5 v + \left[-4i\bar\xi A^0_{3\bar4} - 2i\bar\xi^2 A^0_{2\bar4}\right]|z|^6 v-\xi^2 A^0_{2\bar4}\bar z^4 v^2$$

$$+ 8|\xi|^2 A^0_{2\bar4}\ z\bar z^3 v^2 - 6\bar\xi^2 A^0_{2\bar4}\ z^2\bar z^2 v^2 \Bigg\}.$$

Now h^*N must be put into normal form by a change of coordinates. One finds that if $T(z,w)=(\tilde z,\tilde w)$, where

$$\tilde z=z-\alpha_6-\alpha_7 \ , \ \tilde w=w-\beta_7-\beta_8 \text{ and } \alpha_6,\alpha_7,\beta_7,\beta_8$$

are given by: $\alpha_6 = -2\bar\xi A^0_{4\bar2}\ z^4w+4\xi A^0_{4\bar2}\ z^2w^2 - \frac{4}{3}\ \bar\xi\ A^0_{2\bar4}\ w^3$

$$\beta_7 = \frac{4}{3}\ \xi A^0_{4\bar2}\ zw^3,$$

$$\alpha_7=\left[4\bar\xi^2 A^0_{4\bar2} - 2\bar\xi A^0_{5\bar2}\right]z^5w-8|\xi|^2 A^0_{4\bar2}z^3w^2+\left[4\xi^2 A^0_{4\bar2}-8i\,\mathrm{Im}(\bar\xi A^0_{3\bar4})\right]zw^3$$

$$\beta_8 = -\bar\xi^2 A^0_{4\bar2}\ z^4w^2-i\ \mathrm{Im}(4\bar\xi A^0_{3\bar4}+2\bar\xi^2 A^0_{2\bar4})w^4,$$

then $T^*(h^*N)$ is in normal form through terms of strength 8 and is in fact the normal form whose coefficients are given by Lemma 2.8. It follows that $T^*(h^*N)$ is Nh modulo strength ≥ 9, so that the coefficients given in Lemma 2.8 are those of Nh.

Note that we have now established directly that $A^0_{4\bar4}$ is fully invariant, since $B^0_{4\bar4} = A^0_{4\bar4}$.

To conclude the proof of Theorem 2.1, consider a weight 5 invariant P in $\mathbb{C}^2$. Any cubic expression in the $A^\ell_{p\bar q}$ has weight at least 6, and by Theorem 2.3a) any weight 5 invariant has no linear term, so P must be pure quadratic. The only U(1)-invariant possibilities are linear combinations of $|A^0_{2\bar5}|^2$, $|A^0_{3\bar4}|^2$, $A^0_{2\bar4}\ A^1_{4\bar2}$, $A^0_{4\bar2}\ A^1_{2\bar4}$, so P can be written in the form $\alpha|A^0_{2\bar5}|^2+\beta|A^0_{3\bar4}|^2+\gamma\mathrm{Re}A^0_{4\bar2}\ A^1_{2\bar4} + \delta\,\mathrm{Im}\ A^0_{4\bar2}\ A^1_{2\bar4}$. Now apply Lemma 2.8, first with $\xi=0$, $\eta=i\mu$, $\mu\in\mathbb{R}$. Then $B^0_{2\bar5} = A^0_{2\bar5}$, $B^0_{3\bar4}=A^0_{3\bar4}$, $B^0_{4\bar2}=A^0_{4\bar2}$, $B^1_{2\bar4} = A^1_{2\bar4} - 4\mu A^0_{2\bar4}$, so from the invariance of P it follows that $P(B)=P(A)$, so $\gamma\mathrm{Re}\ A^0_{4\bar2}(A^1_{2\bar4}-4\mu A^0_{2\bar4}) + \delta\,\mathrm{Im}\ A^0_{4\bar2}\ (A^1_{2\bar4}-4\mu A^0_{2\bar4})$ $=\gamma\mathrm{Re}\ A^0_{4\bar2}\ A^1_{2\bar4} + \delta\,\mathrm{Im}\ A^0_{4\bar2}\ A^1_{2\bar4}$, which reduces simply to $\gamma=0$. Next apply

Lemma 2.8 with $\xi\neq 0$, $\mu=0$.

An easy computation shows that P(B)-P(A)

$$=-(2\alpha+5\delta)\ \mathrm{Re}(\bar{\xi}A^{0}_{2\bar{5}}A^{0}_{4\bar{2}}) + (\frac{2\beta}{3} + 3\delta)\ \mathrm{Re}(\xi A^{0}_{3\bar{4}}\ A^{0}_{4\bar{2}}) + (\alpha+\frac{\beta}{9}+3\delta)|\xi|^2|A^{0}_{2\bar{4}}|^2.$$

That this should vanish identically is equivalent to $2\alpha+5\delta=0$, $2\beta+9\delta=0$, $\alpha+\frac{\beta}{9}+3\delta=0$, with unique solution (up to a multiple) $\alpha=5$, $\beta=9$, $\delta = -2$. The above computations prove the invariance of P with this choice of $\alpha,\beta,\gamma,\delta$ concluding finally the proof of Theorem 2.1.

We end this section by indicating how the invariants of Theorem 2.1 arise from the constructions of §1. First consider $\|A^{0}_{2\bar{2}}\|^2$. It is well known and is computed explicitly in [8] that Fefferman's construction of Weyl invariants, when applied to the Kähler invariant $W=\|R\|^2=R_{i\bar{j}k\bar{\ell}}R^{i\bar{j}k\bar{\ell}}$, gives a nonzero multiple of $\|A^{0}_{2\bar{2}}\|^2$. In particular, when n=2, $\|R\|^2=0$ when evaluated at points (z_o,z) with $z\in b\Omega$. Next consider invariants of weight ≥ 3 in $\mathbb{C}^2$. We have already seen in Proposition 2.2 that $A^{0}_{4\bar{4}}$ arises as the first log term coefficient $b\eta_1$ of the Monge-Ampère solution. As for $|A^{0}_{2\bar{4}}|^2$, by one of the criteria given in §1 in the discussion of ambiguity in Fefferman's Weyl invariants, it follows that any quadratic expression in $\nabla^2 R$, e.g. $\|\nabla^2 R\|^2=R_{i\bar{j}k\bar{\ell},p\bar{q}}\ R^{i\bar{j}k\bar{\ell},p\bar{q}}$, gives rise to a well-defined invariant of weight 4 in any dimension $n\geq 2$. So when n=2 any such invariant must be a multiple of $|A^{0}_{2\bar{4}}|^2$. Whether or not a nonzero multiple of $|A^{0}_{2\bar{4}}|^2$ arises this way is not known, but at least $|A^{0}_{2\bar{4}}|^2$ may potentially be a Weyl invariant.

The last invariant of Theorem 2.1b), $5|A^{0}_{2\bar{5}}|^2+9|A^{0}_{3\bar{4}}|^2-2\mathrm{Im}\ A^{0}_{4\bar{2}}\ A^{1}_{2\bar{4}}$, is not one of the invariants constructed from the Monge-Ampère solution or its log term coefficients, however. This can be seen as follows: first this invariant involves $A^{1}_{2\bar{4}}$ which is of weight 3. We saw that in $\mathbb{C}^n$ an invariant involving coefficients of weight n+1 or higher cannot result from Fefferman's construction because of the ambiguity in the Monge-Ampère solution. The other possibility is that this invariant arises as one of the invariants involving the log terms, as described after Theorem 1.6. However this can be ruled out too. Since the invariant is of weight 5, it cannot involve η_k with $k\geq 2$ since these give rise to invariants of weight ≥ 6. As it is quadratic one sees easily that the only possibilities are contr

$(\eta_1\nabla^2 R)$, contr $(\nabla\eta_1\otimes\nabla R)$, contr $(\nabla^2\eta_1\otimes R)$. However all of these vanish, for in any of them one must contract two indices of $R_{i\bar{j}k\bar{l}}$ with each other which gives rise to 0 since the metric $g_{i\bar{j}}$ is Ricci-flat through terms of the appropriate strength.

Actually, one should not really have expected the construction of invariants given in §1 to give all invariants, as the Taylor expansions of the η_k just do not carry enough information.

3. The Bergman Kernel

In this section we show the relationship between the invariant theory of §2 and results of Diederich, Christoffers, and Burns on the asymptotics of the Bergman kernel. The basic idea is described in the introduction: one expands the Bergman kernel K on the diagonal in powers of a defining function, hoping to identify the coefficients. Of course in order to obtain invariant coefficients one must use an invariant defining function, so one expands in terms of one of Fefferman's smooth approximate solutions r to the Monge-Ampère equation, which can be thought of as a locally uniquely determined, canonical defining function, at least to order n+2. Fix some such r; then K can be written $K=\frac{\varphi}{r^{n+1}}+\psi\log r$, $\varphi,\psi\in C^\infty(\bar{\Omega})$, and φ mod r^{n+1}, ψ mod r^∞ are locally determined, and are independent of the choice of r. We will also need to know that φ, ψ depend polynomially on the boundary; precisely, if Ω is a domain containing 0 which is in normal form at 0, then any Taylor coefficient at 0 of φ of order $\le n$, and of ψ of any order, is a universal polynomial in the normal form coefficients $A^{\ell}_{\alpha\bar{\beta}}$. This can be seen from the Boutet-Sjöstrand proof of the existence of the asymptotic expansion. A related argument is given in [11], §2.

First consider $b\varphi=\varphi|_{b\Omega}$. By the transformation laws for K and r under biholomorphisms, $b\varphi$ transforms like an invariant of weight 0; by the observation above it is a polynomial in the normal form coefficients, so $b\varphi$ is an invariant of weight 0 in the sense of Definition 1.3. Hence it must be constant. The constant can be found to be $\frac{n!}{\pi^n}$ by evaluation on the ball.

Now write $\varphi=\frac{n!}{\pi^n}+P_1 r+O(r^2)$, so $P_1=\frac{\varphi-\frac{n!}{\pi^n}}{r}|_{b\Omega}$. The transformation laws show that P_1 transforms like an invariant of weight 1; as before P_1 is a polynomial in normal form coefficients, so it is an invariant of weight 1. By Theorem 2.1 we deduce that $P_1=0$, so $K-\frac{n!}{\pi^n}r^{-n-1}=O(r^{-n+1})$. This result was obtained by Christoffers [8] and Diederich by direct calculation.

Next set $P_2 = \frac{\varphi-\frac{n!}{\pi^n}}{r^2}|_{b\Omega}$. Then P_2 is an invariant of weight 2. If n=2 then by Theorem 2.1 we obtain $P_2=0$. This completes the description of the locally determined part of φ in case n=2: $\varphi=\frac{2}{\pi^2}+O(r^3)$, a result obtained directly by Christoffers [8] without using invariant theory. In case n>2 we get $P_2=c_n\|A^0_{2\bar{2}}\|^2$ for some constant c_n. This was also proved directly by Christoffers, who additionally showed that $c_n\neq 0$.

We restrict attention now to n=2 and consider ψ. The transformation law for K and the observation on polynomial dependence imply that $b\psi=\psi|_{b\Omega}$ is an invariant of weight 3, so must be a multiple of $A^0_{4\bar{4}}$ by Theorem 2.1. Fefferman [9] showed that the multiple is nonzero, and we will see below that $b\psi = -\frac{24}{\pi^2}A^0_{4\bar{4}}$. We would like to continue to determine the next coefficient in ψ. Unfortunately the invariant $A^0_{4\bar{4}}$ is only defined on $b\Omega$ so it is not clear how to invariantly define the next coefficient. This dilemma is resolved by comparison with Proposition 2.2, which states that $b\eta_1=4A^0_{4\bar{4}}$. Thus $\psi+\frac{6}{\pi^2}\eta_1$ is invariantly defined modulo $O(r^3)$ and vanishes at $b\Omega$, so $\frac{\psi+\frac{6}{\pi^2}\eta_1}{r}|_{b\Omega}$ is an invariant of weight 4. By Theorem 2.1 we conclude that it must be a multiple of $|A^0_{2\bar{4}}|^2$, and D. Burns [4] has shown that the multiple is nonzero. Thus we have:

(3.1) $$\psi=-\frac{6}{\pi^2}\eta_1 + k|A^0_{2\bar{4}}|^2 r + O(r^2),\ k\neq 0.$$

Burns also applied (3.1) to obtain

<u>Theorem 3.2</u> (Burns [4]): Let $\Omega\subset\mathbb{C}^2$ and suppose that the log term coefficient ψ of the Bergman kernel satisfies $\psi=O(r^2)$ near $p\in b\Omega$. Then near p, $b\Omega$ is biholomorphically equivalent to the sphere.

<u>Proof:</u> First, by (3.1), $0=b\psi=b\eta_1$ near p. By Proposition 1.8 it follows that $\eta_1=0$ to infinite order near p; in particular $\eta_1=O(r^2)$. Now by (3.1) again one obtains $A^{0}_{2\bar{4}}=0$ near p. However the condition $A^{0}_{2\bar{4}}=0$ characterizes umbilic points on hypersurfaces in $\mathbb{C}^2$, and it is a result of the Chern-Moser theory that a hypersurface consisting entirely of umbilic points must be spherical (see [5]). Thus Theorem 3.2 follows.

We remark that in Theorem 3.2 $\psi=O(r^2)$ cannot be replaced by $\psi=O(r)$. In fact, by Proposition 1.9 and (3.1), there are strictly pseudoconvex domains Ω with $p\in b\Omega$ so that $b\psi=0$ near p, but near p $b\Omega$ is inequivalent to the sphere.

We conclude with a discussion of $b\psi$ in the general case. As above $b\psi$ is seen to be an invariant of weight n+1. Thus by Theorem 2.3b) the linear part of $b\psi$ must be a multiple of $\mathcal{L}$. As mentioned above, it follows from [9] that for n=2 the multiple is nonzero; the same sort of argument can be used to obtain the same result for n>2 too. In fact, Fefferman's method allows one to evaluate directly the linear part of $b\psi$ without recourse to Theorem 2.3. The computation is sketched here.

<u>Theorem 3.3</u>: The linear part of the log term coefficient $b\psi$ of the Bergman kernel is $-\frac{(n+2)!}{\pi^n}\mathcal{L}$, where $\mathcal{L}$ is as in Theorem 2.3.

<u>Proof</u>: We know that the linear part of $b\psi$ is a linear combination of coefficients of weight n+1, so it suffices to evaluate $b\psi$ at 0 for a bounded smooth strictly pseudoconvex domain Ω whose boundary agrees near 0 with $2u-|z|^2=f(z,\bar{z},v)$, where

$$f(z,\bar{z},v)=\sum_{1/2(|\alpha|+|\beta|)+\ell=n+2} A^{\ell}_{\alpha\bar{\beta}}\, z^{\alpha}\bar{z}^{\beta}v^{\ell}.$$

Recall that D is the Siegel domain $2u-|z|^2>0$, and let $\tilde{D}$ be a bounded smooth strictly pseudoconvex domain which agrees with D in some neighborhood of 0 but which agrees with Ω outside of some other neighborhood of 0. The basic idea, in oversimplified form, is as follows: fix $p\in D\cap\Omega$ near 0 and consider the integral $\int_{\tilde{D}} K_{\tilde{D}}(p,q)\, K_{\Omega}(q,p)dq$. Since $K_{\Omega}(\cdot,p)$ is holomorphic, the reproducing property of $K_{\tilde{D}}$ shows that the integral equals $K_{\Omega}(p,p)$. On the other hand it is also $\left[\int_{\tilde D} - \int_{\Omega}\right] K_{\tilde D}(p,q)K_\Omega(q,p)\,dq$ $+ \int_\Omega K_{\tilde D}(p,q)K_\Omega(q,p)\,dq$. But the latter term is $\overline{\int_\Omega K_\Omega(p,q)K_{\tilde D}(q,p)\,dq} = \overline{K_{\tilde D}(p,p)} = K_{\tilde D}(p,p)$. Also modulo much smaller terms, $K_\Omega \approx K_{\tilde D} \approx K_D$, and the last of these is known explicitly. Substituting this into the former integral above then gives $K_\Omega(p,p) \approx K_D(p,p) + \left[\int_{\tilde D} - \int_\Omega\right] |K_D(p,q)|^2\,dq$. The last term can now be asymptotically evaluated as $p \to b\Omega$ to give the asymptotics of K_Ω.

In addition to estimating the errors in this argument one must deal with the fact that the argument is incorrect as stated, for although $K_\Omega(\cdot,p)$ is holomorphic, it need not be holomorphic in all of $\tilde D$. So use of the reproducing property of $K_{\tilde D}$ is not justified, and similarly with Ω and $\tilde D$ interchanged. This difficulty is dealt with by Taylor expanding $K_\Omega(\cdot,p)$ to make everything analytic, modulo more errors. The details are written down carefully as Proposition 3 of [8]. In order to give a statement useful for our purposes, take Ω as above, let p be given by $z=0$, $w=t>0$ in our coordinates (z,w), fix a sufficiently small $\eta>0$ and let $V=\{|z|<\eta,\ |u|<\eta,\ |v|<\eta\}$.

<u>Lemma 3.4</u>: Ω, p as above. Then as $t \to 0$,

$$K_\Omega(p,p) = K_D(p,p) + \left[\int_{D\cap V} - \int_{\Omega\cap V}\right] |K_D(p,q)|^2\,dq + O(1).$$

<u>Proof</u>: As in [8].

Now let $q=(z,w)$ and recall that $K_D(p,q)=\frac{n!}{\pi^n}(t+\bar w)^{-n-1}$, so in particular $K_D(p,p)=\frac{n!}{\pi^n}(2t)^{-n-1}$. Thus writing $p=te_n$, Lemma 3.4 can be restated as:

$$(3.5)\quad K_\Omega(te_n,te_n)= \frac{n!}{\pi^n}(2t)^{-n-1}+\frac{n!^2}{\pi^{2n}}\left[\int_{D\cap V} - \int_{\Omega\cap V}\right] |t+w|^{-2n-2}\,dudvdz + O(1).$$

But $\left[\int_{D\cap V} - \int_{\Omega\cap V}\right] |t+w|^{-2n-2}\,dudvdz$

$$= \int_{|z|<\eta} \int_{|v|<\eta} \int_{1/2|z|^2}^{1/2|z|^2+1/2f} \left[(u+t)^2+v^2\right]^{-n-1} dudvdz$$

$$= \int_{|z|<\eta}\int_{|v|<\eta}\int_{0}^{1/2f} \left[(u+1/2|z|^2+t)^2+v^2\right]^{-n-1} dudvdz$$

$$= \int_{|z|<\eta}\int_{|v|<\eta}\int_{0}^{1/2f} \left[(1/2|z|^2+t)^2+v^2\right]^{-n-1} dudvdz + E, \text{ where}$$

$$E= \int_{|z|<\eta}\int_{|v|<\eta}\int_{0}^{1/2f} \left\{\left[(u+1/2|z|^2+t)^2+v^2\right]^{-n-1} - \left[(1/2|z|^2+t)^2+v^2\right]^{-n-1}\right\} dudvdz$$

One estimates the integrand in E by the mean value theorem. Then easy estimates using the facts that $|f|\le C|z|^4$ so that $u+1/2|z|^2 \approx |z|^2$ on the domain of integration, and $|f|\le C(|z|^2+|v|)^{n+2}$, show that $E=O(1)$ as $t\to 0$.

Hence modulo $O(1)$, the above equals

$$(3.6) \quad \int_{|z|<\eta}\int_{|v|<\eta} 1/2f(z,\bar{z},v)\left[(1/2|z|^2+t)^2+v^2\right]^{-n-1} dvdz.$$

Now $f(z,\bar{z},v)=\sum_{\ell=0}^{n} P_\ell(z,\bar{z})v^\ell$, where $P_\ell(z,\bar{z}) = \sum_{1/2(|\alpha|+|\beta|)=n+2-\ell} A^\ell_{\alpha\bar{\beta}}\, z^\alpha\bar{z}^\beta$.

Clearly when substituted into (3.6) the terms with ℓ odd drop out. So we take $\ell=2j$ and consider

$$1/2 \int_{|z|<\eta}\int_{|v|<\eta} P_{2j}(z,\bar{z})v^{2j}\left[(1/2|z|^2+t)^2+v^2\right]^{-n-1} dvdz.$$

One computes without difficulty that as $t\to 0$ this equals

$$2^{2n-2j}\int_0^\infty \frac{v^{2j}}{(1+v^2)^{n+1}}\, dv \cdot \int_{|z|=1} P_{2j}(z,\bar{z})\, d\sigma\cdot \log\frac{1}{t} + O(1).$$

The only terms in P_{2j} which contribute to $\int_{|z|=1} P_{2j}\, d\sigma$ are those for which $|\alpha|=|\beta|=n+2-2j=p$, and

$$\int_{|z|=1} P_{2j}(z,\bar{z})\, d\sigma = \frac{\omega}{\binom{2n-2j}{n-2}}\, tr^p\, P_{2j}(z,\bar{z}),$$

where ω is the measure of $S^{2n-3}\subset\mathbb{C}^{n-1}$.

Substituting this in, evaluating the constant, and using Moser's normalizations $tr^2 A^\ell_{2\bar{2}} = tr^3 A^\ell_{3\bar{3}}=0$, one obtains that the expression

(3.6) is equal mod O(1) to

$$\frac{(n+2)(n+1)\,\pi^n}{n!}\sum_{j=o}^{\left[\frac{n-2}{2}\right]} 2^{-2j}\,\frac{\binom{n}{j}}{\binom{n+2}{2j}}\,\mathrm{tr}^p\left(A^{2j}_{p\bar{p}}\right)\cdot\log\frac{1}{t}.$$

Returning to (3.5) and recalling the definition of $\mathcal{L}$ gives $K_\Omega(te_n,te_n)=\frac{n!}{\pi^n}(2t)^{-n-1}-\frac{(n+2)!}{\pi^n}\mathcal{L}\log t+O(1)$. Since $\log t=\log r(te_n)+O(1)$, $b\psi(0)$ is just the coefficient of $\log t$ in this expansion, and Theorem 3.3 is proved.

Combining Corollary 2.6 and Theorem 3.3 we see that the first log term coefficients of the Monge-Ampère solution and the Bergman kernel have the same nonzero linear part, up to a multiple. And Theorem 2.3 shows that this property distinguishes these invariants on algebraic grounds. Fefferman's invariant theory predicts that invariants whose weight is roughly equal to the dimension may exhibit unusual behavior, which coincides with our observations. It remains to be seen whether the invariants $b\psi$ and $b\eta_1$ are exceptional in the sense of arising from contractions of natural curvature tensors, as in [11]. Also the relationship, if any, between the full invariants $b\psi$ and $b\eta_1$ is not presently understood if $n>2$.

REFERENCES

[1] M. Beals, C. Fefferman and R. Grossman, Strictly pseudoconvex domains in $\mathbb{C}^n$, Bull. A.M.S. 8 (1983), 125-322.

[2] J. Bland, Local boundary behaviour of the canonical Einstein-Kähler metric on pseudoconvex domains, UCLA PhD. thesis, 1982.

[3] L. Boutet de Monvel and J. Sjöstrand, Sur la singularité des noyaux de Bergman et de Szegö, Astérisque 34-35 (1976), 123-164.

[4] D. Burns, personal communication.

[5] D. Burns and S. Shnider, Real hypersurfaces in complex manifolds, Proc. Symp. Pure Math. 30 (1977), 141-168.

[6] S.-Y. Cheng and S.-T. Yau, On the existence of a complete Kähler metric on noncompact complex manifolds and the regularity of Fefferman's equation, Comm. Pure Appl. Math. 33 (1980), 507-544.

[7] S.S. Chern and J. Moser, Real hypersurfaces in complex manifolds, Acta Math. 133 (1974), 219-271.

[8] H. Christoffers, The Bergman kernel and Monge-Ampère approximations, Chicago PhD. thesis, 1980.

[9] C. Fefferman, The Bergman kernel and biholomorphic mappings of pseudoconvex domains, Invent. Math. 26 (1974), 1-65.

[10] C. Fefferman, Monge-Ampère equations, the Bergman kernel, and geometry of pseudoconvex domains, Ann. Math. 103 (1976), 395-416.

[11] C. Fefferman, Parabolic invariant theory in complex analysis, Adv. in Math. 31 (1979), 131-262.

[12] J. Lee, Higher asymptotics of the complex Monge-Ampère equation and geometry of CR-manifolds, MIT PhD. thesis, 1982.

[13] J. Lee and R. Melrose, Boundary behaviour of the complex Monge-Ampère equation, Acta Math. 148 (1982), 159-192.

Author's Address:

C. Robin Graham
Department of Mathematics
University of Washington
Seattle, WA 98195

BIHOLOMORPHIC SELF-MAPS OF DOMAINS

Robert E. Greene* and Steven G. Krantz*

0. Introduction

In any branch of mathematics, attention naturally focuses on the self-equivalences of whatever structures are under consideration. In complex analysis, the appropriate idea of equivalence among open sets in $\mathbb{C}^n$ or, more generally, among complex manifolds, is the existence of a biholomorphic map. "Self-equivalences" are thus biholomorphic maps of an open set in $\mathbb{C}^n$ or a complex manifold to itself, which we shall call hereinafter automorphisms. The set of all automorphisms of a given domain or complex manifold Ω forms a group with composition of maps as the group operation; this group is called the automorphism group of Ω, denoted by $\mathrm{Aut}(\Omega)$.

The major emphasis of our discussion will be on automorphisms of open sets in $\mathbb{C}^n$. But it is important to realize that many of the results admit immediate generalization to domains in Stein manifolds. In the interest of brevity of exposition, we leave it to the reader to dispose of such generalizations. It is also worth noting that many of the results in this paper that depend only on normal families arguments hold in even greater generality. The information needed to make these generalizations is treated definitively in [WU]

In what follows, theorems, corollaries, and other items are numbered sequentially within each section.

*Supported in part by the National Science Foundation

1. The One Variable Case

The study of the automorphism groups of domains in $\mathbb{C}^n$, like most topics in several complex variables, can be profitably considered as a generalization of the one variable case. Indeed, the one-variable case suggests a number of results that turn out to have direct analogues in several variables. The most efficient methods of proof in the one-variable case are often not generalizable, however; the fastest proof in one-variable usually involve the Uniformization Theorem, which has no analogue in higher dimensions. In higher dimensions, more refined methods are required. These methods often involve the ideas of pseudoconvexity and related topics, such as the $\bar{\partial}$ equation. In one variable, such relatively complicated considerations can usually be avoided, and of course pseudoconvexity itself is not a restriction in the one-variable case.

One important method of the one-variable case can be transferred bodily to the several variables situation. This is the method of normal families. A new feature arises, however, in that it becomes a relatively subtle matter to see that normal families do not degenerate, e.g., that limits do not become constant. In one variable, this is usually easy; in several variables, it is not.

Before beginning the discussion of the results in several variables that are our main subject, we shall review briefly the one-variable situation by way of presenting examples for both their similarity and their contrast with the several variable theory.

a. The Automorphisms of the Simply Connected Riemann Surfaces

The automorphism groups of the simply-connected Riemann surfaces are easily determined by standard one-variable theory. We recall the relevant facts:

Uniformation Theroem: If M is a connected simply connected Riemann surface, then M is biholomorphic to either the unit disc Δ, or the plane $\mathbb{C}$, or the Riemann sphere $S^2 = \mathbb{C}P^1 = \mathbb{C} \cup \{\infty\}$.

Classification of Automoprhism Groups:

(a) $\mathrm{Aut}(\Delta)$ = the set of linear fractional transformations T of the form $T(z) = \omega \dfrac{z-a}{1-\bar{a}z}$ where $|\omega| = 1, \omega \in \mathbb{C}$, $a \in \Delta$.

(b) $\mathrm{Aut}(\mathbb{C})$ = the set of all linear maps T, $T(z)=az+b$, $a,b\in\mathbb{C}$, $a\neq 0$.

(c) $\mathrm{Aut}(S^2)$ = the set of all linear fractional transformation T, $T(z)=(az+b)/(cz+d)$, $a,b,c,d,\in\mathbb{C}$, $ad-bc\neq 0$.

In part (c), S^2 is considered to be $\mathbb{C}\cup\infty$ and T is considered to map to and from ∞ in the usual way.

The Uniformization Theorem is standard, but nontrivial to prove; see [BER]. The classification of automorphisms is easy; but at the risk of being tediously elementary we summarize the proof, because many of the ideas that occur later occur here in embryonic form.

Proof of (a): If $\alpha\in\mathrm{Aut}(\Delta)$ and $\alpha(0)=0$ then the Schwarz Lemma shows that $|\alpha'(0)|\leq 1$. Similarly, $|(\alpha^{-1})'(0)|\leq 1$. Since $(\alpha^{-1})'(0)=1/\alpha'(0)$, it follows that $|\alpha'(0)|=1$ and α is of the form $\alpha(z)=\omega z$, $|\omega|=1$, $\omega\in\mathbb{C}$. Now direct calculation shows that any T of the form indicated in part (a) maps Δ biholomorphically to Δ. If $\alpha\in\mathrm{Aut}(\Delta)$, α arbitrary, then $z\to[\alpha(z)-\alpha(0)]/(1-\bar{\alpha}(0)\alpha(z))$ is in $\mathrm{Aut}(\Delta)$ and takes 0 to 0. Thus for some $\omega\in\mathbb{C}$, $|\omega|=1$,

$$[\alpha(z)-\alpha(0)]/(1-\bar{\alpha}(0)\alpha(z)) \equiv \omega z\ , \quad z\in\Delta\ .$$

Calculation then shows that α has the form required in (a).

Proof of (b): If $\alpha\in\mathrm{Aut}(\mathbb{C})$, then α has a simple pole at ∞ (since α is proper and injective). So α must be a linear function.

Proof of (c): If $\alpha\in\mathrm{Aut}(\mathbb{C}\cup\infty)$ and $\alpha(\infty)=\infty$, then $\alpha|\mathbb{C}\in\mathrm{Aut}(\mathbb{C})$ so that $\alpha(z)=az+b$. Also, the linear fractionals T as described in (c) all belong to $\mathrm{Aut}(\mathbb{C}\cup\infty)$, by a standard calculation. So if $\alpha\in\mathrm{Aut}(\mathbb{C}\cup\infty)$, $\alpha(\infty)\neq\infty$, then

$$z \to 1[\alpha(z)-\alpha(\infty)]$$

is in $\mathrm{Aut}(\mathbb{C}\cup\infty)$ and sends ∞ to ∞; hence this must have the form $z\to az+b$. Direct calculation now shows that α is linear fractional.

b. <u>The Automorphism Groups of Nonsimply Connected Riemann Surfaces</u>

Let M be a Riemann surface and $\pi:\tilde{M}\to M$ be its universal (simply connected) cover so that $\tilde{M}=S^2$, $\mathbb{C}$, or Δ. For each $p\in M$, set $I_p=\{\alpha\in \mathrm{Aut}(M)|\alpha(p)=p\}$. Given a fixed $p\in M$, choose a $q\in\tilde{M}$ with $\pi(q)=p$. Then each element $\alpha\in I_p$ "lifts" to a unique automorphism $\tilde{\alpha}\in\tilde{I}_q \overset{\text{def}}{=} \{B\in \mathrm{Aut}(\tilde{M})|\beta(q)=q\}$. Thus I_p maps injectively into $\tilde{I}_q$, and this map is a group homomorphism. In general, it need not be surjective.

The group $\tilde{I}_q$ is a closed subgroup of $\mathrm{Aut}(\tilde{M})$, closed, that is, in the obvious topology on $\mathrm{Aut}(\tilde{M})$ determined by the linear or linear fractional structure on $\mathrm{Aut}(\tilde{M})$ as discussed in §1.a. In particular, $\tilde{I}_q$ is a Lie group. It is in fact isomorphic to $S^1\subset\mathbb{C}$ if $\tilde{M}=\Delta$, to $\mathbb{C}^*=\{z\in\mathbb{C}|z\neq 0\}$ if $\tilde{M}=\mathbb{C}$ and to $\{az/(cz+d)|a,c,d\in\mathbb{C},d\neq 0\}$ if $\tilde{M}=\hat{\mathbb{C}}$.

The image of I_p in $\tilde{I}_q$ consists exactly of the set of elements β in $\tilde{I}_q$ such that β preserves $\pi^{-1}(x)$ for each $x\in M$. It follows that the image of I_p in $\tilde{I}_q$ is a closed subgroup of $\tilde{I}_q$. Thus I_p is itself a Lie group. General theorems then imply that $\mathrm{Aut}(M)$ must be a Lie group. It is important to note, however, that $\mathrm{Aut}(M)$ does not naturally inject homomorphically into $\mathrm{Aut}(\tilde{M})$. An ambiguity arises in "lifting" the elements of $\mathrm{Aut}(M)$ to $\tilde{M}$; this ambiguity is related to the well known problem of finding a linear faithful representation of the mapping class group.

The situation just described is especially simple in case $\tilde{M}=\Delta$. Then $\tilde{I}_q$ is a compact group isomorphic to $S^1=\{z\in\mathbb{C}|\ |z|=1\}$. Hence I_p, being isomorphic to a closed subgroup of I_q, is also a compact group and in particular is either finite or isomorphic to S^1. In the latter case, it is relatively easy to see that M must be itself biholomorphic to Δ. The proof consists of observing that the elements in the image of I_p in $\tilde{I}_q$ must always map the set $\pi^{-1}(p)$ to itself, since these element fix $p\in M$. But $\pi^{-1}(p)$ is a discrete set, and the only discrete set fixed by all of $\tilde{I}_q$ is the set consisting of q alone, i.e., $\{q\}$. So $\{q\}=\pi^{-1}(p)$ and π is single sheeted, hence biholomorphic from Δ to M.

The Riemann surfaces covered by S^2 and $\mathbb{C}$ (i.e., S^2, $\mathbb{C}$, $\mathbb{C}-\{0\}$, tori) all have the property that their automorphism groups can be explicitly determined quite easily. Henceforth, we therefore consider only Riemann surfaces covered by the unit disc Δ .

If M is a Riemann surface covered by Δ , then Aut(M) is compact if and only if the orbits of Aut(M) are compact. More precisely, if Aut(M) is compact, all orbits are compact. And if any one orbit is compact then so is Aut(M) (and hence all orbits). This follows from the fact that, for any $p\in M$, Aut(M) is a fibre bundle over the orbit θ_p of $p\in M$ with (compact) fibre I_p . Note also that the orbits θ_p are closed sets in M . This follows easily by taking limits of lifts of elements of Aut(M) to elements in $\mathrm{Aut}(\tilde{M})$.

If M is a <u>compact</u> Riemann surface covered by Δ (or, equivalently if M is compact with genus ≥ 2), then Aut(M) is a finite group. This can be seen as follows: Since the orbits of Aut(M) are closed in M , they are compact, so Aut(M) is compact. Hence Aut(M) is finite if it has dimension zero. To rule out the possibility that Aut(M) has positive dimension, not that, if it had, then there would be a vector field on M generating a non-trivial one-parameter subgrup in Aut(M) . Such a vector field would necessarily have a zero, because the Euler characteristic of $M=2-2g\neq 0$. Hence the one-parameter subgroup would fix a point $p_0\in M$. If $\tilde{M}=\Delta$ is the universal cover of M and $\pi:\tilde{M}\to M$ the projection then let $q_0\in\pi^{-1}(p_0)$. It follows that $\tilde{I}_{q_0}$ would also contain a non-trivial one-parameter subgroup and hence would be the full S^1 group, implying as noted that $M=\Delta$, a contradiction. See [HE] for further material on these matters.

In the theory of Riemann surfaces, an explicit bound is determined on the number of elements in Aut(M) when the genus of M is at least 2 : the number of elements in Aut(M) , M compact of genus g , is less than or equal to $84(g-1)$ (Hurwitz). This result does not appear to have a natural generalization to higher dimensions. The Hurwitz theorem is known to be sharp for infinitely many values of g.

The situation is more for noncompact Riemann surfaces covered by Δ . In this case, Aut(M) can be noncompact in nontrivial ways, i.e., even though $M\neq\Delta$. For instance, set $M=\Delta-\{\alpha^n(0)\,|\,n\in\mathbb{Z}\}$ for $\alpha(z)=\left[z+\frac{1}{2}\right]/\left[1+\frac{1}{2}z\right]$ and $\alpha^n=\alpha$ iterated n times. Then $\alpha|M\in$ Aut(M), but the positive powers of $\alpha|M$ have no convergent subsequence. It is also possible for Aut(M) to be positive dimensional; for example, $M=\{z:1<|z|<R\}$, on which S^1 acts in an obvious way. But in fact these annular regions are the only cases in

which Aut(M) is of positive dimension. This can be checked by (1) noting that the Lie algebra of Aut(M) lifts to a subalgebra of the Lie algebra of $\mathrm{Aut}(\tilde{M})$ (2) noting that the only $\neq 0$ vector fields in this subalgebra have no zeroes on Δ (as before, since $M \neq \Delta$): and (3) analyzing the possible such subalgebras.

The description of which dimension zero groups can occur as Aut(M) for some M covered by Δ is an intriguing open problem. ([FK]).

There is one case important to later generalization wherein M is noncompact but the group Aut(M) must be compact.

Theorem 1: If M is a bounded open set in $\mathbb{C}$ with C^1 boundary, then Aut(M) is compact unless M is biholomorphic to Δ .

Corollary 2: Under the hypotheses of the Theorem, Aut(M) is finite unless M is biholomoprhic to Δ or to an annular region $\{z : 1 < |z| < R\}$.

The proof of the theorem is by the following limit argument: Fix $p \in M$. If Aut(M) is noncompact then the orbit of p is noncompact. Suppose $\{\alpha_j(p)\}$ accumulates at $q_0 \in \partial M$. The functions $z \to \alpha_j(z) - q_0$ have no zeroes on M , but have limit 0 at p . By Hurwitz's Theorem, they converge uniformly on compact sets K in M to zero, i.e., $\alpha_{j_k}|K$ goes to q_0 uniformly. If C is a closed curve in M then $\alpha_{j_k}(C)$ lies in a simply connected open subset of M for k large enough because the intersection of M with a small disc around q_0 is simply connected. Hence $\alpha_{j_k}(C)$ is contractible in M and so therefore is C . Thus M is simply connected if Aut(M) is noncompact.

The Theorem as stated can be refined: The C^1 boundary hypothesis can clearly be replaced by a topological condition of local simple connectivity in the neighborhood of boundary points.

Many of the one-variable results here have natural analogues in several variables. These will be discussed in due course, beginning with the next section.

2. Classical Domains

Another avenue of the classical study of automorphism groups of domains is that initiated by E. Cartan. Cartan's program entailed the study of bounded domains $\Omega \subseteq \mathbb{C}^n$ with transitive automorphism group. Such domains are called bounded homogeneous. In the case that such a domain is smoothly bounded it is known that Ω must be biholomorphic to the unit ball (this important result is discussed later in this paper--see also [RO]). Thus the material reviewed in the present section involves domains with non-smooth boundaries (in general, the boundaries are Lipschitz). It should be stressed that the material in the present section is not needed for an understanding of what follows. Since the primary focus of this article is smoothly bounded domains, we discuss only briefly the work of Cartan, plus the work of the Russian school which built on Cartan's. There are two reasons for this brevity:

(i) The methods related to bounded homogeneous domains are purely algebraic in nature, and as such are in contrast to the partial differential equations/differential geometric methods which we use to study smoothly bounded domains.

(ii) The results about bounded homogeneous domains are of a computational nature (these computations are often very difficult) and lead to an explicit classification of domains. This stands in contradistinction to the primarily qualitative results which we obtain for the smoothly bounded case.

a. Cartan's Program

Cartan calculated that, in $\mathbb{C}^2$, any bounded homogeneous domain is biholomorphic to either the ball

$$B^2 = \{(z_1, z_2) \in \mathbb{C}^2 : |z_1|^2 + |z_2|^2 < 1\}$$

or the bidisc

$$D^2 = \{(z_1, z_2) \in \mathbb{C}^2 : |z_1| < 1, |z_2| < 1\}.$$

It can further be calculated that the automorphism group of B^2

consists of those maps generated by the unitary group and by maps of the form

$$\varphi_a : (z_1, z_2) \to \left[\frac{z_1 - a}{1 - \bar{a}z_1} , \frac{(1-|a|^2)^{1/2} z_2}{1 - \bar{a}z_1} \right] , \ a \in \mathbb{C} , \ |a| < 1.$$

In fact, the automorphism group of the vall $B^n = \{z \in \mathbb{C}^n : \Sigma |z_j|^2 < 1\}$ is generated by the unitary group $U(n)$ and by maps of the form

$$\varphi_a : (z_1, \ldots, z_n) \to \left[\frac{z_1 - a}{1 - \bar{\partial} z} , \frac{(1-|a|^2)^{1/2} z_2}{1 - \bar{a}z_1} , \ldots , \frac{(1-|a|^2)^{1/2} z_n}{1 - \bar{a}z_1} \right],$$
$$a \in \mathbb{C} , \ |a| < 1 .$$

Likewise the automorphism group of D^2 is generated by the maps

$$(z_1, z_2) \to \left[\frac{z_1 - a}{1 - \bar{a}z_1} , z_2 \right]$$
$$(z_1, z_2) \to \left[z_1 , \frac{z_2 - a}{1 - \bar{a}z_2} \right]$$
$$(z_1, z_2) \to (z_2, z_1) ,$$

where a ranges over complex values less than unity. (Similarly, the automorphism group of D^n is generated by one variable maps and permutations.) Since the connected component of the identity in the isotropy subgroup of $0 \in \mathrm{Aut}(B^2)$ is non-abelian while that for $\mathrm{Aut}(D^2)$ is abelian, we can recover the result of Poincaré that the ball and the bidisc are biholomorphically inequivalent.

In $\mathbb{C}^3$, Cartan's result is that all bounded homogeneous domains are biholomorphically equivalent to either the ball, the tridisc, or (writing $z_j = x_j + iy_j$) the tube domain

$$\{(z_1, z_2, z_3) : y_3 > [(y_1)^2 + (y_2)^2]^{1/2}\}$$

(While the third of these domains is unbounded, it has a bounded realization.) Because of Cartan's classification, it is clear that in both $\mathbb{C}^2$ and $\mathbb{C}^3$ the bounded homogeneous domains are symmetric. Here a domain Ω is called <u>symmetric</u> if for each $P \in \Omega$ there is a

$\varphi \in \mathrm{Aut}\,\Omega$ such that $\varphi(P)=P$, φ is an involution $(\varphi\circ\varphi=\mathrm{id})$, and 0 is the only fixed point of $d\varphi(P)$. Cartan posed the problem of whether every bounded homogeneous domain in any $\mathbb{C}^n$ is also symmetric (note that an affirmative answer to this question would bring the highly developed and powerful theory of symmetric spaces to bear on the subject of bounded homogeneous domains).

In 1959, Piatetski-Shapiro gave a fairly complete answer to Cartan's question. He showed that in all complex dimensions $n \geq 4$ there exist non-symmetric, bounded homogeneous domains. In fact much more is true: in dimensions 4 ,5 , and 6 , there are only finitely many biholomorphic equivalence classes of homogeneous, bounded, non-symmetric domains. However in dimensions 7 and higher there are a continuum of equivalence classes of such domains. (In <u>any</u> $\mathbb{C}^n$, the set of equivalence classes of bounded <u>symmetric</u> domains is finite, as shown previously by Cartan.)

It was Cartan who calculated the bounded symmetric (hence homogeneous) domains. Call a domain <u>irreducible</u> if it is not a product of bounded homogeneous symmetric domains. Then it is enough to calculate the irreducible bounded symmetric homogeneous domains. These are of two types: there are four "classical" types (which occur in infinitely many distinct dimensions) and two "non-classical" or "exceptional" types (which occur only in dimensions $16=2^4$ and $27=3^3$). We now describe Cartan's classification. For further details on these domains, including calculation of their automorphism groups, one should consult the books [FU] and [HUA].

In what follows, if M is a matrix with complex entries, then ${}^tM$ denotes its transpose, $\overline{M}$ its conjugate, and M^* its conjugate transpose. We let I_m denotes the $m\times m$ identity matrix. Finally, we write $M>0$ if the matrix M is positive definite.

THE CLASSICAL DOMAINS OF CARTAN

<u>Classical Domains of Type I</u>: In $\mathbb{C}^n$, there is a classical domain of Type I for each factorization of n as a product of two natural numbers: $n=p\cdot q$ with $p \geq q>0$. For p , q fixed, the elements of $\mathbb{C}^n$ may be viewed as $p\times q$ matrices of complex numbers. We denote such a matrix with the symbol Z . The resulting classical domain of Type I is

$$\mathfrak{R}_I \equiv \{Z : I_q - Z^* \cdot Z > 0\} .$$

<u>Classical Domains of Type II</u>: These domains occur in $\mathbb{C}^n$ when $n=p(p+1)/2$ for p a positive integer. Under these circumstances we may think of elements of $\mathbb{C}^n$ as $p\times p$ <u>symmetric</u> matrices. Denote such a matrix by the symbol Z. The resulting classical domain of Type II is

$$\mathcal{R}_{II} \equiv \{Z : I_p - Z\cdot\overline{Z} > 0\} .$$

<u>Classical Domains of Type III</u>: These domains occur in $\mathbb{C}^n$ provided $n = p(p-1)/2$ for some integer $p>1$. Under these circumstances we may think of elements of $\mathbb{C}^n$ as skew symmetric matrices $({}^tM = -M)$ of size $p\times p$. Denote such a matrix by the symbol Z. The resulting classical domain of Type III is

$$\mathcal{R}_{III} = \{Z : I_p + Z\cdot\overline{Z} > 0\} .$$

<u>Classical Domains of Type IV</u>: These domains occur in every dimension. We think of $Z=(z_1,\dots,z_n)$ as a $1\times n$ matrix. Then the resulting classical domain of Type IV is

$$\mathcal{R}_{IV} = \{Z : |Z\cdot{}^tZ|^2 - 2\overline{Z}\cdot{}^tZ + 1 > 0 ,\ |Z\cdot{}^tZ| < 1 \} .$$

For clarity, it should be noted that when $n=1$ then classical domains of all four types occur, but they all turn out to be the disc. When $n=2$ then the domains of type I and IV which occur are both the ball, and domains of types II and III do not exist. In dimension 3, the domain of type I is the ball, and those of type II, III, and IV are equivalent to the domain $\{(z_1,z_2,z_3): y_3 > [(y_1)^2+(y_2)^2]^{1/2}\}$. As soon as the dimension is four or greater, then non-trivial choices of p and q are possible in the descriptions of Types I, II, III and genuinely new domains appear.

A description of the exceptional domains (in dimensions 16 and 27) is rather tedious and we omit it. Details may be found in [HEL].

Because of Piatetski-Shapiro's results, we cannot hope to write a list of the bounded, homogeneous, non-symmetric domains. Nevertheless, there is a classification theory, and it is based on the ideas of C.L. Siegel. Call a domain $V \subseteq \mathbb{R}^n$ a (proper) convex cone if

(a) $\lambda x \in V$ whenever $x \in V$ and $\lambda > 0$.

(b) $x + y \in V$ whenever $x, y \in V$.

(c) V contains no entire straight line.

If V is a fixed convex cone then we define the corresponding Siegel domain of the first kind to be

$$\{x + iy \in \mathbb{R}^n \times i\mathbb{R}^n : y \in V\}.$$

Alternatively, such a domain is called a tube domain over the cone V and is denoted by $\mathcal{T}_V$.

More generally, a tube domain may be erected over any base (not necessarily a cone). The theory of tube domains is well developed. It is known that such a domain is a domain of holomorphy if and only if the base is convex (see [HO]). H^p spaces on such domains have been studied [SW] and the automorphism groups of these domains have been considered [YA].

Let $\mathcal{T}_B$ be a tube domain with convex base. Necessary and sufficient for $\mathcal{T}_B$ to be biholomorphically equivalent to a bounded domain is that the base not contain an entire straight line. Thus we might hope to represent every bounded homogeneous domain as a tube domain over a cone. But the Type I domain in $\mathbb{C}^6$ corresponding to $6=3\cdot 2$ has the property that its automorphism group contains no abelian subgroups of real dimension 6, while any tube domain in $\mathbb{C}^6$ certainly would have automorphism group with such a subgroup. Thus this Type I domain could not be realized as a tube domain, and we need a more general type of domain.

Let W be a complex vector space of dimension m. A function

$$F: W \times W \to \mathbb{C}^n$$

is called a V-Hermitian form on W (where V is a convex cone as above) if

(a) $F(u,v)$ is $\mathbb{C}$-linear in u.

(b) $\overline{F(u,v)} = F(v,u)$, where the bar denotes conjugation

(c) $F(u,u) \in \text{closure}(V)$

(d) $F(u,u) = 0$ iff $u = 0$

A Siegel domain of the second kind is defined to be a set of the form

$$\{(x + iy,u) \in (\mathbb{R}^n + i\mathbb{R}^n) \times W : y - F(u,u) \in V\} .$$

Clearly any Siegel domain of the first kind is also a Siegel domain of the second kind (simply let $W=F=0$), but the converse is not true. To see this latter assertion, note that the domain of the second kind corresponding to $V=\mathbb{R}^+$ and F the usual Hermitian inner product is biholomorphic to the ball (see [KR1]); but it can be seen in a number of ways that the ball cannot be written as a tube over a cone. Any Siegel domain of the second kind is biholomorphically equivalent to a bounded domain.

A domain Ω is called affinely homogeneous (over $\mathbb{C}$) if there is a group of complex affine transformations of $\mathbb{C}^n$ which map Ω to itself and which act transitively on Ω . It is elementary to see that a tube domain is affinely homogeneous over $\mathbb{C}$ if its base is affinely homogeneous over $\mathbb{R}$.

The basic result in this subject, due to Vinberg, Gindikin, and Piatetski-Shapiro, is that any bounded homogeneous domain is biholomorphically equivalent to an affinely homogeneous Siegel domain of the second kind. This structure theorem greatly facilitates the study of automorphism groups of bounded homogeneous domains. It gives a canonical form for such a domain, and brings to bear a lot of algebraic machinery on the questions at hand.

b. Circular Domains

What if the domain being considered does not have a transitive group of biholomorphic self-maps? Can something still be said about the structure of the domain or of the group itself?

This circumstance is in fact the generic one (the reasons for this will be discussed later). We now consider this question in the special case of circular domains. Circular domains have by definition a one parameter group of automorphisms, but do not in general have a transitive automorphism group.

A domain Ω is called circular if whenever $z=(z_1,\dots,z_n) \in \Omega$ and $\lambda\in\mathbb{C}$ has modulus one then $\lambda z\equiv(\lambda z_1,\dots,\lambda z_n)\in\mathbb{C}^n$. Of course the ball and polydisc are circular domains, but it is known (see [GK2] and Section 12 below) that even in $\mathbb{C}^2$ the moduli space for the collection of such domains is infinite dimensional. Lempert (personal communication) has constructed a moduli space for these domains. The plethora of circular domains makes the following result particularly striking:

Theorem 1- (Braun, Karp, Upmeier [BKU]): If $\Omega_1, \Omega_2 \subseteq \mathbb{C}^n$ are circular domains, each of which contains the origin, and if Ω_1 and Ω_2 are biholomorphically equivalent, then they are linearly biholomorphically equivalent.

Notice how powerful this result is: If, for example, the ball and polydisc were biholomorphically equivalent, then they would be linearly biholomorphically equivalent. But that is impossible since the ball has smooth boundary while the boundary of the polydisc has corners (note that a linear map, being globally smooth, would preserve smooth boundaries). At least in some cases, the theorem reduces biholomorphic classification to a (not necessarily trivial) problem in linear algebra.

A domain Ω is called complete Reinhardt if whenever $z \in \Omega$ and $\lambda_1, \ldots, \lambda_n$ are complex numbers of modulus not exceeding unity then $(\lambda_1 z_1, \ldots, \lambda_n z_n) \in \Omega$. This is a more stringent condition than that of being circular, and it enabled Sunada [SUN] to obtain a great deal of explicit information about the automorphism groups of such domains. In particular, one can answer the natural question of which complete Reinhardt domains have compact automorphism group and which do not. To put this question in perspective, we note that, for a bounded domain, the property of having a non-compact automorphism group is equivalent to the existence of a point P in the domain such that the orbit of P is non-compact (use normal families considerations). It is known that a smoothly bounded strongly pseudoconvex domain with non-compact automorphism group must be the ball (this will be discussed in some detail later). On the other hand, the smooth (not strongly) pseudoconvex domain

$$\Omega = \{(z_1, z_2) : |z_1|^2 + |z_2|^4 < 1\}$$

has the automorphisms

$$\psi_a : (z_1, z_2) \to \left[\frac{z_1 - a}{1 - \bar{a} z_1}, \frac{[1 - |a|^2]^{1/4} z_2}{(1 - \bar{a} z_1)^{1/2}} \right], \ a \in \mathbb{C}, \ |a| < 1 .$$

These have the property that $\psi_{(-1+1/j)}(P) \to (1,0)$ for any $P\in\Omega$. Some partial results about domains with non-compact automorphism groups will be discussed in later parts of this paper.

3. Smoothly Bounded Domains in Several Complex Variables

It is natural to try to generalize the one-variable results and results for homogeneous domains which we just discussed to more general domains in $\mathbb{C}^n$, and indeed such generalizations are the subject of most of the rest of this paper. It is important to note early, however, that there are immediate obstructions to easy generalization, including most notably:

(1) The failure of the Uniformization Theorem or any analogue thereof. The space of complex structures on the unit ball is infinite dimensional, as is even the space of structures arising from perturbation of the ball in $\mathbb{C}^n$, $n\geq 2$ (see §12 and references therein).

(2) The groups that arise need not be Lie groups. For example, the group $\mathrm{Aut}(\mathbb{C}^2)$ contains all maps of the form $(z,w)\to(z,w+f(z))$, $f:\mathbb{C}\to\mathbb{C}$ holomorphic. This set of maps, being in an obvious sense infinite-dimensional, does not lie in any Lie group.

Difficulty (2) can be eliminated by restricting attention to bounded domains (connected open sets $\Omega\subset\subset\mathbb{C}^n$; as we shall see $\mathrm{Aut}(\Omega)$, Ω a bounded domain, is always a Lie group). But difficulty (1) remains; in fact, the failure of Uniformization is the underlying philosphical reason for the added subtlety and interest of the multi-variable case.

We turn now to considerations of the automorphism groups of bounded domains in $\mathbb{C}^n$. First, the exact definition: $\mathrm{Aut}(\Omega)=\{f:\Omega\to\Omega: f$ holomorphic and bijective$\}$. Note in particular that it is not necessary to assume separately the nonsingularity of f; as in one-variable, a (locally) injective holomorphic map is nonsingular everywhere (cf. [NAR]).

Our first fact about $\mathrm{Aut}(\Omega)$ is a direct generalization of the one-variable situation. Set, as before, for $p \in \Omega$, $I_p = \{f \in \mathrm{Aut}(\Omega) : f(p) = p\}$.

Theorem 1: For each bounded domain Ω, and each $p \in \Omega$, the group I_p is compact in the topology of convergence uniformly on compact subsets of Ω.

The proof of the theorem is a straightforward consequence of normal familier arguments (see §6 for details, in a more general context).

The theorem has as a consequence:

Theorem 2: The map $I_p \to \mathrm{End}(\mathbb{C}^n, \mathbb{C}^n) = GL(n, \mathbb{C})$ determined by $\alpha \to d\alpha|_p$, $\alpha \in I_p$, is a continuous injective homomorphism, homeomorphic onto its image.

The homomorphism property is just an application of the chain rule; continuity follows from the Cauchy estimates. The map is necessarily a homeomorphism since I_p is compact and $GL(n, \mathbb{C})$ a Hausdorff space. To prove injectivity, one uses a classical argument of H. Cartan: It is to be shown that if $d\alpha|_p$=identity then α=identity. If the contrary is assumed and the local power series expansion of α is considered, then it is easy to see that the iterates of α have at p unbounded derivatives of some fixed order, violating Cauchy estimates (see [KR1] for details).

Since the closed subgroups of the Lie group $GL(n, \mathbb{C})$ are Lie groups, it follows that I_p is a Lie group. General results on group actions on finite dimensional spaces then show that $\mathrm{Aut}(\Omega)$ is a Lie group.

Theorem 1 implies that, for fixed $p \in \Omega$, there is a Hermitian inner product on $\mathbb{C}^n$ such that all the $d\alpha|_p$, $\alpha \in I_p$, are unitary transformations relative to this inner product. In particular, all the eigenvalues of the $d\alpha|_p$ are of absolute value 1 and the absolute value of the determinant of $d\alpha|_p$ is 1. These facts are closely related to a generalization of the Schwarz Lemma: If $F: \Omega \to \Omega$ is holomorphic and if, for some $p \in \Omega$, $F(p)=p$, then $d\alpha|_p$ has eigenvalues of absolute value less than or equal to 1 ; moreover, if det $d\alpha|_p$ has absolute value 1 then F if biholomorphic. This generalized Schwarz Lemma in fact holds even more generally than for bounded domains on $\mathbb{C}^n$: see Wu [WU] for the completely general version.

The fact that I_p acts on $\mathbb{C}^n$ as a unitary group (relative to some Hermitian inner product) suggests immediately the possibility of identifying explicitly a Hermitian metric on Ω such that each I_p, or even $\mathrm{Aut}(\Omega)$ altogether, consists of isometries for this metric. Such a metric is provided by the Bergman metric construction discussed in Section 11.

The fact that I_p is invariably a compact group implies, as in the one-variable situation, that $\mathrm{Aut}(\Omega)$ is compact if it has one compact orbit. And of course if $\mathrm{Aut}(\Omega)$ is compact then all its orbits are compact.

The ball $B=\{(z_1,\ldots,z_n)\in\mathbb{C}^n \,|\, \sum_{j=1}^{n} |z_j|^2<1\}$ in $\mathbb{C}^n$ illustrates the observations made so far. The group $I_{(0,\ldots,0)}$ obviously includes the standard unitary group $U(n)$. But $I_{(0,\ldots0)}$ is isomorphic, via $\alpha\to d\alpha|_{(0,\ldots0)}$, to a subgroup of some unitary group; so it follows that $I_{(0,\ldots,0)}=U(n)$. The maps φ_a (see § 2), $a\in\mathbb{C}$, $|a|<1$, can be checked by direct calculation to belong to $\mathrm{Aut}(B)$. The composition of such maps with elements of $U(2)$ suffice to take any $p\in B$ to $(0,\ldots,0)$. Hence B is $\mathrm{Aut}(B)$ homogeneous. In particular $\mathrm{Aut}(B)$ is noncompact. However, in contrast to the one variable case, it is far from true that the ball B (or domains biholomorphic to it) is the only smoothly bounded domain with non-compact automorphism group. For instance, the domain

$$\Omega=\{(z_1,z_2):|z_1|^2+|z_2|^4<1\}$$

has a non-compact automorphism group, as was noted in § 2.

The argument used to prove Theorem 1 of § 1 does in effect extend to yield information in the multivariable case, however. The extension is most easily formulated in terms of the idea of holomorphic support functions.

Definition: Let $\Omega \subseteq \mathbb{C}^n$ be a domain and $x_0 \in \partial\Omega$. A function $f: U\cap\overline{\Omega} \to \mathbb{C}$, U an open neighborhood of x_0 in $\mathbb{C}^n$, is a <u>local holomorphic</u> support function for Ω at x_0 if

(i) f is continous on $U\cap\overline{\Omega}$ and holomorphic on $U\cap\Omega$.
(ii) $|f(z)|>0$ for $z\in U\cap\overline{\Omega}\setminus\{x_0\}$.
(iii) $f(x_0)=0$.

Theorem 3: If Ω is a bounded domain in $\mathbb{C}^n$, if there is a local holomorphic support function at x_0 in $\partial\Omega$, and if there is a sequence $\{\alpha_j\}$, α_j in $\mathrm{Aut}(\Omega)$ such that, for some $z_0\in\Omega$, $\lim_{j\to\infty} \alpha_j(z_0)=x_0$, then $\lim \alpha_j(z)=x_0$ for all $z\in\Omega$, uniformly on compact subsets of Ω. In particular, if $U\cap\Omega$ is contractible (for some open set U with a local holomoprhic support function defined on $U\cap\overline{\Omega}$) then Ω is contractible.

The second conclusion follows from the first as follows: the uniform convergence implies that, for each fixed k-sphere $F:S^k\to\Omega$ it holds that $\alpha_j\circ F:S^k\to\Omega$ is homotopically trivial in $U\cap\Omega$ for all j sufficiently large. Since α_j is biholorphic and in particular homeomorphic, F must be homotopically trivial. Hence $\Pi_k(\Omega)=0$ for all k and Ω is contractible.

The first statement is proved by a normal families argument that is essentially part of the proof of the main theorem of [RO].

Local holomorphic support functions always exist at each boundary point if $x_0\in\partial\Omega$ $\Omega\subset\mathbb{C}$ $(f(z)=z-x_0)$. Thus Theorem 2 is legitimately regardable as a direct generalization of Theorem 1 of Section 1.

The hypothesis that there is a neighborhood U of x_0 such that $U\cap\Omega$ is contractible holds automatically at any x_0 that is a C^1 point of $\partial\Omega$. Thus the following Corollary holds:

Corollary 4: If Ω is a bounded domain with C^1 boundary with a local holomorphic support function at each boundary point and if $\mathrm{Aut}(\Omega)$ is noncompact, then Ω is contractible.

Corollary 4 is illustrated both by the ball B in $\mathbb{C}^2$ and by the domain

$$\Omega=\{(z_1,z_2): |z_1|^2+|z_2|^4<1\} .$$

In the case of B, the sequence $\alpha_j\equiv\varphi_{(-1+1/j)}$ (see § 2) in Aut B has the property that $\lim \alpha_j((0,0))=(1,0)$. It is a straightforward calculation that $\lim \alpha_j((z_1,z_2))=(1,0)$ for all $(z_1,z_2)\in B$, uniformly on compact sets.

For the domain Ω, the sequence $\beta_j\equiv\psi_{(-1+1/j)}$ (again see § 2) belongs to Aut Ω. And again $\lim \beta_j((0,0))=(1,0)$. Calculation shows that $\lim \beta_j((z_1,z_2))=(1,0)$ uniformly on compact sets.

4. Conditions for Biholomorphism to the Ball: Noncompactness of the Automorphism Group

In the previous section, we observed a relationship between topological type and automorphism groups; under reasonably general conditions, nontrivial topology of the domain forces compactness of its automorphism group (see also [BED]). It is natural to ask whether some additional conditions can be imposed under which a domain Ω with $\mathrm{Aut}(\Omega)$ noncompact must be not just homeomorphic to the ball (i.e., contractible) but also biholomorphic to the ball. Such conditions are given by the following striking theorem of J.P. Rosay [RO], generalizing the result of [WO]. (See also [BS], where results similar to those in [WO] are obtained with the technique of Chern-Moser invariants.)

Theorem 1: (Wong-Rosay): Let $\Omega \subset\subset \mathbb{C}^n$ be a bounded domain; suppose there is a point $z_0\in\Omega$, a sequence $\alpha_j\in\mathrm{Aut}(\Omega)$ and a point $x_0\in\partial\Omega$ such that $\lim_{j\to\infty} \alpha_j(z_0)=x_0$ and such that $\partial\Omega$ is C^2 strongly pseudoconvex in a neighborhood of x_0 (see the Appendix for the definition). Then Ω is biholomorphic to the ball B in $\mathbb{C}^n$.

The proof of the Wong-Rosay theorem involves the notion of certain intrinsic volume forms. These can be defined as follows:

Definition: Let Ω be a bounded domain in $\mathbb{C}^n$ and z_0 a point of Ω. The Caratheodory volume element at z_0 relative to Ω, denoted $C_\Omega(z_0)$, is defined to be $C_\Omega(z_0)\equiv\sup\{|\det Jf(z_0)|:f:\Omega\to B,\ f(z_0)=0,\ f$ holomorphic$\}$.

Here Jf = the complex Jacobian of f, B is the unit ball in $\mathbb{C}^n$, det= the determinant.

It is elementary (use Cauchy estimates for the bound above) to see that $0<C_\Omega(z_0)<+\infty$.

Definition: With Ω a bounded domain in $\mathbb{C}^n$ and $z\in\Omega$, define the Eisenmann-Kobayashi volume element at z_0 relative to Ω, denoted $K_\Omega(z_0)$, to be $K_\Omega(z_0)\equiv\inf\{1/|\det Jg(0)|:\ g:B\to\Omega, g(0)=z_0, g$ holomorphic$\}$.

Again it is easy to see that $0<K_\Omega(z_0)<+\infty$.

The volume elements C and K transform in a predictable way under biholomorphic maps: If $F:\Omega_1\to\Omega_2$ is biholomorphic and $f:B\to\Omega_1$ with $f(0)=z_0$ then $F\circ f:B\to\Omega_2$ with $F\circ f(0)=F(z_0)$; moreover, every holomorphic map of B into Ω_2 taking 0 to $F(z_0)$ arises in this way. It follows that $C_{\Omega_1}(z_0)=C_{\Omega_2}(F(z_0))|\det JF(z_0)|$. Similar observations apply to show that $K_{\Omega_1}(z_0)=K_{\Omega_2}(F(z_0))\cdot|\det JF(z_0)|$.

The previous paragraph makes it easy to find C and K for the ball B in $\mathbb{C}^n$. By the generalized Schwarz Lemma (see, for instance, [RU]), $C_B((0,\ldots,0))=1=K_B((0,\ldots,0)$. Calculation with the automorphism group of B then yields

$$C_B((z_1,\ldots,z_n))=K_B((z_1,\ldots,z_n))=1/(1-\Sigma|z_j|^2)^n .$$

The observations on how C and K transform under biholomorphic maps also lead to this important fact:

Proposition 2: The number C_Ω/K_Ω is a biholomoprhic invariant, i.e., if $F:\Omega_1\to\Omega_2$ is biholomoprhic then, for each $z\in\Omega_1$,

$$C_{\Omega_1}(z)/K_{\Omega_1}(z)=C_{\Omega_2}(F(z))/K_{\Omega_2}(F(z)) .$$

If $g:B\to\Omega$ with $g(0,\ldots,0)=z_0$ and $f:\Omega\to B$ with $f(z_0)=(0,\ldots,0)$, then the generalized Schwarz Lemma applied to $f\circ g$ yields that $|\det Jg(0)|\cdot|\det Jf(z_0)|\le 1$. It follows that $C_\Omega(z_0)/K_\Omega(z_0)\le 1$ (all) $z_0\in\Omega$. Rather surprisingly, this inequality admits equality only for the ball, as discovered by B. Wong.

Lemma 3: (Wong [WO]): If for some one $z_0\in\Omega$, Ω a bounded domain, $C_\Omega(z_0)/K_\Omega(z_0)=1$, then Ω is biholomorphic to the ball.

The proof of Wong's Lemma is easy if there is a map $g:B\to\Omega$ with $g(0)=z_0$ and $|\det Jg(0,\ldots,0)|^{-1}=K_\Omega(z_0)$. Note that there <u>is always</u> a map $f:\Omega\to B$ such that $|\det Jf(z_0)|=C_\Omega(z_0)$. The crucial distinction here is that a collection of maps from Ω into the ball is always a normal family (to see this, compose with projection of the ball to the disc). However, the collection of maps from B to Ω need not be normal, even if Ω is smoothly bounded. [In case $\partial\Omega$ has a point P of strict concavity, for example, let ν_P be the unit outward normal at P and μ a unit complex tangent vector at P. For $\epsilon>0$ small the sequence of holomoprhic maps

$$g_j(z)=P-\frac{1}{j}\nu_P+\epsilon z_1\mu$$

satisfies $g_j:B\to\Omega$ and

$$\lim_{j\to\infty} g_j(z)\equiv g(z)=P+\epsilon z_1\mu .$$

Notice the crucial difficulty: the image of g lies partly in $\partial\Omega$ ($g(0)=P$) and partly in Ω.]

To return to Wong's Lemma, in the special case that g exists (for example, if Ω is smooth and pseudoconvex - see [KER]), and if

$C_\Omega(z_0)/K_\Omega(z_0)=1$ then, by the generalized Schwarz Lemma, $f\circ g:B\to B$ and $g\circ f:\Omega\to\Omega$ are biholomoprhic. It follows that f and g are themselves biholomorphic. In the general case (not assuming g exists), an argument similar in flavor, but considerably more technical (see [RO]), is required.

The path from Wong's Lemma to Rosay's result is by way of localization at the boundary. It is a straightforward application of normal families to see that if $x_0\in\partial\Omega$ has a local holomorphic support function and if $\{z_j:z_j\in\Omega\}$ satisfy $\lim_{j\to\infty} z_j=x_0$ then, for any neighborhood A of x_0 in $\mathbb{C}^n$,

$$\lim_{j\to+\infty} K_{A\cap\Omega}(z_j)/K_\Omega(z_j)=1 .$$

[Note: In the literature this argument is usually expressed in terms of local peaking functions. If $P\in\partial\Omega$ then a **local peaking function** at P is a function $\alpha(z)$ on the closure of $U\cap\Omega$, U a neighborhood of P in $\mathbb{C}^n$, such that

(i) α is continuous
(ii) α is holomorphic on $U\cap\Omega$
(iii) $\alpha(P)=1$
(iv) $|\alpha(z)|<1$ if $z\neq P$.

Obviously if α is a local peaking function at P then $1-\alpha$ is a local support function (by our definition).]

Returning to the localization of K_Ω, we note that if x_0 is a C^2 strongly pseudoconvex boundary point, then K_Ω near x_0 behaves like K near the boundary of a strongly pseudoconvex domain, <u>even if</u> Ω is not globally pseudoconvex. [The corresponding statement about the Caratheodary volume form C is false in general, as follows from the Hartogs extension phenomenon applied to a domain of the form $\Omega=B(0,2)\setminus\overline{B}(0,1)\cup B((1,0),1/2)\subseteqq\mathbb{C}^2$: localization of C_Ω fails for this domain at $P=(1/2,0)$. See [KR2] for details on this example.]

In case the local holomorphic support point x_0 is the limit of an orbit of $\mathrm{Aut}(\Omega)$, i.e. there exist $\alpha_j\in\mathrm{Aut}(\Omega)$ and $z_0\in\Omega$ with $\lim \alpha_j(z_0)=x_0$, then the C-localization result also holds. Namely

$$\lim_{j\to+\infty} \frac{C_{A\cap\Omega}(\alpha_j(z_0))}{C_\Omega(\alpha_j(z_0))} = 1 .$$

These localization arguments show that C_Ω/K_Ω at $\alpha_j(z_0)$ behaves like C/K for a sequence converging to the boundary of a strongly pseudoconvex domain. In this case, it is known ([GR]) that C/K converges to 1. [In the particular case at hand, Graham's generalized result can be simplified, using the fact that the sequence here is part of an automorphism group orbit.] Since C_Ω/K_Ω on $\alpha_j(z_0)$ is constantly equal to $C_\Omega(z_0)/K_\Omega(z_0)$ by biholomoprhism invariance, it can have limit 1 only if it equals 1 at z_0. Hence Ω must be the ball, by Wong's Lemma. This completes the proof of the Wong-Rosay Theorem.

In a later section, we shall discuss the question of what can be said if a (noncompact) orbit accumulates at a weakly pseudoconvex point.

It is instructive to compare Rosay's theorem, and its proof, to the statement and proof of the Riemann mapping theorem. Recall that the proof of the Riemann mapping theorem consists in constructing, by a normal families argument, an injective map $\alpha:\Omega\to\Delta$ which maximizes the derivative at a point $P\in\Omega$. The map α turns out to be both injective and surjective. Observe that if $P\in\Omega$ is fixed then C_Ω maximizes, over maps $\varphi:\Omega\to B$ with $\varphi(P)=0$, the derivative $|\det \mathrm{Jac}_{\mathbb{C}}\varphi(P)|$. Likewise K_Ω minimizes, over maps $\psi:B\to\Omega$ with $\psi(0)=P$, the derivative reciprocal $|\det \mathrm{Jac}_{\mathbb{C}}\psi(0)|^{-1}$. For $n>1$, the invariant C_Ω/K_Ω thus measures the extent to which the Riemann mapping theorem fails for Ω.

The following two results about the stability of the C/K invariant for strongly pseudoconvex domains will play a vital role in the arguments that follow. The proofs are quite delicate and cannot be reproduced here (see [GK4] for details). Suffice it to say that stability of K is established by normal families arguments, and that for C is proved by using uniform estimates for the $\bar\partial$-Neumann problem (see § 11).

Theorem 4 (Greene-Krantz [GK4]): Let $\Omega_0\subset\mathbb{C}^n$ be a fixed strongly pseudoconvex domain with C^2 boundary which is biholomorphically inequivalent to the ball. Let p_0 be a fixed point of Ω_0 and let $\delta=1-C_{\Omega_0}(p_0)/K_{\Omega_0}(p_0)$. Then there is a neighborhood $\mathcal{U}$ of Ω_0 in the C^2 topology (see the Appendix for a discussion of this terminology) such that if $\Omega\in\mathcal{U}$ then

$$C_\Omega(p_0)/K_\Omega(p_0) < 1-\delta/2 .$$

In fact $\mathcal{U}$ may be chosen so that the estimate holds uniformly over p_0 lying in a compact subset of Ω_0.

Theorem 5 (Greene-Krantz [GK4]): Let $\Omega\subseteq\mathbb{C}^n$ be a fixed strongly pseudoconvex domain with C^2 boundary (this time we allow Ω_0 to be biholomorphic to the ball). Let $\epsilon>0$. There is a neighborhood W of $\partial\Omega_0$ and a neighborhood $\mathcal{U}$ of Ω_0 in the C^2 topology such that if $\Omega\in\mathcal{U}$ and $z\in\Omega\cap W$ then

$$C_\Omega(z)/K_\Omega(z) > 1-\epsilon .$$

5. Conditions for Biholomorphism to the Ball: Free Mobility

The results of Section 4 can be thought of as a size restriction on $\mathrm{Aut}(\Omega)$, Ω strongly pseudoconvex (see the Appendix for the definition): The group $\mathrm{Aut}(\Omega)$ must be compact unless Ω is biholomorphic to the ball. It is reasonable to inquire whether similar restrictions hold on $I_p=\{\alpha\in\mathrm{Aut}(\Omega) : \alpha(p) = p\}$. As we have observed, I_p is always a compact Lie group, so its maximum size occurs when it is isomorphic to the unitary group $U(n)$, for $\Omega\subset\mathbb{C}^n$. This case is realized for $\Omega=$ the ball $B\subset\mathbb{C}^n$. This is in fact the only case in which this maximum size occurs. Indeed, a much stronger result holds, in which Ω can be replaced by a general noncompact complex manifold. To formulate this result, we need to introduce a definition:

Definition: Let M be a complex manifold and p be a point of M. A subgroup G of I_p <u>acts transitively</u> <u>on real directions</u> **at** p if for any v, w in the real tangent space of M at p, $v \neq 0$, $w \neq 0$, there is an element $g \in G$ with $dg(v) = \lambda w$ for some $\lambda \in \mathbb{R}^+$.

Theorem 1 (Greene-Krantz [GK5]): If M is a noncompact (connected) complex manifold, and if, for some $p \in \Omega$, there is a compact subgroup G of I_p that acts transitively on real tangent directions at p then Ω is biholomorphic to B in $\mathbb{C}^n$ or to $\mathbb{C}^n$ itself, $n = \dim_{\mathbb{C}} \Omega$.

We note that in the paper [BDK], Bland, Duchamp, and Kalka treat the case of M compact. They also weaken the hypothesis "transitive on real directions" to "transitive on complex directions". At least in the compact case, questions of the type discussed here have been treated using Lie group methods by Huckleberry and Oeljeklaus [HUC]. Moreover Morimoto and Nagano [MNA] prove a theorem about Lie group actions which, in effect, contains Theorem 1. We thank E. Oeljeklaus for bringing this to our attention.

If M is a bounded domain then I_p is automatically compact so the compact subgroup G can be just replaced by I_p. However, in the general case, I_p may act transitively on real directions without M being $\mathbb{C}^n$ or B^n, so the compactness of G is essential. To illustrate this possibility, consider $M = \mathbb{C}^2 \setminus \{(1,0)\}$ and $p = (0,0)$. The quadratic maps

$$(z,w) \to (z, w + \alpha z^2)$$
$$(z,w) \to (z + \beta w^2, w)$$

belong to $\mathrm{Aut}(\mathbb{C}^n)$ and compositions of these act transitively on $\mathbb{C}^n \setminus \{(0,0)\}$. Hence for any element $A \in U(n)$, there is a composition B of the quadratic maps indicated such that $B \circ A(1,0) = (1,0)$. Clearly $d(B \circ A)\big|_{(0,0)} = dA\big|_{(0,0)}$. Hence $I_{(0,0)}$ of $M = \mathbb{C}^n \setminus \{(1,0)\}$ acts transitively on real directions, but M is biholomorphically neither the plane or the ball. Similar arguments can be used to show that $I_{(0,0)}$ acts transitively on real directions for

$M=\mathbb{C}^n\setminus\{p_1,\ldots,p_k\}$, $p_j\neq(0,0)$. The obvious point for consistency with the theorem here is that I_p is not compact in these cases.

The role of the compactness of G in the proof of the theorem is that G then acts as isometries relative to some Hermitian metric on M . The noncompactness of M and the transitivity of G on real directions implies easily that M is real-diffeomorphic to $\mathbb{R}^{2n}$. This completes the proof in case $n=1$. In case $n\geq 2$, the proof then proceeds by showing that the balls about the distinguished point p in the G-invariant Hermitian metric are strongly pseudoconvex. Then it is shown that these balls are each biholomorphic to the (standard) ball in $\mathbb{C}^n$ and that these biholomorphisms fit together to provide a biholomorphic map of M to the ball or to $\mathbb{C}^n$. For further details, see [GK5] .

Results related to those described in this section, formulated in terms of mobility at two points of the domain, appear in [PA].

6. Normal Families of Biholomorphic Maps

The basic approach to comparing automorphism groups of different domains is the use of normal families. The essential idea is that the limit of a sequence of biholomorphic maps must be either again a biholomorphic map or a map that has image contained in the boundary. Included in this limiting process is the formation of limits of both the domain and the image domains themselves. Thus to formulate a precise statement we need a definition of convergence of domains.

Definition: Let $\{\Omega_i : i = 1,2,\ldots\}$ be a sequence of bounded open sets in $\mathbb{C}^n$ and Ω_0 be a bounded open set. The sequence Ω_i is said to converge to Ω_0 , written

$$\lim \Omega_i = \Omega_0$$

if and only if

(a) For every compact set $K\subseteqq\Omega$ there is a $J=J(K)$ such that $i\geq J$ implies $K\subseteqq\Omega_i$

and

(b) If K is a compact set which is contained in Ω_i for all sufficiently large i then $K \subseteq \Omega_0$.

We can now formulate our basic normal families result.

Theorem 1 (The Normal Families Theorem): Suppose $\{\Omega_i : i=1,2,\ldots\}$ and $\{A_i : i=1,2,\ldots\}$ are sequences of bounded domains in $\mathbb{C}^n$ with $\lim \Omega_i = \Omega_0$ and $\lim A_i = A_0$ for some (uniquely determined) bounded domains Ω_0 , A_0 in $\mathbb{C}^n$. Suppose also that $\{f_i : A_i \to \Omega_i : i=1,2,\ldots\}$ is a sequence of biholomorphic maps and that $f_i : A_i \to \mathbb{C}^n$ converges uniformly on compact subsets of A_0 to a limit $f_0 : A_0 \to \mathbb{C}^n$. Then one of two mutually exclusive conditions holds: Either f_0 maps A_0 biholomorphically onto Ω_0 <u>or</u> f_0 maps A_0 into $\partial\Omega_0$, the boundary of Ω_0 .

Before undertaking the proof of this theorem, we offer some explanatory remarks: First, notice that the concept of the $\{f_i\}$ sequence converging uniformly on compact subsets of A_0 makes sense. If K is a compact subset of A_0 then it follows from the definition of convergence that, for all sufficiently large i , $K \subset\subset A_i$ so that f_i is in fact defined on K . Second note that it is inevitably and obviously the case that $f_0(A_0) \subseteq \bar{\Omega}_0$. To see this note that for each point p it must be that $f_i(p) \in \bar{\Omega}_0$. If $\lim f_i(p) \notin \bar{\Omega}_0$ then $\lim \Omega_i = \Omega_0$ is contradicted.

It is important to notice that the second alternative in the conclusion of the theorem can actually occur. The most obvious example is $\Omega_j = \Omega_0 = A_j = A_0 = \{z \in \mathbb{C} : |z| < 1\}$ and the sequence

$$f_j(z) = \frac{z-(-1+1/j)}{1-(-1+1/j)z}$$

with $\lim f_j(z) \equiv 1$. But there are more subtle examples in which not only is $f_0 : A_0 \to \overline{\Omega}_0$ not a biholomorphic map but Ω_0 and A_0 are not biholomorphic at all.

To construct such a sequence, let $\rho(z_1, z_2)$ be a smooth, real valued function which is a small C^2 perturbation of $|z_1|^2 + |z_2|^2 - 1$ and which satisfies

(a) $\rho(z_1, z_2) = \rho(|z_1|, |z_2|)$;

(b) $\rho(z_1, z_2) = |z_1|^2 + |z_2|^2 - 1$ outside a small neighborhood U of $(1,0)$;

(c) $\rho(z_1, z_2) > |z_1|^2 + |z_2|^2 - 1$ in a smaller neighborhood $V \subseteqq U$ of $(1,0)$;

(d) $\nabla\rho \neq 0$ on $\{z : \rho(z) = 0\}$.

Let $A_0 = \{z \in \mathbb{C}^2 : \rho(z) < 0\}$. It can be shown (see [GK4]) that A_0 is not biholomorphic to the unit ball B . Define $f_j(z_1, z_2) = \varphi_{(-1+1/j)}(z_1, z_2)$ (see Section 2). Finally, let $A_j = A_0$, all j , and let $\Omega_j = f_j(A_j)$. It is easy to see that $\lim \Omega_j = B \equiv \Omega_0$. But, by construction, A_0 is not biholomorphic to Ω_0 .

This example is consistent with the theorem because the image of $f_0 \equiv \lim f_j$ is the singleton $\{(1,0)\}$, which lies in the boundary of Ω_0 .

Outline of the Proof of Theorem 1

Suppose that f_0 maps some point p_0 of A_0 into Ω_0 . Then we will show that f_0 is a biholomorphism of A_0 onto Ω_0 . Let $q_0 = f_0(p_0)$. For each i , denote $(f_i)^{-1}$ by g_i . By normal families, there is a subsequence $\{g_{i_j}\}$ which converges uniformly on compact subsets of Ω_0 to a map $g_0 : \Omega_0 \to \overline{A}_0$. Now

$$g_0(q_0) = g_0(f_0(p_0)) = \lim_{j\to\infty} g_{i_j}(f_0(p_0)) = \lim_{j\to\infty} g_{i_j}(f_{i_j}(p_0)) = p_0 ,$$

where the next to last identity is by uniform convergence. Hence

$g_0(q_0)=p_0$. But again by uniform convergence this allows us to conclude that if W is a sufficiently small neighborhood of q_0 then there is a relatively compact neighborhood U of p_0 such that $g_{i_j}(W) \subseteqq U$ for all j sufficiently large. Observe that $f_{i_j} \to f_0$ uniformly on U and $g_{i_j} \to g_0$ uniformly on W . Therefore, for $w \in W$, we have

$$f_0(g_0(w))=\lim_{j\to\infty} f_{i_j}(g_{i_j}(w))=w$$

Passing to derivatives yields

$$\text{Jac}_{\mathbb{C}} f_0(g_0(w)) \circ \text{Jac}_{\mathbb{C}} g_0(w)=\text{Id} .$$

In particular, $\text{Jac}_{\mathbb{C}} f_0(z)$ is non-singular for points z in $g_0(W)$. This is our first main partial result.

Consider now the functions

$$m_i(z) = \det \text{Jac}_{\mathbb{C}} f_i(z) .$$

By hypothesis, each of these functions is non-vanishing on A_i . Moreover, they converge uniformly on compacta to $m_0 \equiv \det \text{Jac}_{\mathbb{C}} f_0(z)$. We know that m_0 is not identically zero by the result of the first paragraph. By Hurwitz's theorem, it follows that m_0 never vanishes. As a result, f_0 is an open map. Since f_0 takes at least one point p_0 into Ω_0 , it follows that $f_0(A_0) \subseteqq \Omega_0$.

Of course this entire argument may be repeated to see that the function g_0 of the first paragraph is an open mapping and hence satisfies $g_0(\Omega_0) \subseteqq A_0$. But then uniform convergence allows us to conclude that for all $z \in A_0$ it holds that

$$g_0 \circ f_0(z)=\lim_{j\to\infty} g_{i_j} \circ f_{i_j}(z)=z$$

and likewise for all $w \in \Omega_o$ it holds that

$$f_0 \circ g_0(w) = \lim_{j\to\infty} f_{i_j} \circ g_{i_j}(w) = w .$$

This proves that f_0 and g_0 are each one-to-one and onto, hence in particular that f_0 is a biholomorphic map. □

7. Closure of Biholomorphic Equivalence Classes

Theorem 1 of Section 6 is most interesting in situations where it is guaranteed that the limit mapping is interior, i.e. that $f_0(A_0) \subseteq \Omega_0$. A consideration of the examples in Section 6 of the alternative possibility (that $f_0(A_0) \subseteq \partial\Omega_0$) suggests that the topology (that is, the number of derivatives) on which convergence of the boundaries of the domains takes place is playing a vital role. (See the Appendix for a detailed discussion of these various topologies.) In particular, in the last example of Section 6, the Ω_j, which have C^∞ boundaries, converge only in the C^0 sense to a C^∞ boundary domain Ω. But the convergence fails to take place on the C^2 level. The following theorem states in effect that convergence in C^2 does guarantee that the limit mapping is interior and hence biholomorphic, at least in the case that the domains in question are strongly pseudoconvex (see the Appendix for the definition of this term).

Of course, C^2 convergence is quite a natural condition for strongly pseudoconvex domains because strong pseudoconvexity is itself defined in terms of two boundary derivatives.

Theorem 1(Greene-Krantz [GK4]): Suppose $\{A_j : j=1,2,...\}$ and $\{\Omega_j : j = 1,2,...\}$ are sequences of C^2 strongly pseudoconvex bounded domains and that $f_j : A_j \to \Omega_j$ is a sequence of biholomorphic maps. Suppose that $\{A_j\}$ and $\{\Omega_j\}$ converge in the C^2 topology (see the Appendix for the definition) to C^2 strongly pseudoconvex domains A_0 and Ω_0 respectively. Finally suppose that A_0 is not biholomorphic to the ball B. If the sequence $\{f_j\}$ converges uniformly on compact subsets of A_0 to $f_0 : A_0 \to \overline{\Omega}_0$ then f_0 is a biholomorphic map from A_0 to Ω_0 .

This theorem follows from combining the Normal Families Theorem of Section 6 with the results of Section 4 on the stability of the C/K invariant. In more detail, we proceed as follows:

Choose a point p_0 in A_0 . Since A_0 is not the ball, Lemma 2 of Section 4 guarantees that $Q_0 \equiv C_{A_0}(p_0)/K_{A_0}(p_0) < 1$. Set $\delta = 1 - Q_0$. Choose i_0 such that $i \geq i_0$ implies that

$$C_{A_i}(p_0)/K_{A_i}(p_0) < 1 - \delta/2$$

(using Theorem 4 of §4). Choose also, by Theorem 5 of §4, $\epsilon > 0$ and an i_1 such that if $i \geq i_1$ and $q \in \Omega_i$ with $\mathrm{dis}(q, \partial\Omega_i) < \epsilon$ then

$$C_{\Omega_i}(q)/K_{\Omega_i}(q) > 1 - \delta/4 \ .$$

With these choices, it follows from the biholomorphic invariance of C/K that, if $i > \max\{i_0, i_1\}$ then

$$\mathrm{dis}(f_i(p_0), \partial\Omega_i) \geq \epsilon \ .$$

Hence $\mathrm{dis}(f_0(p_0), \partial\Omega_0) \geq \epsilon$ so $f_0(p_0) \in \Omega_0$. By Theorem 1 of Section 6, f_0 is a biholomorphic map from A_0 to Ω_0 . □

The only role of the C/K invariant in the proof just given is to insure that the limit map is nondegenerate in the sense that it takes A_0 into Ω_0 , rather than into $\partial\Omega_0$. For this purpose, it would would suffice to consider any biholomorphic invariant that would bound $f_j(p_0)$ away from $\partial\Omega_j$. A geometrically interesting case of this observation is the Bergman metric curvature, which we discuss in detail in Section 12.

8. Rigid Domains: Openness

As noted earlier, a domain Ω is called <u>rigid</u> if $\mathrm{Aut}(\Omega)=\{\mathrm{identity}_\Omega\}$. It is plausible by analogy with properties of geometric figures to guess that rigidity should be preserved under sufficiently small perturbations of the boundary. If an object has no symmetries, then a sufficiently similar object should also have none. By contrast, a symmetric object can be made asymmetric by arbitrarily small changes. Both these geometric intuitions have analogues in the complex domain, for strongly pseudoconvex domains and automorphism groups (as symmetry groups). In this section, we shall discuss the preservation of rigidity by small perturbations; in Section 10, we shall discuss briefly the destruction of symmetry (i.e. the behavior of non-rigid domains under perturbation).

Let $\mathcal{U}$ denote the set of all C^2 (bounded) strongly pseudoconvex domains in $\mathbb{C}^n$, $n\geq 1$. We impose on $\mathcal{U}$ the C^2 topology (see the Appendix for details).

Theorem 1 (Greene-Krantz [GK2]): The set of rigid C^2 strongly pseudoconvex domains is an open set in $\mathcal{U}$ (in the C^2 topology).

Proof: It is most convenient to express the setup in terms of sequences. In other words, we want to show that if a sequence $\{\Omega_i\in\mathcal{U}\}$ of non-rigid domains with automorphism groups $\mathrm{Aut}(\Omega_i)$ converges to a domain Ω_0 then Ω_0 is also non-rigid. If Ω_0 is biholomorphically a ball, we are done, so we henceforth assume that Ω_0 is not biholomorphic to the ball.

Any subsequence $\{\Omega_{i_k}\}$ and $f_{i_k}\in\mathrm{Aut}(\Omega_{i_k})$ has itself a subsequence that convergerges uniformly on compact subsets of Ω_0 to a map $f_0\in\mathrm{Aut}(\Omega_0)$. That there is a subsequence that converges normally follows by the usual normal families argument (Montel's theorem). That the limit is non-degenerate follows as in the proof of Theorem 1 in Section 7.

We now proceed by contradiction. The proof of the theorem is complete if there is some such f_0 which is not the identity mapping on Ω_0 . Such an f_0 will exist <u>unless</u> the groups $\mathrm{Aut}(\Omega_i)$ converge to the identity in the sense that every sequence $\{f_i : f_i \in \mathrm{Aut}(\Omega_i)\}$ converges to the identity, uniformly on compact subsets of Ω_0. So it suffices to show that the supposition that $\mathrm{Aut}(\Omega_i)$ converges to the identity leads to a contradiction. (The careful reader will realize that the logical convolutions here are simply a device for avoiding too many passages to subsequences.)

Suppose now that $\mathrm{Aut}(\Omega_i)$ converges to the identity. Choose a point $p \in \Omega_0$. We claim the following:

(*) There is an i_0 and a sequence of points $q_i \in \Omega_0$ such that for all $i \geq i_0$ and $f_i \in \mathrm{Aut}(\Omega_i)$ it holds that $f_i(q_i) = q_i$ and $\lim q_i = p$.

The proof of statement (*) is a bit tricky, so we assume its validity for now and complete the proof of the theorem. The proof of (*) is at the end of the section.

Assuming (*), we have for each $i \geq i_0$ that the differentials $\{df_i(q_i) : f_i \in \mathrm{Aut}(\Omega_i)\}$ form a group of linear endomorphisms, a group that is in fact isomorphic to $\mathrm{Aut}(\Omega_i)$--cf. Theorem 2 of § 3. Since $\{p\} \cup \{\cup\{q_i\}\}$ is a compact set, it follows that the differentials $\{df_i(q_i)\}$ converge to the identity. We thus have that, for every $\epsilon > 0$, there is an i_1 such that if $i \geq i_1$ then $df_i(q_i)$ is within ϵ of the identity in a fixed one of the usual metrics on the space of linear endomorphisms of $\mathbb{C}^n$. From this it follows that in fact for all i sufficiently large, the group $\{df_i(q_i) : f_i \in \mathrm{Aut}(\Omega_i)\}$ is the group consisting of the identity alone (here we use the fundamental fact that $GL(n, \mathbb{C})$ does not contain arbitrarily small subgroups). But then $\mathrm{Aut}(\Omega_i) = \{\text{identity}\}$, contradicting the fact that the Ω_i were all non-rigid. □

Proof of statement (*): It is enough to show that, for each $\epsilon > 0$, there is an i_ϵ such that for all $i \geq i_\epsilon$ there is a fixed point q_i of $\mathrm{Aut}(\Omega_i)$ with $\mathrm{dis}(q_i, p) \leq \epsilon$. For this purpose, set $\varphi(z) = \mathrm{dis}^2(z, p)$. Then φ is a C^∞ strongly convex function, i.e., φ has positive second derivative along every line.

For all i sufficiently large, it holds that for every $f \in \mathrm{Aut}(\Omega_i)$ the function $z \to \varphi \circ f(z)$ is also C^∞ strongly convex on the closed ball of radius ϵ around p. Moreover, it also holds that $\lim_{i\to\infty} \varphi \circ f_i \geq \epsilon^2$ on the boundary of this closed ball and $\lim(\varphi \circ f_i)(p)=0$. These statements are an immediate consequence of the fact that f_i converges C^2 to the identity as $i \to +\infty$.

Now define (for all i sufficiently large)

$$\varphi_i(z) = \int_{f \in \mathrm{Aut}(\Omega_i)} (\varphi \circ f)(z)\,df$$

where integration is with respect to Haar measure on the compact Lie group $\mathrm{Aut}(\Omega_i)$, normalized to have total mass 1. (Because Ω_0 is not biholomorphic to the ball, it holds for all i sufficiently large that Ω_i is not biholomorphic to the ball by Theorem 1 of Section 6. Thus $\mathrm{Aut}(\Omega_i)$ is compact for all i sufficiently large.) For i sufficiently large, we have

(i) φ_i is C^∞ strongly convex on the closed ball of radius ϵ around p.

(ii) φ_i converges uniformly to φ on the closed ball of radius ϵ as $i \to +\infty$.

Statement (i) implies that there is a unique point q_i in the closed ball such that $\varphi_i(q_i)$ minimizes $\varphi_i(z)$ over the closed ball (strong convexity of φ_i implies the uniqueness). The convergence of φ_i to φ implies that $\lim_{i\to+\infty} q_i = p$. In particular, for i large, it holds for all $f \in \mathrm{Aut}(\Omega_i)$ that $f(q_i)$ is in the closed ϵ-ball about p because q_i is close to p and $f(q_i)$ is close to q_i (by the convergence of $\{f_i\}$ to the identity). But $\varphi_i \circ f(q_i) = \varphi_i(q_i)$ so by uniqueness $f(q_i) = q_i$. □

The argument just given to establish statement (*) is a standard rocess from differential geometry for finding fixed points, applied in an ad hoc version suitable for the present situation.

9. Existence and Density of Rigid Domains

A detailed discussion of the existence and density (in the C^∞ topology) of rigid strongly pseudoconvex domains requires extended development of the boundary invariant theory of Chern-Moser-Tanaka ([CM],[TA]). Such a development is beyond the scope of this article, so we shall content ourselves with outlining a few basic ideas and supplying references to more extended discusssions which appear elsewhere.

The boundary invariant theory is perhaps best approached by contrast with the one-variable situation. Consider a domain Ω in $\mathbb{C}$ with real analytic boundary near $0\in\partial\Omega$ given by $y=f(x)$ where $f(x)$ is a real analytic function of the real variable x (in $z=x+iy$) with (for normalization purposes) $f(0)=f'(0)=0$. It is natural to ask whether we can find a local biholomorphic map $w=F(z)$ that transforms the real axis $\{x+iy : y=0\}$ into $\{x+iy:y=f(x)\}$, in other words to ask whether $\partial\Omega$ is locally equivalent to the real axis. It is easy to see that such a local biholomorphic map exists: Just take $w=\hat{f}(z)$ where $\hat{f}$ is the extension of f to be a (complex) holomorphic function in a neighborhood of 0. In this case, we can also arrange (by taking $\hat{f}(-z)$ for F if necessary) that the upper half plane $\{x+iy:y>0\}$ maps locally to Ω. We summarize this situation by saying that every pair $(\Omega,\partial\Omega)$ is locally equivalent at each point of $\partial\Omega$ to the pair (upper half plane, real axis) at 0.

The situation just described is not particularly surprising since the Riemann Mapping Theorem yields that $\Omega\cap B(0,\epsilon)$, ϵ small, is biholomorphic to the unit disc Δ, and by Schwarz Reflection this biholomorphic map extends across 0 holomorphically. This construction yields a local equivalence of $(\Omega,\partial\Omega)$ at 0 to $(\Delta,\partial\Delta)$ at some point of $\partial\Delta$ and of course this is the same as a local equivalence to the half plane situation (compose with a linear fractional transformation).

In $\mathbb{C}^n$, $n\geq 2$, the Riemann Mapping Theorem fails, so there is no hope for there to be an analogue of the construction just described. The question of local boundary equivalence has thus to be investigated _ab initio_ as in our first observations about the one variable case. Such an investigation leads rapidly to the observation that not all possible combinations $(\Omega, \partial\Omega)$ at $p_0 \in \partial\Omega$ are equivalent in the sense indicated. We now elaborate on this point.

The most obvious feature of the situation in several variables is that strong pseudoconvexity is preserved under local biholomorphic equivalence. Thus if p_0 is a strongly pseudoconvex boundary point then $(\Omega, \partial\Omega)$ at p_0 cannot be equivalent to, for instance, the pair

$$\left[\{(z_1, \ldots, z_n): y_n > 0\} , \{(z_1, \ldots, z_n): y_n = 0\}\right]$$

This is clear by calculation and was very well known classically. Note that in one complex variable the distinguished role of strong pseudoconvexity does not arise because every smooth (C^2) boundary point is vacuously strongly pseudoconvex.

What is less obvious is that not all (real analytic) strongly pseudoconvex domains in $\mathbb{C}^n$, $n \geq 2$, are locally equivalent. To see this, one counts parameters. The set of k-jets (power series up to and including order k) of local biholomorphic maps at a point p_0 operates on the k-jets of strongly pseudoconvex hypersurfaces at p_0. If all (real analytic) strongly pseudoconvex hypersurfaces were locally equivalent, this action would clearly have to be transitive. This is true if $n=1$, as we have already noted. But for $n \geq 2$, it fails (for k large enough: large enough turns out to be $k \geq 6$ if $n=2$, $k \geq 4$ if $n > 2$). To see this it suffices to note that the real dimension of the space of k-jets of local biholomorphic maps fixing p_0 (which dimension is equal to $2n$ times the number of complex coefficients in a power series in n variables up to k^{th} order) is less than the dimension of the space of all k-jets of real hypersurfaces. Hence, since the action is differentiable, indeed algebraic, we cannot have the smaller dimensional set acting transitively on the larger.

The inequalities of dimension just noted increase with k, i.e., as k goes to infinity, the difference of the two dimensions does also. This means that _a priori_ there must exist infinitely many differential invariants that have to agree at p_0 in $\partial\Omega_0$ and $p_1 \in \partial\Omega_1$ if $(\Omega_0, \partial\Omega_0)$ is to be equivalent to $(\Omega_1, \partial\Omega_1)$ at p_0 and

p_1 respectively. Moreover some at least of these differential invariants exist on higher and higher levels of differentiability. In particular, given ℓ, it must be possible for any given p_0, Ω_0, $\partial\Omega_0$ and p_1, Ω_1, $\partial\Omega_1$, to make a perturbation of $(\Omega_0,\partial\Omega_0)$ that is arbitrarily small in C^ℓ but nonetheless suffices to insure that $(\Omega_0,\partial\Omega_0)$ at p_0 is not locally biholomorphically equivalent to $(\Omega_1,\partial\Omega_1)$ at p_1.

The local boundary invariant theory is not a priori related to the question of the existence of automorphisms or biholomorphic maps, because the maps could in principle be very irregular in behavior as the boundary of the domain is approached. The missing link is supplied by Fefferman's smoothness-to-the-boundary theorem ([FE]). This result, which has been simplified, localized, and generalized in [BE], [BEL], and [LI], asserts the following:

If Ω_0 and Ω_1 are C^∞ strongly pseudoconvex domains in $\mathbb{C}^n$ and if $F:\Omega_0 \to \Omega_1$ is a biholomorphic map from Ω_0 to Ω_1 then there is a (unique) smooth extension $\hat{F}$ of F which is a diffeomorphism of the closure of Ω_0 onto the closure of Ω_1.

Unless Ω_0 and Ω_1 are real analytic, the map $\hat{F}$ will not be in general complex analytic (i.e. holomorphic) on the closure of Ω_0, as one can see by easy one-variable examples. Nonetheless, it is true (and indeed obvious) that $\bar{\partial}\hat{F}$ vanishes to infinite order on $\partial\Omega_0$. And a moment's thought will show that this suffices to insure that the local invariant theory will apply. Namely, any boundary differential invariant defined at a finite jet level k will be preserved by $\hat{F}$.

The existence of a large colleciton of preserved differential invariants immediately suggests the possibility of constructing rigid domains by imitating the corresponding "bumpy metric" construction of Riemannian geometry (see [AB]). In other words, one sees in effect that generically a small localized perturbation of the boundary suffices to make a given point inequivalent to all other points of the boundary, and by iteration of this perturbation process a rigid domain should be obtainable. In fact, rigid domains should be generic. Perhaps somewhat surprisingly, carrying out this construction in detail is far from easy. The details are given in [BSW] (see also [BDD]).

It is worth noting that once we are in possession of a C^∞ rigid domain, then Theorem 1 of Section 8 shows us how to obtain a real analytic rigid domain--by approximation of the given one. The same theorem also shows us how to obtain a rigid domain that has boundary of no greater regularity than C^2, again by approximation. To the authors' knowledge, Theorem 1 of Section 8 is in fact the only known method for treating the real analytic and C^2 cases. The former is otherwise intractable because it disallows localized perturbations; the latter because there is no local invariant theory at the C^2 level.

10. Semicontinuity of Automorphism Groups

The openness of rigid domains in the set of C^2 strongly pseudoconvex domains is a special case of a more general concept that we shall call semicontinuity of automorphism groups. For this idea, think of the openness of rigid domains as saying that if Ω_0 is a C^2 strongly pseudoconvex rigid domain then domains that are C^2 close to Ω_0 do not have larger automorphism groups than Ω_0 itself--and, since $\mathrm{Aut}(\Omega_0)=\{\text{identity}\}$, that these nearby domains have only the identity automorphism. We can consider this situation without making the assumption that Ω_0 is rigid. That is, suppose that Ω_0 is a (general) C^2 strongly pseudoconvex domain; then is it the case that, for each Ω that is C^2 sufficiently close to Ω_0, the group $\mathrm{Aut}(\Omega)$ is not larger than $\mathrm{Aut}(\Omega_0)$?

Of course, the precise meaning of this question depends on the sense in which the two groups are compared, i.e., the sense of the phrase "not larger than" as applied to Lie groups. It turns out that the answer to the question is "yes" in the strongest reasonable sense of the phrase "not larger than". Namely, we consider a Lie group A to be "not larger than" a Lie group B if A is isomorphic to a subgroup of B. This implies, among other things, that the dimension of A does not exceed the dimension of B and that if the dimension of A equals the dimension of B then the cardinality of {components of A} does not exceed the cardinality of {components of B}

The semicontinuity statement for C^2 strongly pseudoconvex domains is the following.

Theorem 1 (Greene-Krantz [GK3]): If Ω_0 is a C^2 strongly pseudoconvex domain in $\mathbb{C}^n$ not biholomorphic to the ball, then there is a neighborhood $\mathcal{U}$ of Ω_0 in the C^2 topology (see the Appendix for this terminology) such that, for each $\Omega \in \mathcal{U}$, the Lie group $\mathrm{Aut}(\Omega)$ is isomorphic to a subgroup of $\mathrm{Aut}(\Omega_0)$.

If Ω_0 is a C^2 strongly pseudoconvex domain not biholomorphic to the ball, then it is relatively straightforward to construct a sequence of subdomains $\{\Omega_j \subseteq \Omega_0 : j=1,2,\ldots\}$ such that

(a) Each Ω_j has real analytic boundary;

(b) Each Ω_j is invariant under the action of $\mathrm{Aut}(\Omega_0)$ on Ω_0;

(c) The sequence $\{\Omega_j\}$ converges to Ω_0 in the C^2 topology.

Such a sequence is constructed as sublevel sets of suitable $\mathrm{Aut}(\Omega_0)$-invariant functions; see [GK3], [GK4] for details, and also the discussion to be given momentarily of the proof of the theorem. For such a sequence $\{\Omega_j\}$, the theorem implies that $\mathrm{Aut}(\Omega_j)=\mathrm{Aut}(\Omega_0)$ for all j sufficiently large; this follows because then $\mathrm{Aut}(\Omega_j)$ is isomorphic to a subgroup of $\mathrm{Aut}(\Omega_0)$ while at the same time $\mathrm{Aut}(\Omega_j)$ contains $\mathrm{Aut}(\Omega_0)$, since Ω_j is $\mathrm{Aut}(\Omega_0)$-invariant. Thus one obtains the following striking corollary:

Corollary 2 (Greene-Krantz [GK3]): If Ω_0 is a C^2 strongly pseudoconvex domain not biholomorphic to the ball then there is a strongly pseudoconvex domain Ω in C^n with real analytic boundary with $\mathrm{Aut}(\Omega)$ isomorphic to $\mathrm{Aut}(\Omega_0)$. Indeed, there is such a Ω in any given C^2 neighborhood of Ω_0.

The rest of this section contains a sketch of the proof of the theorem, together with discussion of related topics. We assume once and for all that the domain Ω_0 is not biholomorphic to the ball.

The proof of the theorem begins with the same basic observation that began the proof of the openness of rigid domains. Namely, suppose $\{\Omega_j\}$ is a sequence of C^2 domains converging in the C^2 topology to Ω_0 ; then for each (sub)sequence $\{\alpha_{j_k}\}$, $\alpha_{j_k} \in \mathrm{Aut}(\Omega_{j_k})$, there is a subsequence of the α_{j_k} that converges uniformly on compact subsets of Ω_0 to an element of $\mathrm{Aut}(\Omega_0)$. The reasoning to establish this is the same as in Section 9: nondegeneracy of normal families, etc.

It follows that if Ω is a C^2 domain that is C^2 very close to Ω_0 then $\mathrm{Aut}(\Omega)$ is element-by-element close to $\mathrm{Aut}(\Omega_0)$ on any fixed compact subset of Ω_0 . By this is meant that for each element of $\mathrm{Aut}(\Omega)$ there is an element α_0 of $\mathrm{Aut}(\Omega_0)$ that is C^0 close to α on the given compact set. This situation is very suggestive because of the following result from Lie group theory.

Theorem 3 (Montgomery-Zippin [MZ]): If G is a compact Lie group and H is a closed subgroup of G , then there is a neighborhood V of H in G such that if K is a subgroup of G with $K \subseteqq V$ then K is isomorphic to a subgroup of H ; indeed, there is an $x \in G$ such that $x^{-1}Kx \subseteqq H$.

Thus inside a fixed (perhaps large) Lie group, a group K being element-by-element close enough to a fixed group H does imply that K is isomorphic to a subgroup of H . This fact does not apply to our automorphism group situation directly, however, because we have no way of knowing <u>a priori</u> that the groups $\mathrm{Aut}(\Omega_j)$ and $\mathrm{Aut}(\Omega_0)$ lie together in any compact Lie group, or even that they are isomorphic to groups lying in a fixed compact Lie group.

To treat this difficulty, a considerable detour is necessary. The idea is to reduce the problem to the case of compact manifolds. For compact manifolds, the following result follows immediately from [EB] and Theorem 3:

Theorem 4: Suppose M is a compact manifold and $\{G_j\}$ is a sequence of compact Lie groups in the diffeomorphism group of M, equipped with the C^∞ topology. Suppose G_0 is a compact Lie group acting faithfully on M with the property that, for each subsequence $\{G_{j_k}\}$

of subgroups and $\gamma_{j_k} \in G_{j_k}$, there is a (sub)subsequence of the $\{\gamma_{j_k}\}$ that converges in the C^∞ topology (of maps from M to M) to an element of G_0 . Then: for all j sufficiently large, G_j is isomorphic to a subgroup of G_0 .

Notice that the diffeomorphism group of M , being infinite dimensional, is not a Lie group. It is instructive to think of the $\{G_j\}$ as a sequence of compact Lie groups each armed with a faithful action on M . The convergence of γ's statement is the same as saying that for all j sufficiently large G_j belongs to any fixed C^∞ neighborhood of G_0 . We call this "element-by-element" convergence of $\{G_j\}$ to G_0 .

Actually, Theorem 4 is not proved point-blank in [EB]; rather, the results there are stated in terms of isometry groups of metrics. The result we state is obtained from that in [EB] by noting that, under the hypotheses indicated, there is a sequence g_j of C^∞ G_j-invariant Riemannian metrics that converges in the C^∞ topology to a C^∞ G_0-invariant Riemannian metric g_0 . Such a sequence is constructed by first finding a G_0-invariant metric by a standard integration (averaging) process over G_0 , and then setting g_j equal to the G_j-average of g_0 . Since G_j converges element-by-element to G_0 , it follows that $\{g_j\}$ converges to g_0 . The results in [EB] then show that, for j large,

$$G_j \subseteq [\text{Isometry group of } g_j] \subseteq [\text{Isometry group of } g_0] ,$$

where the second $\subseteq$ is after an isomorphism which can be taken close to the identity for j large. Now the isometry group of g_0 may be larger than G_0 . But the previous result from Lie group theory now applies to show that the isometry group of g_j is isomorphic to a subgroup of G_0 . (See [GK7] for further details.)

To relate these results for compact manifolds to the case of domains in $\mathbb{C}^n$ requires two steps. The first step is to show that the compact-manifold results extend to compact manifolds with boundary. The second step, which will be discussed further later, is to convert the domain case to the manifold-with-boundary case by restricting the automorphism groups to suitable smoothly-bounded invariant subdomains.

To carry out the first step, note that a group action on a compact manifold M with boundary can be extended to an action of the same group on the double of the manifold, by an obvious reflection process, the reflection being taken relative to a group-invariant metric. Pursuing this idea in detail (cf. [GK7]) yields the following result, which is Theorem 4 for manifolds with boundary:

Theorem 5 (Greene-Krantz): Suppose M is a compact manifold with boundary, and that $\{G_j\}$ is a sequence of faithful compact Lie group actions on M. Suppose that G_0 is a compact Lie group acting faithfully on M with the property that, for each subsequence G_{j_k} and sequence $\{\gamma_{j_k}\}$, $\gamma_{j_k} \in G_{j_k}$, there is a (sub)subsequence of the γ_{j_k} that converges in the C^∞ topology to an element of G_0. Then, for all j sufficiently large, each G_j is isomorphic to a subgroup of G_0.

To apply Theorem 5 to the automorphism groups of domains, we proceed as follows: First, construct a C^∞ function $\psi_0 : \Omega_0 \to \mathbb{R}$ on Ω_0 with $\psi_0 < 0$ and $\lim \psi_0 = 0$ at $\partial\Omega_0$. Averaging ψ_0 with respect to the group action of $\mathrm{Aut}(\Omega_0)$ will produce a nonconstant C^∞ $\mathrm{Aut}(\Omega_0)$-invariant function $\hat{\psi}_0$ on Ω_0, with again the property that $\hat{\psi}_0 < 0$ on Ω_0 and $\lim \hat{\psi}_0 = 0$ at $\partial\Omega_0$. Choose $\alpha < 0$ such that $\phi \neq \{x \in \Omega_0 : \hat{\psi}_0(x) \leq \alpha\} \subset\subset \Omega_0$ and such that α is a noncritical value of $\hat{\psi}_0$ (this choice is possible by Sards's theorem). Set $M_0 = \{x \in \Omega_0 : \hat{\psi}_0(x) \leq \alpha\}$. Then $\mathrm{Aut}(\Omega_0)$ acts faithfully on the C^∞ compact manifold with boundary M_0; the faithfulness is a consequence of the fact that an automorphism is determined by its values on a nonempty open set, and M_0 has nonempty interior.

Now let

$$\hat{\psi}_j = [\text{the average over } G_j \text{ of } \hat{\psi}_0] .$$

Of course ψ_j is defined only at those points x of Ω_0 such that the G_j orbit of x lies in Ω_0. But for j very large, the set of such points x will include any fixed compact subset of Ω_0, since G_j converges element-by-element to G_0. Set

$$M_j \equiv \{x \in \Omega_0 : \hat{\psi}_j(x) \text{ is defined and } \hat{\psi}_j(x) \leq \alpha\} .$$

For j large, M_j will be a compact manifold with boundary that is diffeomorphic to M_0. Furthermore, diffeomeophisms $F_j : M_0 \to M_j$ can be chosen to converge to the identity of M_0 in the C^∞ topology. These statements follow from observing that the $\hat{\psi}_j$ converge C^∞ on compact subsets of Ω_0 to $\hat{\psi}_0$. This latter convergence is a consequence of the facts that $\hat{\psi}_0$ is $\mathrm{Aut}(\Omega_0)$ invariant and that, for j large, each element of $\mathrm{Aut}(\Omega_j)$ is C^∞ close on compact sets to an element of $\mathrm{Aut}(\Omega_0)$.

We now consider the faithful action of $\mathrm{Aut}(\Omega_j)$ on M_j transferred to a faithful action on M_0 by $\alpha \to F_j^{-1} \circ \alpha \circ F_j$. Because $\{\mathrm{Aut}(\Omega_j)\}$ converges to $\mathrm{Aut}(\Omega_0)$ in the by now usual sense and because $\{F_j\}$ converges to the identity, it follows that for j large the action of $\mathrm{Aut}(\Omega_j)$ on M_0 is isomorphic to a subgroup of the action of $\mathrm{Aut}(\Omega_0)$ on M_0. Since all actions are faithful, $\mathrm{Aut}(\Omega_j)$ is isomorphic to a subgroup of $\mathrm{Aut}(\Omega_j)$, as desired. This completes the proof of Theorem 1.

It is important to realize that the argument just given to prove Theorem 1 proves in fact more. Namely, this argument applies to any sequence of domains $\{\Omega_j\}$ converging in a C^0 sense to Ω_0 and having the property that any sequence $\{f_j : \Omega_j \to \Omega_j\}$ of automorphisms, j ranging over infinitely many values, must have a subsequential nondegenerate limit $f_0 : \Omega_0 \to \Omega_0$, $f_0 \in \mathrm{Aut}(\Omega_0)$. The only role of the C^2 and strong pseudoconvexity hypotheses of the theorem was to guarantee the nondegeneracy of normal families via the C/K invariant. Once this nondegeneracy is guaranteed the argument proceeds in complete generality (indeed, see Section 12 for an instance of this greater generality). This generality is sometimes useful in practice (in [BDD], for instance, nondegeneracy is guaranteed by topological considerations in circumstances where C^2 convergence is not available). We state the general result formally:

Theorem 6 (Greene-Krantz): Suppose $\{\Omega_j\}$ is a sequence of bounded domains in $\mathbb{C}^n$ converging C^0 to a bounded domain Ω_0. Suppose also that for every subsequence Ω_{j_k} of Ω_j and every sequence $\{f_{j_k}\}$, $f_{j_k} \in \mathrm{Aut}(\Omega_{j_k})$, there is a subsequence that converges uniformly on compact subsets of Ω_0 to an element of $\mathrm{Aut}(\Omega_0)$. Then for all j sufficiently large $\mathrm{Aut}(\Omega_j)$ is isomorphic to a subgroup of $\mathrm{Aut}(\Omega_0)$.

Here C^0 convergence means, as usual, that any compact set in Ω_0 is in Ω_j for all j sufficiently large and any compact set in Ω_j for all sufficiently large j is also in Ω_0. The somewhat peculiar looking condition for convergence of the f's is needed here and earlier because if one simply assumed the existence of a nondegenerate subsequential limit for $\{f_j\}$, $j = 1,2,\ldots$, and if $\mathrm{Aut}(\Omega_j)=\{\text{identity}\}$ for infinitely many j then no information could be obtained about the remaining Ω_j.

In case one assumes high enough differentiability of the boundaries and convergence of boundaries with many derivatives (C^k topology, k large enough), one can obtain a more refined conclusion in Theorem 6. Namely, it can be shown that $\mathrm{Aut}(\Omega_j)$ is isomorphic to a subgroup of $\mathrm{Aut}(\Omega_0)$ via conjugation by a diffeomorphism from the closure of Ω_0 to the closure of Ω_j ([GK3]). (These two closures are in this case themselves compact manifolds with boundary.) Moreover, the sequence of conjugating diffeomeorphisms can be taken to converge to the identity on the closure of Ω_0, in the C^ℓ topology, where the k required above depends on the ℓ desired here.

Related to the previous paragraph is the fact that, in case the (C^2 strongly pseudoconvex) domain Ω_0 has a sufficiently smooth boundary and is still biholomorphically distinct from the ball then one can find a special metric g_0 on Ω_0 which is C^∞ on the closure of Ω_0 and is $\mathrm{Aut}(\Omega_0)$ invariant. In particular, the existence of this special metric interprets geometrically the smoothness to the boundary of the elements of $\mathrm{Aut}(\Omega_0)$. Namely, $\mathrm{Aut}(\Omega_0)$ then acts as isometries on the boundary of Ω_0, these isometries being necessarily C^∞ by the Myers-Steenrod theorem. And then this boundary action extends smoothly to a neighborhood of $\partial\Omega_0$ in Ω_0 by noting that geodesics perpendicular to the boundary must be taken to geodesics perpendicular to the boundary. This viewpoint, although derived from known smoothness-to-the-boundary results, is both geometrically appealing and, on occasion, technically useful; cf. Section 12 and the forthcoming [GK8].

Theorem 1 can be extended to the case where Ω_0 is biholomorphic to the ball, at least in the situation of the C^∞ topology on domains. See Section 12 for further details.

It is natural to inquire which compact Lie groups can occur as automorphism groups of C^2 strongly pseudoconvex domains. Rather surprisingly, the answer is that given any compact Lie group G there is a C^2 strongly pseudoconvex domain Ω in $\mathbb{C}^n$ (for some n) such that $\mathrm{Aut}(\Omega)\cong G$. (As we noted earlier, Ω can in fact be taken to have real analytic boundary). The fact that all compact Lie groups so occur was proved independently by Saerens/Zame [SZ] and by a shorter method by Bedford/Dadok [BD]. Both these sets of authors also showed that a corresponding result holds for isometry groups of compact manifolds, i.e., every compact Lie group is the exact isometry group of some compact Riemannian manifold.

11. The Bergman Kernel and Stability Questions

In Section 12 we shall consider differential invariants which arise naturally from the geometry of the Bergman metric. In the present section we therefore give an introduction to the Bergman kernel and Bergman metric. We include some results about stability of estimates for the $\bar{\partial}$-problem and $\bar{\partial}$-Neumann problem which are needed for the study of the Bergman kernel and in other parts of this paper as well.

a. <u>The Bergman Kernel</u>

If $\Omega \subseteq \mathbb{C}^n$ is a bounded domain, we define

$$A^2(\Omega)=\{f \text{ holomorphic on } \Omega : \int_\Omega |f(z)|^2 \, dV(z) < \infty\} .$$

Here dV is volume measure on $\mathbb{C}^n \approx \mathbb{R}^{2n}$. The expression

$$\|f\| \equiv \left[\int_\Omega |f|^2 \, dV\right]^{1/2}$$

is a norm on the complex vector space $A^2(\Omega)$. We may equip the space $A^2(\Omega)$ with the inner product

$$\langle f,g\rangle \equiv \int_\Omega f(\xi)\overline{g(\zeta)} \, dV(\zeta) ,$$

which makes A^2 into a Hermitian inner product space. The norm is compatible with the inner product in the usual sense.

The following lemma is the key to the Bergman space theory:

Lemma 1: If E is compact in Ω then there is a constant $C=C(E,\Omega)$ such that for any $f \in A^2(\Omega)$ and all $z \in E$ it holds that

$$|f(z)| \le C\|f\| .$$

Proof: By the compactness of E , there is an $r>0$ such that for all $e \in E$ the ball $B(e,r)$ is contained in Ω. By the mean value property for holomorphic functions we then have that if $e \in E$ then

$$f(e) = \frac{1}{V(B(e,r))} \int_{B(e,r)} f(\zeta) \, dV(\zeta) .$$

As a result, we may use Schwartz's Inequality to obtain that

$$|f(e)| \le \frac{1}{V(B(e,r))} \int_{B(e,r)} |f(\zeta)| \, dV(\zeta)$$

$$\le \frac{1}{V(B(e,r))^{1/2}} \int_{B(e,r)} |f(\zeta)|^2 \, dV(\zeta)^{1/2}$$

$$= C(r) \cdot \|f\|$$

as desired. □

The lemma shows that norm convergence in the space $A^2(\Omega)$ implies uniform convergence on compact sets. Since the uniform limit on compact sets of a sequence of holomorphic functions is itself holomorphic, we see that $A^2(\Omega)$ is complete; that is, $A^2(\Omega)$ is a Hilbert space.

For each fixed $z \in \Omega$ consider the linear functional $\eta_z : A^2(\Omega) \to \mathbb{C}$ defined by

$$\eta_z(\varphi) = \varphi(z) .$$

By the lemma, with $E=\{z\}$, we see that this functional is continuous. By the Riesz representation theorem, we then conclude that there is an element $k_z \in A^2(\Omega)$ such that

$$\eta_z(\varphi) = \langle \varphi, k_z \rangle$$

for all $\varphi \in A^2(\Omega)$. Defining

$$K(z,\zeta) = \overline{k_z(\zeta)} ,$$

we see that for all $\varphi \in A^2(\Omega)$ and all $z \in \Omega$ it holds that

$$\varphi(z) = \int_\Omega \varphi(\zeta) K(z,\zeta) \, dV(\zeta) .$$

The function K is called the <u>Bergman kernel</u> for the domain Ω.

It can be shown by a purely formal argument that $K(z,\zeta) = \overline{K(\zeta,z)}$. As a result,

$$K(z,\cdot) \in \overline{A^2}(\Omega) \quad \text{and} \quad K(\cdot,\zeta) \in A^2(\Omega)$$

for any fixed z, ζ in Ω. Therefore

$$K(z,z) = \int_\Omega K(\zeta,z)\cdot K(z,\zeta)\, dV(\zeta) = \int_\Omega |K(z,\zeta)|^2\, dV(\zeta) \geq 0 .$$

The last expression is the square of the norm of the functional η_z and hence is not zero. We may thus consider the function $\log K(z,z)$ and we may define

$$g_{ij}(z) = \frac{\partial^2}{\partial z_i \partial \bar{z}_j} \log K(z,z)$$

for $z\in\Omega$. The matrix $(g_{ij}(z))$ defines a hermitian, indeed a Kähler, metric on Ω. This metric is called the Bergman metric. The key property of the Bergman metric, which may be checked by a formal calculation, is that if Ω_1 and Ω_2 are bounded domains and if $\phi:\Omega_1\to\Omega_2$ is a biholomorphic map then ϕ is an isometry of Bergman metrics. Details on this and other matters relating to the Bergman metric may be found in [KR1].

b. The $\bar\partial$-Neumann Problem

On $\mathbb{C}^n$, we define the differential operators

$$\frac{\partial}{\partial z_j} = \frac{1}{2}\left[\frac{\partial}{\partial x_j} - i\frac{\partial}{\partial y_j}\right]$$

and

$$\frac{\partial}{\partial \bar{z}_j} = \frac{1}{2}\left[\frac{\partial}{\partial x_j} + i\frac{\partial}{\partial y_j}\right].$$

It can be checked directly that a continuously differentiable function u on a domain $\Omega\subseteq\mathbb{C}^n$ satisfies the Cauchy-Riemann equations in each of the variables $z_1,\ldots,z_n$ if and only if

$$\frac{\partial}{\partial \bar{z}_j} u \equiv 0 , \quad j = 1,\ldots,n .$$

In order to study these equations, it is convenient to introduce the notation

$$dz_j = dx_j + idy_j$$

and

$$d\bar{z}_j = dx_j - idy_j .$$

Notice that we have the pairings

$$\left\langle \frac{\partial}{\partial z_j} , dz_k \right\rangle = \delta_{jk} , \left\langle \frac{\partial}{\partial \bar{z}_j} , d\bar{z}_k \right\rangle = \delta_{jk}$$
$$\left\langle \frac{\partial}{\partial \bar{z}_j} , dz_k \right\rangle = \left\langle \frac{\partial}{\partial z_j} , d\bar{z}_k \right\rangle = 0 ,$$

by direct calculation using known properties of real vectors and covectors. Here δ_{jk} is the Kronecker delta. Now we define the exterior differential operators

$$\partial u \equiv \sum_{j=1}^{n} \left[\frac{\partial}{\partial z_j} f \right] dz_j$$

and

$$\bar{\partial} u \equiv \sum_{j=1}^{n} \left[\frac{\partial}{\partial \bar{z} j} f \right] d\bar{z}_j .$$

We see that for a continuously differentiable u , $\bar{\partial}u \equiv 0$ if and only if u is holomorphic.

For maximum flexibility in constructing functions, we will consider the equation

$$\bar{\partial} u = f , \quad (*)$$

for $f = f_1(z)d\bar{z} + \ldots + f_n(z)d\bar{z}_n$ a given form. For such an f we define

$$\bar{\partial} f = \sum_{j>k} \left[\frac{\partial f_k}{\partial \bar{z}_j} - \frac{\partial f_j}{\partial \bar{z}_k} \right] d\bar{z}_j \wedge d\bar{z}_k .$$

One calculates that for any twice continuously differentiable function u we have

$$\bar{\partial}(\bar{\partial} u) = 0 .$$

As a result, a necessary condition for the solvability of the equation (*) is that

$$\bar\partial f = 0 . \quad (**)$$

A form f satisfying $\bar\partial f=0$ is called $\bar\partial$-closed. On domains of holomorphy, the condition (**) is sufficient for solvability of (*). The theory of solutions of this equation is highly developed (see [FK] for a survey), and we shall only touch on a few high points here.

Notice that the operator $\bar\partial$ acting on functions has as its kernel the class $\mathcal{H}$ of all holomorphic functions. Thus the solution of

$$\bar\partial u = f ,$$

when the compatibility condition (**) is satisfied, will be a coset of the additive group of holomorphic functions. Following a cue from Hodge theory, we concentrate on finding a solution of (*) which is orthogonal (in some metric) to $\mathcal{H}$. This is called the solution of the "$\bar\partial$-Neumann" problem. The method for studying this solution is to consider the symmetrized operator

$$\bar\partial\bar\partial^* + \bar\partial^*\bar\partial \equiv \square .$$

The calculation of this operator, in particular the consideration of $\bar\partial$, involves degenerate boundary conditions which must be adjoined to the problem. As a result, the system

$$\square v = f$$

plus

(boundary conditions)

is extremely difficult to solve (see [FK] for a thorough treatment). It turns out that if f is $\bar\partial$-closed and if v is the "canonical solution" to $\square v=f$ gotten using Hodge theory then $\bar\partial^* Nv$ is the (unique) solution to the equation $\bar\partial u=f$ which is orthogonal to ker $\bar\partial$.

c. <u>Stability Questions for the L^2 Topology</u>

It will be crucial for the geometric results to be studied in later sections that we understand how the Bergman kernel and metric change when the domain Ω is perturbed slightly. A closely related question, which will also be needed to understand the Bergman metric situation, is the stability of solutions to the $\bar{\partial}$-Neumann problem. A careful consideration of these matter is necessarily very tedious and technical (details are contained in [GK2]). In the interest of clarity, we use this space only to summarize the key results.

We consider a C^∞ strongly pseudoconvex domain $\Omega=\{z:\rho(z)<0\}$ (see the Appendix for the definition). The question that we now want to discuss is this: if the defining function ρ for a given smoothly bounded strongly pseudoconvex domain is perturbed slightly in the C^∞ topology then how are solutions of the $\bar{\partial}$-Neumann problem affected and, concomitantly, how is the Bergman metric affected?

In order to answer this question, it is convenient to introduce another way to think about boundary perturbations. Let Ω_0 be a fixed, smoothly bounded domain in $\mathbb{C}^n$ and let $\iota:\Omega_0\to\mathbb{C}^n$ be the (identity) injection. Then a domain $\Omega\subseteqq\mathbb{C}^n$ will be considered a (small) C^∞ perturbation of Ω_0 if Ω is the image of an injection $\jmath:\Omega_0\to\mathbb{C}^n$ such that $\iota-\jmath$ is small in the C^∞ topology. (The Appendix contains a discussion of this topology). Alternatively, $\Omega=\{\rho(z)<0\}$ is a C^∞ small perturbation of $\Omega_0=\{\rho_0(z)<0\}$ if the C^∞ defining functions ρ and ρ_0 for Ω and Ω_0 respectively satisfy $\|\rho-\rho_0\|_{C^\infty(\mathbb{C}^n)}$ is small. Now our stability results are these:

BASIC STABILITY RESULT FOR THE $\bar{\partial}$-NEUMANN PROBLEM:

If Ω_0 is a fixed strongly pseudoconvex domain and if $\epsilon>0$ then there is a $\delta>0$ such that if Ω is a smooth perturbation of Ω_0 and

(i) If $f^0=\Sigma f_j^0$ on Ω_0 and $f=\Sigma f_j d\bar{z}_j$ on Ω are $\bar{\partial}$-closed forms;

(ii) If there is an injection $\jmath:\Omega_0\to\mathbb{C}^n$ such that (image $\jmath$)=Ω and $\|\iota-\jmath\|_{C^\infty(\Omega_0)}<\delta$ and $\Sigma\|f_j^0\circ\jmath-f_j\|_{C^\infty(\Omega_0)}<\delta$, ι being the identity injection of Ω_0 into $\mathbb{C}^n$;

THEN

The $\bar{\partial}$-Neumann solutions to the equations

$$\bar{\partial}u^0 = f^0$$

and

$$\bar{\partial}u = f$$

on Ω_0 and Ω respectively satisfy

$$\|u^0 - u\circ j\|_{C^\infty(\Omega_0)} < \epsilon\cdot\sum_j \|f_j^0 - f_j\circ j\|_{C^\infty(\Omega_0)} .$$

BASIC STABILITY RESULT FOR THE BERGMAN KERNEL AND METRIC:

Fix a smoothly bounded strongly pseudoconvex domain $\Omega_0=\{z\in\mathbb{C}^n:\psi_0(z)<0\}$. According to Fefferman's theorem [FE] there are $C^\infty(\bar{\Omega})$ functions ϕ_{Ω_0} , $\tilde{\phi}_{\Omega_0}$, and $\tilde{\tilde{\phi}}_{\Omega_0}$ such that, on the diagonal, the Bergman kernel $K_{\Omega_0}(z,z)$ has an expansion of the form

$$K_{\Omega_0}(z,z)=\phi_{\Omega_0}(z)\cdot\psi_{\Omega_0}^{-(n+1)}(z)+\tilde{\phi}_{\Omega_0}(z)\cdot\log\ \psi_{\Omega_0}(z)+\tilde{\tilde{\phi}}_{\Omega_0}(z) \quad (*)$$

on Ω .

Now let $\epsilon>0$. There exists a $\delta>0$ such that if Ω is another smoothly bounded strongly pseudoconvex domain and if there is an injection $j:\Omega_0\to\mathbb{C}^n$ such that (image j)$=\Omega$ and $\|\iota-j\|_{C^\infty(\Omega_0)}<\delta$ (ι being the identity injection of Ω_0 into $\mathbb{C}^n$) then, on the diagonal, the Bergman kernel $K_\Omega(z,z)$ for Ω has an asymptotic expansion

$$K_\Omega(z,z) = \phi_\Omega(z)\cdot\psi_\Omega^{-(n+1)}(z)+\tilde{\phi}_\Omega(z)\cdot\log\ \psi_\Omega(z)+\tilde{\tilde{\phi}}_\Omega(z) \quad (**)$$

with

$$\|\phi_{\Omega_0}-\phi_\Omega\circ j\|_{C^\infty(\Omega\cup\Omega_0)}<\epsilon$$

$$\|\tilde{\phi}_{\Omega_0}-\tilde{\phi}_\Omega\circ j\|_{C^\infty(\Omega\cup\Omega_0)}<\epsilon$$

$$\|\approx\!\!\!\phi_{\Omega_0}-\approx\!\!\!\phi_\Omega\circ j\|_{C^\infty(\Omega\cup\Omega_0)}<\epsilon$$

The Bergman metric tensor is calculated by taking second order derivatives of $K(z,z)$ and the holomorphic sectional curvature tensor R is calculated by taking two more derivatives. The stability of the asymptotic expansion enables us to obtain a stability estimate for the holomorphic sectional curvature tensor of the Bergman metric. This will be discussed in the next section.

d. Uniform Estimates for the $\bar{\partial}$-Problem

Crucial in the construction of holomorphic functions is the use of uniform estimates for the $\bar{\partial}$ equation. Thus we need to formulate separately a stability result for such estimates.

STABILITY RESULT FOR UNIFORM ESTIMATES FOR THE $\bar{\partial}$ PROBLEM:

Let Ω_0 be a fixed strongly pseudoconvex domain with twice continuously differentiable boundary and let $\epsilon>0$. Then there is a $\delta>0$ such that if Ω is another domain with twice continuously differentiable boundary and

(i) If $f^0=\Sigma f_j^0\, d\bar{z}_j$ and $f=\Sigma f_j\, d\bar{z}_j$ are $\bar{\partial}$-closed forms with bounded coefficients on Ω_0 and Ω respectively;

(ii) If there is an injection $j:\Omega_0\to C^n$ such that (image j)$=\Omega$ and $\|i-j\|_{C^2(\Omega_0)}<\delta$ and $\Sigma\|f_j\circ j-f_j\|_{C^2(\Omega_0)}<\delta$, i being the identity injection of Ω_0 into C^n;

THEN

There is a construction (due to Henkin) of solutions to the equations

$$\bar{\partial}u^0 = f^0$$
$$\bar{\partial}u = f$$

on Ω_0 and Ω respectively satisfying

$$\|u\circ\jmath - u^0\|_{L^\infty(\Omega_0)} < \epsilon\left[\sum\|f_j - f_j^0\|_{L^\infty}\right] .$$

Proof of the first stability result involves detailed checking of the proof of Kohn's solution ([GK2]) to the $\bar{\partial}$-Neumann problem. Proof of the second stability result involves detailed checking of Fefferman's asymptotic expansion [GK2] ; the first stability result is needed here because of the fact that the Bergman kernel is the solution of a $\bar{\partial}$-Neumann problem: if z is a fixed point of Ω and δ_z is the point mass measure at z then

$$K(z,w) = \delta_z - \bar{\partial}^* N\bar{\partial}(\delta_z) .$$

Proof of the third stability result involves detailed checking of Henkin's construction, together with the first stability result as a vital step in a part of the Henkin construction.

12. The Curvature Viewpoint on Nondegeneracy of Normal Families

Throughout our discussion so far of automorphism groups, a central role has been played by nondegeneracy of normal families. We have seen that nondegeneracy could be insured by topological conditions (see § 1 and, for instance, [BED]). Also, nondegeneracy could be insured under very general conditions (essentially just C^2 strong pseudoconvexity) by considering the C/K invariant. In this section, we shall discuss another mechanism of insuring nondegeneracy of normal families, namely, by the curvature of the Bergman metric. As it turns out, this curvature method is generally less powerful technically than the C/K method: more boundary derivatives are required. However, the curvature method yields sometimes more detailed information, and it also has the advantage of putting the normal families considerations into a general geometric framework.

The Bergman metric is a Kähler metric, and it is a standard theorem of Kähler geometry that the whole Riemannian curvature tensor at a point of a Kähler manifold is completely determined by the Riemannian sectional curvature of the J-invariant 2-planes at that point. Here J is the almost-complex structure tensor of the manifold so that J is an endomorphism of each real tangent space to itself with $J^2=-id$. In euclidean space, the only case we shall consider in detail, J is the map $(x_1,y_1,\dots,x_n,y_n) \to (-y_1,x_1,\dots,-y_n,x_n)$. Thus J is multiplication by $\sqrt{-1}$, expressed in real coordinates. The sectional curvatures of J-invariant 2-planes are called <u>holomorphic sectional curvatures</u>.

The Bergman metric of the ball has constant holomorphic sectional curvature. This is clear because the biholomorphic maps, which act as isometries, act transitively on the J-invariant 2-planes. The holomorphic sectional curvature of the Bergman metric of the ball in $\mathbb{C}^n$ is $-4/(n+1)$, with our choice of normalization for the Bergman metric. This negative constant will be denoted by c_n throughout the remainder of this section.

The Bergman metric of the ball is a complete Riemannian metric. This follows from the fact that its isometry group is transitive. The following result characterizes the ball as the only complete, constant-holomorphic-sectional-curvature Bergman metric.

Theorem 1 (Lu Qi-Keng [LK]): If a bounded domain Ω in $\mathbb{C}^n$ has complete Bergman metric of constant holomorphic sectional curvature, then the constant must be c_n and Ω must be biholomorphic to the ball.

This theorem is remarkable for the fact that no hypothesis of simple connectivity of Ω is required. The corresponding result in the geometry of abstract Kähler metrics (not necessarily Bergman metrics) definitely <u>does</u> require simple connectivity; consider for example the fact that compact Riemann surfaces of genus at least two admit Kähler metrics of constant negative curvature.

The following result, which is obtained by calculation from C. Fefferman's asymptotic expansion of the Bergman kernel [FE], is a useful supplement to the theorem of Lu Qi-Keng:

Theorem 2 (P. Klembeck [KL]): If Ω is a C^∞ strongly pseudoconvex domain then given $\epsilon>0$ there is a $\delta>0$ such that, for each J-invariant 2-plane p at each point $x\in\Omega$ with $\mathrm{dis}(x,\partial\Omega)<\delta$, it holds that

$$|\kappa(p)-c_n| < \epsilon .$$

Here $\kappa(p)$ is the Riemannian sectional curvature of p and "dis" denotes Euclidean space distance.

This result should be thought of as saying that any C^∞ strongly pseudoconvex domain has asymptotically constant holomorphic sectional curvature c_n at the boundary of the domain. This result combined with Lu Qi-Keng's theorem immediately yields (in the case of C^∞ strongly pseudoconvex domains) a new proof of Bun Wong's compactness-of-automorphism groups result. The argument is as follows:

Suppose Ω is a C^∞ strongly pseudoconvex domain with noncompact automorphism group. Choose an arbitrary point $y\in\Omega$ and a J-invariant 2-plane p at y. Now, given $\delta>0$, there is an element $\alpha\in\mathrm{Aut}(\Omega)$ such that $\mathrm{dis}(\alpha(Q),\partial\Omega)<\delta$ because if the orbit of Q were bounded from $\partial\Omega$ then the orbit would be compact and hence the group would be compact (normal families reasoning as before--see also Section 1). Hence, for any $\epsilon>0$, there is an element $\alpha\in\mathrm{Aut}(\Omega)$ such that

$$|\kappa(\alpha_*(p))-c_n| < \epsilon .$$

This is an application of Klembeck's result. But $\kappa(\alpha_*(p))=\kappa(p)$. Hence

$$|\kappa(p)-c_n| < \epsilon$$

for all $\epsilon>0$ and we conclude that $\kappa(p)=c_n$. Since p was arbitrary, Lu Qi-Keng's theorem implies that Ω must be the ball.

This reasoning is geometrically satisfying, but it is not easily extended to Rosay's generalization of Bun Wong's theorem. Moreover, it requires C^∞ boundary. More precisely, it requires enough derivatives to make Fefferman's expansion valid--more than two (see

[GK2] for exact specifications of the finite number of derivatives needed here and throughout. For simplicity we shall formulate all results here for C^∞ boundary.)

For understanding the variation of domains, it is important to know that the estimates in Klembeck's results are stable with respect to small changes in the boundary of the domain.

The relevant results are in two parts. First, it is necessary to know that small boundary perturbations make small changes (in the C^∞ sense) in the Bergman kernel restricted to a fixed compact subset. This is a consequence of standard approximation theorems for square integrable holomorphic functions and extremal properties of K (see [GK2] for details). Note that this interior stability is considerably easier than the stability, up to the boundary, of the asymptotic expansion which was discussed in the last section.

The second result that is needed is in fact the one from Section 11 involving the stability of Fefferman's expansion with respect to boundary perturbations. This is needed in order to deduce that the δ for given ϵ in Klembeck's result can be taken to be stable with respect to small variations of the boundary.

The exact result that comes from these considerations and that suffices for the desired normal families constructions is the following:

Theorem 3 (Greene and Krantz [GK2]): Let Ω_0 be a C^∞ strongly pseudoconvex domain, C a compact subset of Ω_0 and ϵ a positive number. Then there is a neighborhood $\mathcal{U}$ of Ω_0 in the C^∞ topology on C^∞ domains and a $\delta>0$ such that:

(a) If $\Omega\in\mathcal{U}$, then $C\subset\Omega$ and for each 2-plane p at each point of C it holds that

$$|\kappa_\Omega(p)-\kappa_{\Omega_0}(p)| < \epsilon$$

where κ_Ω and κ_{Ω_0} are the sectional curvature functions of the Bergman metrics of Ω and Ω_0 respectively.

(b) If $\Omega\in\mathcal{U}$ and p is a J-invariant 2-plane at a point $x\in\Omega$ with $\mathrm{dis}(x,\partial\Omega)<\delta$ then

$$|\kappa_\Omega(p)-c_n| < \epsilon .$$

If Ω_0 is a C^∞ strongly pseudoconvex domain that is not biholomorphic to the ball, then there is a J-invariant 2-plane p at some point x with $\kappa_{\Omega_0}(p)\neq c_n$. If we set $\epsilon=(1/4)|\kappa_{\Omega_0}(p)-c_n|$ and $C=\{x\}$ in Theorem 3 then it follows that the orbit of x under $\mathrm{Aut}(\Omega_0)$ has all its points at euclidean distance at least δ from $\partial\Omega_0$. This follows by the biholomorphic invariance of sectional curvature. But note also that the same observation holds for the orbit of x under $\mathrm{Aut}(\Omega)$ for any Ω which is C^∞ sufficiently close to Ω_0. From this, we can obtain the kind of nondegeneracy of normal families that we derived before from the C/K invariant. In particular, the reader will find it easy to reprove the C^∞ cases of Theorem 1 of Section 4, Theorem 1 of Section 7, Theorem 1 of Section 8 and Theorem 1 of Section 10 using these curvature estimates in lieu of the C/K estimates.

Theorem 3 has interesting applications in case Ω_0 is taken to be the unit ball in $\mathbb{C}^n$. In particular, the two points (a) and (b) combine to show that:

Corollary 4: Let $\epsilon>0$ be given. Then there is a C^∞ neighborhood $\mathcal{U}$ of the unit ball $B\subseteq\mathbb{C}^n$ such that, if $\Omega\in\mathcal{U}$ then, for every J-invariant 2-plane p at each point $x\in\Omega$, it holds that

$$|\kappa_\Omega(p)-c_n| < \epsilon .$$

It is a well known result in Kähler geometry that if a Kähler manifold has nearly constant holomorphic sectional curvatures then its Riemannian sectional curvatures are almost the same as those of the space of exactly constant holomorphic sectional curvature. This is in fact true point by point ("pointwise") on a Kähler manifold ([BE]). In our particular case, this yields the following result:

Corollary 5: Let $\epsilon>0$ be given. There is a C^∞ neighborhood $\mathcal{U}$ of the ball $B\subseteq\mathbb{C}^n$ such that, for each $\Omega\in\mathcal{U}$, the Riemannian sectional curvatures κ of the Bergman metric of Ω lie between $c_n-\epsilon$ and

$(c_n/4)+\epsilon$. In particular, there is a C^∞ neighborhood $\mathcal{U}$ of the ball B in $\mathbb{C}^n$ such that if $\Omega\in\mathcal{U}$ then the Bergman metric of Ω is a (complete, Kähler) metric of negative Riemannian sectional curvature.

The completeness is assured by the strong pseudoconvexity of Ω, since every C^∞ strongly pseudoconvex domain has a complete Bergman metric.

The corollary can be combined with a classical theorem of E. Cartan to yield a result about fixed points of automorphism groups. The result of Cartan states that if M is a complete, simply connected Riemannian manifold of nonpositive sectional curvature then every compact group of isometries of M has a fixed point. Combining this result with the previous corollary we obtain:

Theorem 6 (Greene-Krantz [2]): There is a C^∞ neighborhood $\mathcal{U}$ of the ball $B \subseteq \mathbb{C}^n$ such that every $\Omega\in\mathcal{U}$ either (a) is biholomorphic to B or (b) has automorphism group with a fixed point in Ω.

In case (b), $\mathrm{Aut}(\Omega)$ is compact by the B. Wong-Rosay theorem so that the Cartan theorem and Corollary 5 yield the conclusion.

The theorem just stated yields a particularly interesting result when applied to circular domains, i.e. domains invariant under the map $(z_1,\ldots z_n)\to(e^{i\theta}z_1,\ldots,e^{i\theta}z_n)$ for all $\theta\in\mathbb{R}$. Clearly, only $(0,\ldots,0)$ can be a fixed point of $\mathrm{Aut}(\Omega)$ if Ω is a circular domain. It follows that if $\Omega_1,\Omega_2\in\mathcal{U}$ are as in the theorem, if both Ω_1 and Ω_2 contain the origin and if neither Ω_1 nor Ω_2 is biholomorphic to the ball, then a biholomorphic map (if any) from Ω_1 to Ω_2 must take $(0,\ldots,0)$ to $(0,\ldots,0)$.

It is a standard theorem of H. Cartan that if $F:\Omega_1\to\Omega_2$ is a biholomorphic map of circular domains and if $F((0,\ldots,0))=(0,\ldots,0)$, then F must be linear. From this, and the fact that $\mathrm{Aut}(B)$ is transitive we see that:

Corollary 7: With $\mathcal{U}$ as in Theorem 6, let $\Omega_1,\Omega_2\in\mathcal{U}$ be circular domains both of which contain the origin. Suppose that Ω_1 and Ω_2

are biholomorphic and that $F:\Omega_1 \to \Omega_2$ is a biholomorphic map. Then either

(a) Ω_1 and Ω_2 are both linearly equivalent to the ball and hence to each other;

or

(b) $F:\Omega_1 \to \Omega_2$ is a linear map.

If $n \geq 2$, then it is obvious that the space of all circular domains is infinite dimensional. On the other hand, it is also clear that the set of linear equivalences is finite dimensional. Thus the set of biholomorphically distinct circular domains is infinite dimensional if $n \geq 2$.

Theorem 6 has been generalized to all strongly convex domains in $\mathbb{C}^n$ by L. Lempert, using the special properties of the Kobayashi and Caratheodory metrics that hold for convex domains ([LEM]). This generalization implies of course that the corollaries of the theorem are valid for strongly convex (circular) domains. Lempert has also constructed a (necessarily infinite dimensional) moduli space for the strongly convex domains in $\mathbb{C}^2$.

The fixed point result, Theorem 6, provides the tool needed to extend Theorem 1 of Section 10 to the case where Ω_0 is the ball, provided we replace the C^2 topology by the C^∞ topology. If Ω is C^∞ close to a C^∞ domain Ω_0 biholomorphic to the ball, then either (a) Ω is itself biholomorphic to the ball, in which case the conclusion of the theorem is obvious; or (b) Ω is not biholomorphic to the ball. In the latter case, $\mathrm{Aut}(\Omega)$ has a fixed point (by Theorem 6) and hence $\mathrm{Aut}(\Omega)$ is isomorphic to a subgroup of $U(n)$. Since $U(n)$ is a subgroup of $\mathrm{Aut}(B)$, the theorem holds in case (b) as well.

13. Weakly Pseudoconvex Domains

The theorem of B. Wong and Rosay (see Section 4) does not admit direct extension to weakly pseudoconvex domains. As we noted earlier (Section 2), the domain $\Omega^*=\{(z_1,z_2):|z_1|^2+|z_2|^4<1\}$ has a noncompact automorphism group; but it is not biholomorphic to the ball B^2 in $\mathbb{C}^2$. If Ω is a C^2 weakly pseudoconvex domain with $\mathrm{Aut}(\Omega)$ not compact, then the orbits of $\mathrm{Aut}(\Omega)$ are noncompact (in Ω) since if they were compact then the group itself would be compact (see Section 1). All the orbits of the automorphism group of any bounded domain are closed in that domain, by an easy normal families argument. So the noncompactness of the orbits of $\mathrm{Aut}(\Omega)$ guarantees that each of these orbits accumulates at some boundary point. (This could of course be seen in the case of Ω^* by direct calculation, as in Section 2). If Ω is not biholomorphic to the ball, then the orbits can accumulate only at weakly pseudoconvex boundary points, since accumulation at a strongly pseudoconvex boundary point would imply biholomorphism to the ball via the Bun Wong-Rosay result.

Suppose that Ω is a C^2 weakly pseudoconvex domain and that an orbit of $\mathrm{Aut}(\Omega)$ accumulates at a particular boundary point $p\in\partial\Omega$. The Wong-Rosay result can be thought of in the following terms: If, at p, the boundary $\partial\Omega$ of Ω agrees after a local biholomorphic change of coordinates with the boundary ∂B^n of the ball up to (and including) second order, then Ω is biholomorphic to B^n. Here we have used the fact that p is a strongly pseudoconvex boundary point if and only if there is a local biholomorphic change of coordinates in a neighborhood of p after which $\partial\Omega$ agrees with ∂B^n up to (and including) second order.

This seemingly circuitous view of the Wong-Rosay theorem has the purpose of suggesting a possible mode of generalization. Let us choose as our "model" not B^n, but rather a weakly pseudoconvex domain with non-compact automorphism group such as (to be specific) $\Omega^*=\{(z_1,z_2):|z_1|^2+|z_2|^4<1\}$. Suppose that Ω, another weakly pseudoconvex domain, has an orbit of $\mathrm{Aut}(\Omega)$ accumulating at a boundary point p. Suppose also that $\partial\Omega$, in a neighborhood of p, agrees with $\partial\Omega^*$ in a neighborhood of $(1,0)$ to a high order (after a biholomorphic change of coordinates). The role of $(1,0)$ here is that the orbits of $\mathrm{Aut}(\Omega^*)$ accumulate there. It is now natural to ask, as a possible generalization of the Wong-Rosay result, whether in this situation it might necessarily follow that Ω is biholomorphic to Ω^* .

The general philosophy of this question is that the accumulation of an orbit at a particular boundary point makes the structure of the boundary near that boundary point carry complete information about the domain. This is of course exactly what the Wong-Rosay theorem guarantees in the case that the boundary structure is C^2 strongly pseudoconvex.

The question raised in the last two paragraphs has an affirmative answer. In fact, somewhat more general conclusions hold.

Theorem 1 (Greene-Krantz [GK6]): Define

$$\Omega_m = \{(z_1,z_2) \in \mathbb{C}^2 : \rho_m(z_1,z_2) = |z_1|^2+|z_2|^{2m}-1 < 0\} \quad .$$

Let $\Omega \subseteq \mathbb{C}^2$ be a bounded domain with C^6 boundary, and p a point of $\partial\Omega$. Suppose that, after a local biholomorphic change of coordinates near p , Ω near p is given by $\{z : \rho'(z)<0\}$ where

$$\rho'(z) = \rho_m(z)+0\Big[|z-(1,0)|\Big]^{2m+\epsilon} \quad ,$$

some $\epsilon>0$. Finally, assume that there is a point $q\in\Omega$ and elements $\alpha_j \in \mathrm{Aut}(\Omega)$ such that $\alpha_j(q)\to p$ as $j\to\infty$. Then Ω is biholomorphic to Ω_m .

Although we do not know a true generalization of this result to n dimensions, there is the following partial result:

Theorem 2 (Greene-Krantz [GK6]): Define

$$\Omega_m = \{(z_1,z_2,\ldots,z_n)\in\mathbb{C}^n : \rho_m(z_1,z_2,\ldots,z_n)$$

$$\equiv |z_1|^2+|z_2|^2+\ldots+|z_{n-1}|^2+|z_2|^{2m}-1 < 0\} \ .$$

Let $\Omega\subset\mathbb{C}_n$ be a bounded domain with C^{2n+2} boundary, and p a point of $\partial\Omega$. Suppose that, after a local biholomorphic change of coordinates near p , Ω near p is given by $\{z : \rho'(z)<0\}$ where

$$\rho'(z)=\rho_m(z)+0\Big[|z-(1,0,\ldots,0)|\Big]^{2m+\epsilon} \quad ,$$

some $\epsilon>0$. Finally, assume that there is a point $q\in\Omega$ and elements $\alpha_j\in \mathrm{Aut}(\Omega)$ such that $\alpha_j(q)\to p$ as $j\to\infty$. Then Ω is biholomorphic to Ω_m .

We shall summarize the proof of the theorems first for the special case that the boundary of Ω near $(1,0)$ actually coincides with the boundary of the model Ω_m near $(1,0,\ldots,0)\equiv\mathbf{1}$. The modifications necessary to treat the general case will be explained later.

If $\alpha_j\in \mathrm{Aut}(\Omega)$ is a sequence of automorphisms with $\lim \alpha_j(q)=\mathbf{1}$ for some point q in Ω , it is easy to see by a local holomorphic support function argument that $\lim \alpha_j=\mathbf{1}$ uniformly on compact subsets of Ω (cf. the corresponding reasoning in [RO]). Choose $\epsilon>0$ such that $B(\mathbf{1},\epsilon)\cap\Omega=B(\mathbf{1},\epsilon)\cap\Omega_m$. Define a holomorphic map G_j on the set of all $z\in\Omega$ such that $\alpha_j(z)\in B(\mathbf{1},\epsilon)$ by $G_j=\beta_j^{-1}\circ\alpha_j$ where $\beta_j\in \mathrm{Aut}(\Omega_m)$ will be chosen later.

Now the idea is to choose the β_j in such a way that the points $G_j(P_0)$, j large, all lie together in a compact subset of Ω_m or, equivalently, that they are bounded away from the boundary of Ω_m . If such β_j can be chosen, then it is easy to see that the sequence of maps G_j has a subsequence that converges uniformly on compact subsets of Ω and that the limit of such a convergent subsequence must be a biholomorphic map from Ω to Ω_m .

The proof of the theorems is thus reduced to showing that it is possible to choose the β_j as specified. Since $\mathrm{Aut}(\Omega_m)$ acts transitively on $\{(z_1,\ldots,z_n)\in\Omega_m : z_n=0\}$, we would be finished if $\alpha_j(q)$ always belonged to this set. It is of course unreasonable to expect this to actually happen. A more reasonable expectation is that $\{\alpha_j(q)\}$ converges to $\mathbf{1}$ "approximately" along the normal at $\mathbf{1}$ in the sense, say, that if we could choose β_j to make the z_n-component of $\beta_j^{-1}\circ\alpha_j(q)$ equal to zero then the sequence $\{\beta_j^{-1}\circ\alpha_j(q)\}$ would lie in a compact subset of $\{(z_1,\ldots,z_n)\in\Omega_m : z_n=0\}$. Why is this expectation reasonable? Because it is true for $\Omega=\Omega_m$! This is easily verified by explicit calculation with the automorphisms ψ_a of Ω_m described in Section 2.

In order to give a proof that $\{\alpha_j(q)\}$ converges to **1** approximately along the normal we return to our considerations of C/K. First note that $C_\Omega(q)/K_\Omega(q)<1$ since Ω is not biholomorphic to the ball. (Actually, a little argument is necessary to see, under the exact hypotheses of our theorems, that Ω could not be biholomorphic to the ball; see [GK6] for the details. On the other hand, if Ω were a ball, with hence transitive automorphism group, then we could certainly have chosen $\alpha_j \in \mathrm{Aut}(\Omega)$ with $\{\alpha_j(q)\}$ converging precisely along the normal to the point **1** .) Now if x_0 is a boundary point of Ω_m with the z_n component of x_0 unequal to zero then x_0 is a strongly pseudoconvex point of Ω_m . Consequently

$$\lim_{\substack{z\to x_0\\ z\in\Omega}} C_\Omega(z)/K_\Omega(z) = 1 .$$

Since $C/K < 1$ at q , each such point x_0 has a "forbidden" neighborhood into which the orbit of q under $\mathrm{Aut}(\Omega)$ cannot penetrate. If these forbidden neighborhoods can be estimated to have sufficient size (diminishing of course as x_0 moves toward a weakly pseudoconvex boundary point), then we could deduce information about the way in which $\{\alpha_j(q)\}$ converges to **1** . In particular, we could show that, up to rotations in the z_n variable, the convergence was almost along the normal at **1** in the sense already indicated.

The necessary estimates of C/K can be obtained only by a detailed analysis of the localization results of [GR]. In [GR], it is shown that localization holds at every strongly pseudoconvex boundary point. But for the purposes at hand, localization has to be estimated explicitly. The technical details of this process are outside the scope of informal summary; the interested reader should consult [GK6]. Included in these technical details is the fact that the localization requires matching of the boundaries of Ω and Ω_m only up to a fixed finite order (after a biholomorphic change of coordinates). In this generality the maps G_j have to be modified slightly by dilating a small amount so that Ω lies locally (near **1**) inside Ω_m .

Theorems 1 and 2 open up the possibility of much more general results. Namely, one might hope in principle to be able to make a list of "good" points, say of finite type in the sense of [DA], generalizing [KO], with the property that an automorphism orbit could, in principle, accumulate at the point (i.e., there are no infinitesimal obstructions). Each such finite-type point of this sort would, one hopes, come with an explicit "model" bounded domain having a boundary of the sort specified and actually having an automorphism orbit accumulating there. An example would be the point **1** in Theorem 1 as the finite type point and Ω_m as the global model. Then one might hope to prove the analogue of Theorems 1 and 2, namely that if Ω were an arbitrary (weakly pseudoconvex) domain containing a boundary point x_0 agreeing with one of the "good" points up to a high enough order, and if the domain Ω had an $\mathrm{Aut}(\Omega)$ orbit accumulating at the point x_0 , then Ω would have to be biholomorphic to the "model" domain associated with the particular "good" point.

This optimistic program can be carried out for more cases than those of Theorems 1 and 2 by using the same type of normal family argument combined with C/K estimates. But progress toward a general theorem seems for the moment blocked by the fact that the C/K analysis must be carried out separately in each case. (See also the discussion in [GK6]).

Viewing the matter a bit differently, we might also speculate that there should be a complementary list of "bad" points of finite type with the property that if a boundary point x_0 of a (weakly pseudoconvex) domain Ω agreed to a high enough order with a "bad" point, then x_0 could <u>not</u> be an accumulation point of an $\mathrm{Aut}(\Omega)$ orbit. (The order of agreement would of course depend on the "bad" point in question, just as the order of agreement in the "good" case depends on the good point in question.)

For example, the automorphism group of $\Omega_0 \equiv \{(z_1,z_2)\in\mathbb{C}^2: |z_1|^4+|z_2|^4<1\}$ is compact, consisting as it does of permutations of the two variables composed with rotations in each of the two variables separately. Thus no orbit accumulates at $\mathbf{1}=(1,0)$. These facts are most easily seen by global reasoning (each element of $\mathrm{Aut}(\Omega_0)$ extends smoothly to $\overline{\Omega}_0$ and hence takes each weakly pseudoconvex point to another; this in turn puts severe restrictions on possible orbit structures. For details, see [GK6]).

Even though the compactness of $\mathrm{Aut}(\Omega_0)$ is ascertained from global considerations, we might hope that the point 1 in $\partial\Omega_0$ is "bad" in the sense described above. In other words, we might hope that the following is possible: that any weakly pseudoconvex domain Ω with a boundary point x_0 near which $\partial\Omega$ coincides with $\partial\Omega_0$ cannot have any orbits accumulating at x_0.

What would the "badness" of 1 in $\partial\Omega_0$ amount to? It would amount to there being a truly local obstruction to the existence of automorphisms with orbit accumulating at 1. Could we have deduced the nonexistence of such orbits for Ω_0 not from global considerations but from the local geometry of $\partial\Omega_0$ near 1? If so, the deduction could presumably be transferred to Ω near x_0. And, if the local obstructions occurred at an estimatable finite jet level, then we would need only agreement of $\partial\Omega$ at x_0 with $\partial\Omega_0$ at 1 up to a known finite order to be sure that orbits of $\mathrm{Aut}(\Omega)$ could not accumulate at x_0.

One can imagine carrying out some such program with Chern-Moser-Tanaka invariants in connection with the Wong-Rosay result; such an approach was indeed carried out in [BS] (see also [GK6]). The difficulty with the present case is that no satisfactory local invariant theory exists for the finite-type weakly pseudoconvex case. A second related difficulty arises with the idea of listing the weakly pseudoconvex "good" and "bad" cases, and with hoping that each finite type point is necessarily good or bad. This difficulty is simply that no systematic method is known for listing the finite type points in $\mathbb{C}^n (n>2)$, even ignoring the question of their "goodness" or "badness". This is a subject of ongoing investigation.

APPENDIX
Definitions of Special Terms

The results discussed in this paper use a certain amount of technical language. While their spirit can be appreciated without a detailed understanding of these technical concepts, their exact meaning and, in particular, their proofs, cannot. We therefore isolate a concise but careful discussion of the necessary terminology here.

A domain with C^k boundary in $\mathbb{C}^n$ is defined to be a bounded, connected open set Ω of the form

$$\Omega = \{z \in \mathbb{C}^n : \rho(z) < 0\}$$

for some C^k function $\mathbb{C}^n \to \mathbb{R}$ satisfying $\nabla\rho \neq 0$ on $\partial\Omega$. Domains in this paper will generally have at least C^2 boundary, although sometimes it will be convenient to assume only that the boundary is C^2 near a point $P \in \partial\Omega$ (the obvious modification of the definition is left to the reader).

A vector $w = (w_1, \ldots, w_n)$ is a real tangent vector to $\partial\Omega$ at $P \in \partial\Omega$ if

$$\mathrm{Re} \sum_{j=1}^{n} \frac{\partial\rho}{\partial z_j}(P) w_j = 0 .$$

An easy calculation shows that this is simply the usual advanced calculus definition of tangent vector in new notation. We call w a complex tangent vector to $\partial\Omega$ at P if

$$\sum_{j=1}^{n} \frac{\partial\rho}{\partial z_j}(P) w_j = 0 .$$

Observe that all complex tangent vectors are also real tangent vectors, and that if w is a complex tangent vector then iw is also a complex tangent vector.

When it is convenient we write

$$z = (z_1, \ldots, z_n) = (x_1 + iy_1, \ldots, x_n + iy_n) = (x_1, y_1, \ldots, x_n, y_n)$$
$$= (t_1, \ldots, t_{2n})$$

and

$$w = (w_1, \ldots, w_n) = (u_1+iv_1, \ldots, u_n+iv_n) = (u_1, v_1, \ldots, u_n, v_n)$$
$$= (c_1, \ldots, c_{2n}) .$$

A point P in the boundary of $\Omega=\{z\in\mathbb{C}^n : \rho(z)<0\}$ is called <u>convex</u> if $\partial\Omega$ is C^2 at P and

$$\sum_{j,k=1}^{2n} \frac{\partial^2 \rho}{\partial t_j \partial t_k}(P)\, c_j c_k \geq 0$$

for each real tangent vector c at P. The point P is called <u>strongly convex</u> if strict inequality obtains whenever $c\neq 0$. A domain Ω is called convex (resp. strongly convex) when all its boundary points are convex (resp. strongly convex). If Ω has C^2 boundary near P and if P is a point of strong convexity, then it can be shown that nearby boundary points are also points of strong convexity.

Convexity is very useful in complex analysis, but it is not a biholomorphic invariant; so the more technical notion of (Levi) pseudoconvexity has evolved. We call a point P in the boundary of Ω <u>pseudoconvex</u> if $\partial\Omega$ is C^2 near P and

$$\sum_{j,k=1}^{n} \frac{\partial^2 \rho}{\partial z_j \partial \bar{z}_k}(P) w_j \bar{w}_k \geq 0$$

for all complex tangent vectors w at P. The point P is called <u>strongly pseudoconvex</u> if strict inequality obtains whenever $w\neq 0$. A domain Ω is called pseudoconvex (resp. strongly pseudoconvex) if each boundary point is such. If Ω has C^2 boundary near P and if P is a point of strong pseudoncovexity then it can be shown that nearby boundary points are strongly pseudoconvex.

Among C^2 domains, the pseudoconvex ones are precisely the domains of holomorphy--this is the classical solution of the Levi problem. Any weakly pseudoconvex domain can be exhausted by an increasing union of C^∞ strongly pseudoconvex domains. So strongly pseudoconvex domains are in some sense generic among domains of holomorphy, and are reasonable objects of study.

It can be checked that the definitions just given for convexity and pseudoconvexity are independent of the choice of defining function

ρ . For some purposes it is convenient, however, to choose a particular defining function. When Ω has C^k boundary, $k \geq 2$, then the function

$$\varphi_\Omega(z) = \begin{cases} -\operatorname{dist}(z, \partial\Omega) , & z \in \Omega \\ \operatorname{dist}(z, \partial\Omega) , & z \in \mathbb{C} \setminus \Omega \end{cases}$$

is C^k near $\partial\Omega$ (see [SK1]). Using a patching argument, a defining function may be created which agrees with φ_Ω near $\partial\Omega$ and each of whose derivatives is bounded on all of $\mathbb{C}^n$. For the rest of this Appendix, when Ω is a given domain then it will be supposed that Ω is equipped with such a defining function $\rho = \rho_\Omega$.

If f is any C^k function on $\mathbb{C}^n$ then we define

$$\|f\|_{C^k} = \sum_{|\alpha| + |\beta| \leq k} \frac{1}{\alpha!\beta!} \left| \frac{\partial^\alpha}{\partial z^\alpha} \frac{\partial^\beta}{\partial \bar{z}^\beta} f(z) \right| .$$

We may also define, for $f \in C^\infty$,

$$\|f\|_{C^\infty} = \sum_{k=0}^{\infty} \frac{1}{2^k} \frac{\|f\|_{C^k}}{1 + \|f\|_{C^k}} .$$

Next we wish to define the C^k and C^∞ topology on <u>domains</u>. Fix k (here we allow k to be ∞) and consider the collection of all C^k bounded strongly pseudoconvex domains. Each such domain Ω is equipped with the defining function ρ_Ω as described above. A sub-base for the C^k topology on domains is given by the sets

$$S_k(\Omega_0, \epsilon) \equiv \{\Omega : \|\rho_\Omega - \rho_{\Omega_0}\|_{C^k} < \epsilon\}$$

for all possible choices of C^k domain Ω_0 and all $\epsilon > 0$. Many of the results in this paper are formulated in terms of C^k perturbations of a given domain Ω_0 . A domain Ω will be called a C^k perturbation of Ω_0 if $\Omega \in S_k(\Omega_0, \epsilon)$ for some small ϵ .

REFERENCES

[AB] R. Abraham, Bumpy metrics, Proc. Symp. Pure Math.- XIV, Amer. Math. Soc., Providence, 1970, 1-4..

[BED] E. Bedford, Invariant forms on complex manifolds with applications to holomorphic mappings, Math. Ann. 265 (1983), 377-396.

[BDD] E. Bedford and J. Dadok, Bounded domains with prescribed automorphism groups, preprint.

[BE] S. Bell, Biholomorphic mappings and the $\bar{\partial}$ problem, Ann. Math. 114 (1981), 103-112.

[BEL] S. Bell and E. Ligocka, A simplification and extension of Fefferman's theorem on biholomorphic mappings, Invent. Math. 57 (1980), 283-289.

[BER] L. Bers, Riemann Surfaces, Stevens, New York, 1959.

[BERG] M. Berger, Pincement riemannien et pincement holomorphe, Ann. Scuola Norm. Sup. Pisa 14 (1960), 151-159.

[BDK] J. Bland, T. Duchamp, and M. Kalka, A characterization of $\mathbb{CP}^n$ by its automorphism group, Proceedings of the Complex Analysis Week at Penn State, Springer Verlag, to appear.

[BKU] R. Braun, W. Karp, and H. Upmeier, On the automorphisms of circular and Reinhardt domains in complex Banach spaces, Manuscripta Mat. 25 (1978), 97-133.

[BS] D. Burns and S. Shnider, Geometry of hypersurfaces and mapping theorems in $\mathbb{C}^n$, Comment. Math. Helv. 54 (1979), 199-217.

[BSW] D. Burns, S. Shnider, and R. Wells, On deformations of strictly pseudoconvex domains, Inventiones Math. 46(1978), 237-253.

[CM] S. Chern and J. Moser, Real hypersurfaces in complex manifolds, Acta Math. 133(1974), 219-271.

[DA] J. D'Angelo, Hypersurfaces, orders of contact and applications, Ann. Math. 115 (1982), 615-638.

[EB] D. Ebin, The manifold of Riemannian metrics, Proc. Symp. Pure Math. XV, Am. Math. Soc., Providence, 1970.

[FK] H. Farkas and I. Kra, Riemann Surfaces, Springer, New York, 1974.

[FE] C. Fefferman, The Bergman kernel and biholomorphic mappings of pseudoconvex domains, Invent. Math. 26(1974), 1-65.

[FK] G. B. Folland and J. J. Kohn, The Neumann Problem for the Cauchy-Riemann Complex, Princeton University Press, Princeton, 1972.

[FU] B. A. Fuks, Special Chapters in the Theory of Analytic Functions of Several Complex Variables, Trans. Amer. Math. Soc., Providence, 1965.

[GR] I. Graham, Boundary behavior of the Caratheodory and Kobayashi metrics on strongly pseudoconvex domains in $\mathbb{C}^n$ with smooth boundary, Trans. Am. Math. Soc. 207(1975), 219-240.

[GK1] R. E. Greene and S. G. Krantz, Stability of the Bergman kernel and curvature properties of bounded domains, in Recent Developmemts in Several Complex Variables, Annals of Math. Studies 100, Princeton University Press, Princeton, 1981.

[GK2] R. E. Greene and S. G. Krantz, Deformations of complex structures, estimates for the $\bar{\partial}$ equation, and stability of the Bergman kernel, Adv. Math 43(1982), 1-86.

[GK3] R. E. Greene and S. G. Krantz, The automorphism groups of strongly pseudoconvex domains, Math. Ann. 261(1982), 425-446.

[GK4] R. E. Greene and S. G. Krantz, Stability of the Caratheodory and Kobayashi metrics and applications to biholomorphic mappings, in Complex Analysis of Several Variables, Proc. Symp. Pure Math. 41, Amer. Math. Soc., Providence, 1984.

[GK5] R. E. Greene and S. G. Krantz, Characterization of complex manifolds by the isotropy subgroups of their automorphism groups, Indiana Univ. Math. Jour. 34(1985), 865-879.

[GK6] R. E. Greene and S. G. Krantz, Characterization of certain weakly pseudoconvex domains with non-compact automorphism groups, Proceedings of the Complex Analysis Week at Penn State, Springer Verlag, to appear.

[GK7] R.E. Greene and S. G. Krantz, Normal families and the semicontinuity of isometry and automorphism groups, Math. Z. 190(1985), 455-567.

[GK8] R. E. Greene and S. G. Krantz, A new invariant metric in complex analysis and some applications, to appear.

[HE] M. Heins, On the number of 1-1 directly conformal maps which a multiply-connected plane region of finite connectivity $p(>2)$ admits onto itself, Bull. Am. Math. Soc. 52(1946), 454-457.

[HEL] S. Helgason, Differential Geometry and Symmetric Spaces, Academic Press, New York, 1962.

[HO] L. Hörmander, Introduction to Complex Analysis in Several Variables, North Holland, Amsterdam, 1973.

[HUA] L. K. Hua, Harmonic Analysis of Functions of Several Complex Variables in the Classical Domains, Am. Math. Soc., Providence, 1963.

[HUC] A. Huckleberry and E. Oeljeklaus, Classifications Theorems for Almost Homogeneous Spaces, Institut Elie Cartan, Equipe Associée au C.N.R.S., n° 839.

[KER] N. Kerzman and J. P. Rosay, Fonctions plurisousharmoniques d'exhaustion borneeé et domaines taut, Math. Ann. 257(1981), 171-184.

[KL] P. Klembeck, Kähler metrics of negative curvature, the Bergman metric near the boundary, and the Kobayashi metric on smooth bounded strictly pseudoconvex sets, Indiana Univ. Math. J. 27(1978), 275-282.

[KO] J. J. Kohn, Boundary behavior of $\bar{\partial}$ on weakly pseudoconvex manifolds of dimension two, J. Diff. Geom. 6(1972), 523-542.

[KR1] S. Krantz, Function Theory of Several Complex Variables, John Wiley and Sons, New York, 1982.

[KR2] S. Krantz, Integral formulas in complex analysis, The Beijing Lectures, Ann. of Math. Studies, Princeton, 1986.

[LEM] L. Lempert, Intrinsic distances and holomorphic retracts, Complex Analysis and Applications 81(1984), 341-364.

[LI] E. Ligocka, preprint.

[LK] Lu Qi-Keng, On Kähler manifolds with constant curvature, Acta. Math. Sinica 16(1966), 269-281,(Chinese);(= Chinese Math. 9(1966), 283-298.)

[MZ] D. Montgomery and L. Zippin, Topological Transformation Groups, Interscience, New York, 1955.

[MNA] A Morimoto and T. Nagano, On pseudo-conformal transformations of hypersurfaces, J. Math. Japan 15 (1963), 289-300.

[NAR] R. Narasimhan, Several Complex Variables, University of Chicago Press, Chicago, 1971.

[PA] G. Patrizio, Characterization of strongly convex domains biholomorphic to a circular domain, Bull. Am. Math. Soc. 9(1983), 231-233.

[RO] J. P. Rosay, Sur une characterization de la boule parmi les domains de $\mathbb{C}^n$ par son groupe d'automorphismes, Ann. Inst. Four. Grenoble XXIX(1979), 91-97.

[RU] W. Rudin, Function Theory in the Unit Ball of $\mathbb{C}^n$, Springer Verlag, New York, 1980.

[SW] E. M. Stein and G. Weiss, Introduction to Fourier Analysis on Euclidean Space, Princeton University Press, Princeton, 1971.

[SUN] T. Sunada, Holomoprhic equivalence problems for bounded Reinhardt domains, Math. Ann. 235 (1978), 111-128.

[SY] N. Suita and A. Yamada, On the Lu Qi-Keng conjecture, Proc. Am. Math. Soc. 59(1976), 222-294.

[TA] N. Tanaka, On generalized graded Lie algebras and geometric structures I, J. Math. Soc. Japan 19(1967), 215-254.

[WO] Bun Wong, Characterizations of the ball in $\mathbb{C}^n$ by its automorphism group, Invent. Math. 41(1977), 253-257.

[WU] H. H. Wu, Normal families of holomorphic mappings, Acta Math. 119(1967), 193-233.

[YA] P. Yang, Automorphisms of tube domains, Am. Jour. Math. 104 (1982), 1005-1024.

Authors' addresses:

Department of Mathematics
UCLA
405 Hilgard Avenue
Los Angeles, California 90024 U.S.A.

and

Department of Mathematics
Washington University
St. Louis, Missouri 63130 U.S.A.

SOME $\mathbb{C}^N$ CAPACITIES AND APPLICATIONS

J. Korevaar

Abstract. Homogeneous Siciak-type capacities will be derived from extended Green functions with pole at infinity and from polynomial approximation numbers. The latter describe how well monomials can be approximated on a bounded set E in $\mathbb{C}^N$ by linear combinations of other monomials of the same degree. Such polynomial approximation numbers lead to a precise form of a lemma by Wiegerinck and the author on the estimation of mixed derivatives in terms of directional derivatives of the same order. A number of other applications will be surveyed, including the Sibony-Wong theorem on the growth of entire functions, a result on real-analyticity and a simple edge-of-the-wedge theorem, Siciak's convergence theorem for polynomial series and Wiegerinck's results on the Radon transformation and N-dimensional holomorphic extension.

1. Introduction.

An important theme in complex analysis is the estimation of the size or growth of holomorphic functions f in $\mathbb{C}^N$.

For $N = 1$, ordinary planar potential theory is an efficient tool for the estimation problem. The relevant facts are:

(i) $v = \log|f|$ is subharmonic (SH);

(ii) On every bounded domain, a subharmonic function v is majorized by the harmonic function u with the same boundary values (maximum principle for the SH function $v - u$).

Subharmonicity of a function $v: D \to \mathbb{R} \cup \{-\infty\}$ is characterized locally by the mean value inequality, combined with suitable continuity. At each point a in D, the value $v(a)$ must be majorized by the mean values $\overline{v}(a;r)$ of v over the small circles $C(a,r)$ (or the spheres $S(a,r)$ if we are in $\mathbb{R}^k$). In view of this inequality, the usual condition of upper semicontinuity simplifies to $v(a) = \limsup v(z)$ for $z \to a$.

For $N \geq 2$, $\log|f|$ is much more special than just subharmonic. Indeed, for every a in D and every b in $\mathbb{C}^N$ different from 0, the restriction $f(a+wb)$ will be holomorphic as a function of the single complex variable w around $w = 0$. Thus $\log|f(a+wb)|$ will be subharmonic and $v = \log|f|$ is then said to be plurisubharmonic (PSH). By definition, plurisubharmonicity of $v: D \to \mathbb{R} \cup \{-\infty\}$ means that for every a in D and $b \neq 0$, the restriction $v(a+wb)$ is subharmonic around $w = 0$ (possibly identically $-\infty$ for special a and b). Moreover, $v(a) = \limsup v(z)$ for $z \to a$. We usually require of a PSH function that it not be identically $-\infty$ on its domain.

Potential theory for PSH functions in $\mathbb{C}^N$ is relatively recent, and quite different from the classical potential theory related to SH functions on $\mathbb{R}^{2N}$ $(N \geq 2)$. The applications go beyond the estimation of holomorphic functions and are important for, among others, algebraic geometry and real analysis.

In many situations, holomorphic functions can be adequately approximated by polynomials. Thus it is perhaps not surprising that a useful potential theory may be based on a study of <u>polynomials</u>

$$(1.1) \qquad p(z) = p_n(z) = \Sigma c_k z^k = \Sigma c_{k_1 \ldots k_N} z_1^{k_1} \ldots z_N^{k_N} \quad (n \geq \deg p_n).$$

It is convenient to consider two <u>basic problems</u>: Given an upper bound for

$$(1.2) \qquad \|p\|_E = \sup_E |p(z)|$$

on some compact set E in $\mathbb{C}^N$,

(i) Find a bound on $|p(z)|$ for arbitrary z in $\mathbb{C}^N$;

(ii) Obtain bounds for the coefficients c_k.

We begin with a brief discussion of the case $N = 1$ (Sections 2-3); some references may be found at the end of Section 3.

2. <u>Green functions on $\mathbb{C}$ with pole at infinity</u>.

Let D_∞ denote the unbounded component of the complement $\mathbb{C} - E$. The estimation problem for the value $|p(z)|$ leads to the consideration of the Green function $g(z)$ for D_∞ with pole at ∞. For compact sets E with "good" outer boundary $\partial_0 E = \partial D_\infty$ the Green function is well-defined; it is harmonic on D_∞, has boundary values 0 on ∂D_∞ and behaves like $\log|z| + O(1)$ for $|z| \to \infty$. On the other hand, the function

$$v(z) = \frac{1}{n}\log|p_n(z)|/\|p_n\|_E$$

will be subharmonic, ≤ 0 on ∂D_∞ and $\leq \log|z| + O(1)$ for $|z| \to \infty$. Hence by the maximum principle for SH functions (applied to $v - (1+\varepsilon)g$), $v(z)$ is majorized by $g(z)$ throughout D_∞, hence

(2.1) $$|p_n(z)| \le \|p_n\|_E \exp\{ng(z)\}.$$

The upper bound $g(z)$ for $v(z)$ is sharp in the following asymptotic sense: there exist special polynomials P_n of degree n such that

$$\lim_{n\to\infty} \frac{1}{n}\log|P_n(z)|/\|P_n\|_E = g(z) \quad \text{throughout } D_\infty .$$

One may for example take P_n equal to a monic polynomial vanishing at n^{th} order Fekete extreme points for E. The latter are points, $z_1,\dots,z_n$ on E such that

$$\prod_{1\le j<k\le n} |z_j - z_k|$$

is maximal for the given n.

The preceding considerations motivate the following definition of an <u>extended pre-Green function</u> g_E with pole at ∞ which applies to arbitrary nonempty compact sets E in $\mathbb{C}$:

(2.2) $$g_E(z) \stackrel{\text{def}}{=} \sup_{n\ge 1} \sup_{\deg p_n \le n} \frac{1}{n}\log\frac{|p_n(z)|}{\|p_n\|_E}, \quad z\in\mathbb{C}.$$

It is easy to see that $g_E(z) = 0$ on E and on any domain enclosed by E (take $p_1 \equiv 1$). If $E \subset F$ where F is the closed disc $\bar{B}(0,R)$, then $\|p_n\|_E \le \|p_n\|_F$ and hence

(2.3) $$g_E(z) \ge g_F(z) = \log^+|z|/R .$$

Pre-Green functions g_E may have annoying discontinuities where E is thin, for example at isolated points. Thus for $E = \{0\}$ one has $g_e(0) = 0$ and $g_E(z) = +\infty$ for $z \ne 0$. In order to remove this sort of discontinuity one applies upper <u>regularization</u> as follows:

<u>Definition 2.4</u>. (<u>extended Green function</u> with pole at ∞):

$$g_E{}^*(z) = \limsup_{\zeta\to z} g_E(\zeta) = \lim_{\delta\downarrow 0} \sup_{B(z,\delta)} g_E(\zeta), \quad z\in\mathbb{C}.$$

<u>Principal Fact 2.5</u>. <u>Depending on</u> E, <u>one either has</u>

$$g_E{}^*(z) \equiv +\infty \text{ (for "thin" } E),$$

<u>or</u>

$$g_E{}^*(z) < +\infty \ \underline{\text{everywhere}},$$

<u>and then</u> $g_E{}^*$ <u>is subharmonic on</u> $\mathbb{C}$. (In the present case of E in $\mathbb{C}$, $g_E{}^*$ is moreover equal to g_E and harmonic outside E.)

The basic step in the proof is the following observation: if there is some point a where $g_E{}^*(a) < +\infty$, then by 2.4 there is a closed disc $\bar{B}(a,r)$ on which $M = \sup g_E(\zeta) < +\infty$. It now follows from the maximum principle that for all polynomials p_n with $\|p_n\|_E = 1$,

(2.6) $$\frac{1}{n}\log|p_n(z)| \le M + \log^+\frac{|z - a|}{r}, \quad z \in \mathbb{C}.$$

The functions $g_E(z)$ and $g_E{}^*(z)$ will have the same upper bound on $\mathbb{C}$.

<u>Remark 2.7</u>. It can be shown that E is "thin" in the sense of 2.5 if and only if it is <u>polar</u>, that is, if it is contained in the "$-\infty$" set of an SH function on $\mathbb{C}$. For nonpolar E, there is a unique probability measure λ on E of finite minimal potential energy $\gamma = \gamma(E)$. This λ is an equilibrium mass distribution in the following sense: the associated logarithmic potential U^λ is constant and equal to γ everywhere on E outside a polar subset (which is empty for "good" $\partial_0 E$). There is a close connection between the Green function $g_E{}^*$ and the equilibrium potential U^λ:

$$g_E{}^*(z) = \gamma - U^\lambda(z) \quad \text{throughout } \mathbb{C}.$$

In particular $g_E{}^*$ satisfies the (distributional) Poisson equation

(2.8) $$\Delta g_E{}^* = -\Delta U^\lambda = 2\pi\lambda \quad \text{on } \mathbb{C}.$$

3. <u>Two capacities for sets</u> E <u>in</u> $\mathbb{C}$.

The <u>logarithmic capacity</u> cap E of a compact set E in $\mathbb{C}$ may be obtained from the Greenfunction via the Robin constant $\gamma = \gamma(E)$:

(3.1) $$\begin{cases} \gamma = \limsup\limits_{|z|\to\infty} \{g_E{}^*(z) - \log|z|\} = \limsup\limits_{|z|\to\infty} \{g_E(z) - \log|z|\}, \\ \operatorname{cap} E = e^{-\gamma}. \end{cases}$$

In the present case of $\mathbb{C}$, the lim sup for γ is actually a limit. Indeed, either $g_E{}^*(z) \equiv +\infty$ and then $\gamma = +\infty$ and $\operatorname{cap} E = 0$, or $g_E{}^*(z) - \log|z|$ is a bounded harmonic function around $z = \infty$ which thus has a removable singularity at ∞, cf. (2.3), (2.6). In the latter case γ is finite and cap E positive. In particular

$$\operatorname{cap} \bar{B}(a,r) = r.$$

it turns out that a compact set is polar if and only if it has capacity zero.

There are other ways of obtaining cap E. We mention the

Chebyshev problem to determine the best uniform approximation to z^n on E by polynomials of lower degree. Specifically, one may define the nth Chebyshev number for E as

$$(3.2)\qquad \tau_n(E) = \inf_b \left\| z^n - \sum_{j=0}^{n-1} b_j z^j \right\|_E^{1/n}, \quad n = 1,2,\ldots .$$

As $n\to\infty$, the numbers $\tau_n(E)$ tend to a limit, the Chebyshev constant $\tau(E)$. More precisely,

$$(3.3)\qquad \begin{cases} \tau_n(E)\to\tau(E) \overset{\text{def}}{=} \inf_m \tau_m & \text{for } n\to\infty, \\ \tau(E) = \text{cap } E. \end{cases}$$

The author has shown that a more general polynomial approximation constant likewise equals a (generalized) capacity. In connection with the coefficient problem mentioned at the end of Section 1, it is natural to introduce "polynomial approximation numbers". For given n, we consider the best uniform approximation to z^k $(0 \le k \le n)$ on our compact $E \subset \mathbb{C}$ by polynomials of degree $\le n$ which do not involve z^k. As in the Chebyshev problem, we take the nth root of the minimal distance. Furthermore, we minimize over k:

$$(3.4)\qquad \alpha_n(E) \overset{\text{def}}{=} \inf_{0\le k\le n} \inf_b \left\| z^k - {\sum_{j=0}^{n}}' b_j z^j \right\|_E^{1/n}, \quad n = 1,2,\ldots,$$

where Σ' means a summation over $j \ne k$.

For the polynomial p_n of (1.1) with $N = 1$, definition (3.4) gives

$$\|p_n\|_E = |c_k| \left\| z^k + {\sum_{j=0}^{n}}' (c_j/c_k) z^j \right\|_E \ge |c_k| \alpha_n(E)^n$$

whenever $c_k \ne 0$, hence

$$(3.5)\qquad \max_k |c_k| \le \|p_n\|_E / \alpha_n(E)^n .$$

Question: How do the numbers $\alpha_n(E)$ behave for $n\to\infty$? In particular, when will the polynomial approximation constant

$$(3.6)\qquad \alpha(E) \overset{\text{def}}{=} \inf_n \alpha_n(E)$$

be greater than zero?

The author has shown that $\alpha_n(E)\to\alpha(E)$ as $n\to\infty$ and that the constant $\alpha(E)$ is equal to a generalized capacity $c(E)$ of the kind introduced by Siciak [21] for the case of $\mathbb{C}^N$:

(3.7) $c(E) \overset{\text{def}}{=} \exp\{-\sup_{|z|<1} g_E^*(z)\} = \exp\{-\sup_{|z|<1} g_E(z)\}$.

In particular $\alpha(E) > 0$ unless $g_E^* \equiv +\infty$ or equivalently, unless E is polar or cap E = 0.

For references to Sections 2-3, see Korevaar [10], where various properties of g_E, g_E^*, $c(E)$ and $\alpha(E)$ are proved in detail. The contents of that paper owe much to the $\mathbb{C}^N$ theories of Siciak [19,20,21] and Bedford and Taylor [3].

4. Most of the preceding can be extended to $\mathbb{C}^N$!

Thus for compact sets E in $\mathbb{C}^N$, one may define a pre-Green function g_E and a Green function g_E^* with pole at infinity as in Definitions (2.2) and 2.4, this time using polynomials in $z_1,\ldots,z_N$. The Principal Fact 2.5 remains valid, with SH replaced by PSH and "thin" now meaning pluripolar; pluripolar sets are defined in analogy to Remark 2.7.

One may define logarithmic capacity in $\mathbb{C}^N$ via a Robin constant as in (3.1), but because γ is not a limit in general, it is not known if cap E has the regularity properties of a generalized capacity when $N \geq 2$. However, the Siciak capacity $c(E)$ given by Definition (3.7) is a good generalized capacity also in $\mathbb{C}^N$: it has the same regularity properties as $\rho(E)$ in 9.2 below, cf. Siciak [21].

Certain things are very different with $N \geq 2$: the Green function g_E^* need not be smooth outside E, let alone (pluri)harmonic. We know from the fundamental work of Bedford and Taylor ([3] and earlier papers) that the "right" generalization of the Laplace operator in (2.8) is the highly nonlinear complex Monge-Ampère operator $\det[\partial^2/\partial z_i \partial \bar{z}_j]$.

An equivalent of the Green function on $\mathbb{C}^N$ with pole at infinity was introduced by Siciak exactly 25 years ago, cf. [19]. His extremal functions Φ_E and Φ_E^* correspond to $\exp g_E$ and $\exp g_E^*$. However, outside Eastern Europe, Siciak's original work long failed to attract the attention which it deserved, even through he kept adding to it. Today, there is much interest in the area of extremal PSH functions and $\mathbb{C}^N$ capacities. This is partly due to the development of complex analysis in several variables generally, and partly to the results of Bedford and Taylor and the surprising growth theorem of

Sibony and Wong [18], cf. Section 6 below. The latter developments were followed by extensive new work of Siciak [20,21]. The area has attracted many other mathematicians. It is not possible to be complete here, but we mention H. Alexander, Cegrell, Demailly, Josefson, Kiselman, Levenberg, Molzon, Ronkin, Sadullaev, Shiffman, Wiegerinck, Zakharyuta and Zériahi. Most of these authors have written more papers on the subject than are listed in the bibliography, cf. the surveys by Siciak [21] and Sadullaev [17a,b] and Mathematical Reviews or the Zentralblatt.

In the sequel we will discuss some important multidimensional applications that are related to homogeneous polynomials

$$(4.1) \qquad q(z) = q_n(z) = \sum_{|k|=n} c_k z^k = \sum_{k_1+\ldots+k_N=n} c_{k_1\ldots k_N} z_1^{k_1} \ldots z_N^{k_N} .$$

Here k stands for a multi-index $(k_1,\ldots,k_N)$ of nonnegative integers and $|k|$ denotes its height $k_1+\ldots+k_N$; the polynomial q_n is homogeneous of degree n.

In the discussion we will again be guided by our two basic problems: Given an upper bound for

$$\|q\|_E = \sup_E |q(z)|$$

where E is (for the time being) some compact set,

(i) Find a bound on $|q(z)|$ for arbitrary z in $\mathbb{C}^N$;

(ii) Obtain bounds for the coefficients c_k.

5. Logarithmically homogeneous Green functions on $\mathbb{C}^N$.

For the first problem above, an appropriate pre-Green function may be defined as follows, cf. [10] and Siciak [21] whose extremal function Ψ_E equals $\exp h_E$:

Definition 5.1. For E in $\mathbb{C}^N$ compact, nonempty and $\neq \{0\}$,

$$h_E(z) = \sup_{n \geq 1} \sup_{q_n \text{ hom,deg } n} \frac{1}{n} \log \frac{|q_n(z)|}{\|q_n\|_E} , \quad z \in \mathbb{C}^N,$$

where q_n runs over the homogeneous polynomials of degree n with nonzero sup norm $\|q_n\|_E$.

Remark 5.2. The first sup in the Definition may be replaced by lim as $n \to \infty$. Indeed, suppose that the homogeneous polynomial Q_s provides a good approximation to h_E at a, where $a_\nu \neq 0$. Then, decomposing $n = ms + r$, $0 \le r < s$, the homogeneous polynomials

$$q_n(z) = Q_s(z)^m z_\nu{}^r$$

will provide nearly as good an approximation to h_E at a as $n \to \infty$:

$$\frac{1}{n} \log \frac{|q_n(a)|}{\|q_n\|_E} \ge \frac{ms}{ms + r} \cdot \frac{1}{s} \log \frac{|Q_s(a)|}{\|Q_s\|_E} + \frac{r}{ms + r} \log \frac{|a_\nu|}{\|z_\nu\|_E} .$$

The function h_E will be ≤ 0 on E and $h_E \ge h_F$ whenever $E \subset F$. The function h_E is <u>logarithmically homogeneous</u> on $\mathbb{C}^N$:

(5.3) $$h_E(wb) = h_E(b) + \log|w|,\ b \in \mathbb{C}^N - \{0\},\ w \in \mathbb{C},$$

because the functions behind the second sup in 5.1 have that property. Clearly $h_E = h_{E_c}$, where E_c is the <u>circular set</u> generated by E, that is, the set of all points $z = e^{i\theta}\zeta$ with $\zeta \in E$ and $\theta \in \mathbb{R}$.

<u>Examples 5.4</u>. First let E be the closed <u>ball</u> $\bar{B}(0,r)$. For any point $b \in \partial B(0,r)$ the homogeneous polynomial $\bar{b}_1 z_1 + \ldots + \bar{b}_N z_N$ takes on its maximum absolute value (on E) at b, hence $h_E(b) = 0$. Thus by (5.3), writing $z = wb$ with $b \in \partial B(0,r)$,

$$h_{\bar{B}(0,r)}(z) = \log|w| = \log\frac{|z|}{r} .$$

For the closed <u>equiradial polydisc</u>

$$\bar{\Delta}(0,r) = \{z \in \mathbb{C}^N: \ |z_\nu| \le r,\ \nu = 1,\ldots,N\}$$

one similarly finds that

$$h_{\bar{\Delta}(0,r)}(z) = \max_\nu \{\log|z_\nu|/r\}.$$

The examples show that always $h_E(z) \to +\infty$ as $|z| \to \infty$, more precisely:

(5.5) $$E \subset \bar{B}(0,R) \text{ implies } h_E(z) \ge \log|z|/R.$$

<u>Characterization of the important set</u>

(5.6) $$V = \{z \in \mathbb{C}^N: \ h_E(z) \le 0\}.$$

It is clear that V contains the circular set E_c. It will also contain the polynomially convex hull $\hat{E}_c$ of E_c, that is, the set of all points a in $\mathbb{C}^N$ such that $|p(a)| \le \|p\|_{E_c}$ for every polynomial p. Indeed, if $a \in \hat{E}_c$ then in particular $|q(a)| \le \|q\|_{E_c} = \|q\|_E$ for all homogeneous polynomials q, hence $h_E(a) \le 0$.

One can also go in the other direction. Suppose $h_E(b) \le 0$ and let p be any polynomial in N variables such that $\|p\|_{E_c} = 1$ and

$s = \deg p \geq 1$. Writing p^m as a sum of homogeneous polynomials q_n of different degrees n, the Cauchy inequalities applied to $p(w\zeta)^m = \Sigma q_n(\zeta)w^n$ with $|w| \leq 1$ show that $|q_n(\zeta)| \leq 1$ for all $\zeta \in E$ and every $n \geq 0$. Hence Definition 5.1 gives $|q_n(b)| \leq \exp nh_E(b) \leq 1$ and thus

$$|p(b)|^m \leq \Sigma_0^{ms} |q_n(b)| \leq ms + 1.$$

Conclusion for $m \to \infty$: $|p(b)| \leq 1$. It follows that b belongs to $\hat{E}_c$. We have thus proved

<u>Proposition 5.7</u> (Siciak [21]). <u>For compact</u> E <u>in</u> $\mathbb{C}^N$,

$$V \overset{\text{def}}{=} \{z \in \mathbb{C}^N: \ h_E(z) \leq 0\} = \hat{E}_c.$$

This proposition may be used to show that for circular sets E,

$$g_E(z) = \max\{h_E(z), 0\}.$$

<u>Definition 5.8</u> (logarithmically homogeneous <u>Green functions</u>):

$$h_E^*(z) = \limsup_{\zeta \to z} h_E(\zeta), \ z \in \mathbb{C}^N.$$

<u>Principal Fact 5.9</u> <u>Depending on</u> E, <u>one either has</u>

$$h_E^*(z) \equiv +\infty \quad (\text{for "thin" } E),$$

<u>or</u>

$$h_E^*(z) < +\infty \ \underline{\text{everywhere}},$$

<u>and then</u> h_E^* <u>is plurisubharmonic on</u> $\mathbb{C}^N$.

The proof is similar to the one for 2.5: if a PSH function $v(z)$ on $\mathbb{C}^N$ is majorized by M on $\bar{B}(a,r)$ and by $\log|z| + O(1)$ at infinity, then it follows from the subharmonicity of $v(a + wb)$ on $\mathbb{C}$ that

$$v(z) \leq M + \log^+\frac{|z - a|}{r} \quad \text{throughout } \mathbb{C}^N.$$

In the present context, E is "thin" iff it is LH-polar, that is, if it is contained in the "$-\infty$ set" of a logarithmically homogeneous PSH function. An equivalent statement would be that E_c is <u>pluripolar</u>, that is, contained in the "$-\infty$ set" of a PSH function on $\mathbb{C}^N$, cf. [10]. It follows from the work of Bedford and Taylor [3] that the set where h_E^* is actually larger than h_E is always pluripolar. The second of the Examples 5.4 shows that (a finite function) h_E^* need not be smooth outside E, let alone pluriharmonic.

6. Siciak's LH-capacity and the Sibony-Wong growth theorem.

The standard capacity associated with the logarithmically homogeneous Green function h_E^* is given by

Definition 6.1 (LH-capacity, cf. Siciak [21]):

$$\rho(E) = \exp\left\{-\limsup_{|z|\to\infty} [h_E^*(z) - \log|z|\right\}$$

$$= \exp\{-\sup_B h_E^*(z)\} = \exp\{-\sup_B h_E(z)\}$$

$$= \inf_{n \ge 1} \inf_{q_n \text{hom,deg } n} \left(\|q_n\|_E/\|q_n\|_B\right)^{1/n}, \quad B = B(0,1).$$

Clearly $\rho(E) = \rho(E_c)$. A (compact) set E in $\mathbb{C}^N$ will be LH-polar if and only if $\rho(E) = 0$; a circular set E will be pluripolar relative to $\mathbb{C}^N$ if and only if $\rho(E) = 0$. Any closed ball $\bar{B}(a,r)$ will have LH-capacity r. The closure E of a nonempty open subset of the real unit sphere S^{N-1} in $\mathbb{R}^N + i0 \subset \mathbb{C}^N$ has positive LH-capacity $\rho(E)$; in particular $\rho(S^{N-1}) = 1/\sqrt{2}$. Cf. Examples 7.7 and [10].

If one restricts oneself to circular subsets $E = E_c$ of the unit sphere ∂B in $\mathbb{C}^N$, the function $\rho(E)$ gives rise to a capacity on projective space $\mathbb{P}^{N-1}$. Somewhat different projective capacities (for which the same sets have capacity zero) have been introduced and studied by Alexander [1] and by Molzon, Shiffman and Sibony [14].

The LH-capacity has an important geometric description, cf. [1,2a]:

Proposition 6.2. For compact E, the quantity $\rho(E) = \rho(E_c)$ is equal to the radius $r(E)$ of the largest ball $\bar{B}(0,r)$ contained in the polynomially convex hull $\hat{E}_c = \{h_E(z) \le 0\}$.

Indeed, one has $\rho(E) = 0$ iff $h_E^* \equiv +\infty$ iff $r(E) = 0$. Suppose now that $\rho(E) > 0$. Then by (5.3) and 6.1,

$$\sup_{B(0,r)} h_E(z) = \sup_{B(0,1)} h_E(z) + \log r = \log r/\rho(E), \quad r > 0,$$

hence $B(0,r) \subset \hat{E}_c$ or $\bar{B}(0,r) \subset \hat{E}_c$ iff $r \le \rho(E)$.

Proposition 6.2 is equivalent to a sharp form of the Sibony-Wong theorem on the growth of entire and PSH functions on $\mathbb{C}^N$. Let E be any compact subset of the unit sphere ∂B in $\mathbb{C}^N$ for which $\rho(E)$ is equal to some positive number ρ or equivalently, for which the complex cone $\mathbb{C}\cdot E$ is nonpluripolar. The polynomially convex hull $\hat{E}_c$ will contain the closed ball $\bar{B}(o,\rho)$, and the polynomially convex hull $(rE_c)^\wedge$ will contain the ball $\bar{B}(0,\rho r)$. We conclude:

Theorem 6.3. For every polynomial and hence for every entire function F on $\mathbb{C}^N$,

$$\sup_{|z| \le \rho r} |F(z)| \le \sup_{z \in rE_c} |F(z)| = \sup_{z \in \mathbb{C}\cdot E,\ |z| \le r} |F(z)|.$$

Thus the growth of $|F|$ on $\mathbb{C}^N$ is circumscribed by its growth on any nonpluripolar closed complex cone. One may for example take $E = S^{N-1} \subset \mathbb{R}^N + i0$ so that $\mathbb{C}\cdot E$ becomes the cone of "semireals"; in this case $\rho = 1/\sqrt{2}$. The inequalities in Theorem 6.3 are actually valid for arbitrary PSH functions u on $\mathbb{C}^N$ instead of $|F|$. This is so because the polynomially convex hull of a compact set E is also its PSH convex hull relative to $\mathbb{C}^N$, cf. Hörmander [7a].

The important discovery of a theorem of the form 6.3 is due to Sibony and Wong [18]; a sharper form was obtained by Alexander [1] and the precise constant $\rho = \rho(E)$ was given by Siciak [21].

7. Homogeneous polynomial approximation numbers and another LH-capacity.

We now address problems (i) and (ii) from the end of Section 4: the estimation of $|q_n(z)|$ and the coefficients c_k, given an upper bound for

$$\|q_n\|_E = \|q_n\|_{\hat{E}} .$$

We start with compact E. In terms of the pre-Green function h_E of 5.1 we have the obvious inequality

(7.1) $$|q_n(z)| \le \|q_n\|_E \exp\{nh_E(z)\},\ z \in \mathbb{C}^N.$$

The coefficients c_k of q_n are estimated most simply with the aid of the Cauchy inequalities for the unit polydisc $\Delta = \Delta(0,1)$:

(7.2) $$\max_k |c_k| \le \|q_n\|_\Delta .$$

It is thus natural to introduce an LH-capacity involving $\sup h_E(z)$ on Δ instead of B.

Definition 7.3:

$$\sigma(E) = \exp\{-\sup_\Delta h_E{}^*(z)\} = \exp\{-\sup_\Delta h_E(z)\}$$
$$= \lim_{n\to\infty} \inf_{q_n \text{ hom,deg } n} (\|q_n\|_E/\|q_n\|_\Delta)^{1/n} .$$

For the last equality, cf. Remark 5.2 and Definition 6.1. Clearly $\sigma(E) = \sigma(E_c) = 0$ iff $h_E^* \equiv +\infty$ iff $\bar{E}_c$ is pluripolar, cf. Section 6. For the close

the Examples 5.4 gives $\sigma(E) = r$. The constant $\sigma(E)$ is always equal to the radius of the largest equiradial polydisc $\bar{\Delta}(0,r)$ contained in $\hat{E}_c$.

Formulas (7.1), (7.2) and Definition 7.3 give the coefficient inequality

$$(7.4) \qquad \max_k |c_k| \le \|q_n\|_E \exp\{n \sup_\Delta h_E(z)\} = \|q_n\|_E/\sigma(E)^n .$$

There is, however, a more direct way to estimate $\max |c_k|$ in terms of $\|q_n\|_E$, cf. Section 3. To this end, we introduce homogeneous polynomial approximation numbers; from here on, we also allow noncompact sets E.

Definition 7.5. For nonempty bounded E in $\mathbb{R}^N$ or $\mathbb{C}^N$ we set

$$\beta_n(E) = \inf_{|k|=n} \inf_b \Big\| z^k - \sum_{|j|=n}{}' b_j z^j \Big\|_E^{1/n}, \quad n = 1,2,\dots .$$

For given $k = (k_1,\dots k_N)$, j runs over all multi-indices $j \ne k$ of height $|j| = j_1+\dots+j_N$ equal to $n = |k|$.

Clearly $\beta_n(E) = \beta_n(\bar{E})$. For our polynomials q_n (4.1), we have

$$\|q_n\|_E = |c_k|\cdot\Big\|z^k + \sum_{|j|=n}{}' (c_j/c_k) z^j\Big\|_E \ge |c_k|\beta_n(E)^n$$

whenever $c_k \ne 0$. Thus we obtain the following

Basic Inequality 7.6.

$$\max_k |c_k| \le \|q_n\|_E/\beta_n(E)^n .$$

Question: when will

$$\beta(E) \overset{\text{def}}{=} \inf_n \beta_n(E) > 0?$$

Examples 7.7. For the case where E is the equiradial polydisc $\Delta(0,r)$ or its distinguished boundary, the equiradial torus $T(0,r) = \{z \in \mathbb{C}^N: |z_\nu| = r, \nu = 1,\dots,N\}$, the best approximation to z^k by sums $\Sigma' b_j z^j$ is obtained when all b's are zero (Cauchy inequalities!), hence

$$\beta_n(\Delta(0,r)) = \beta_n(T(0,r)) = r.$$

Calculations by Wiegerinck and the author show that $\beta(E) > 0$ for any nonempty open subset E of the unit sphere S^{N-1} in $\mathbb{R}^n$ [11].

How does Inequality 7.6 compare to (7.4) (where E was compact)? Inequality 7.6 is sharp for each $n \geq 1$ and appropriate q_n, cf. Definition 7.5. It follows that

$$\beta_n(E) = \beta_n(E) \geq \sigma(E), \ n = 1,2,\ldots \ . \tag{7.8}$$

We will now prove

Theorem 7.9. *For arbitrary bounded* E *in* $\mathbb{R}^N$ *or* $\mathbb{C}^N$,

$$\beta_n(E) \to \beta(E) = \inf_m \beta_m(E) \ \text{as} \ n \to \infty \ .$$

The polynomial approximation constant $\beta(E)$ *is equal to the* LH-*capacity* $\sigma(E)$. *In particular* $\beta(E) > 0$ *unless* E *is pluripolar in* $\mathbb{C}^N$.

A somewhat related result may be found in Siciak [21, section 9].

Proof. We may assume that E is compact. By (7.8)

$$\liminf_{n\to\infty} \beta_n(E) \geq \inf_m \beta_m(E) = \beta(E) \geq \sigma(E)$$

hence we need only show that

$$\limsup_{n\to\infty} \beta_n(E) \leq \sigma(E). \tag{7.10}$$

It follows from the last line of Definition 7.3 that we can find a sequence of homogeneous polynomials $\{q_n\}$ such that

$$\|q_n\|_E \leq 1, \ \lim_{n\to\infty} \|q_n\|_\Delta{}^{1/n} = 1/\sigma(E). \tag{7.11}$$

We now apply the Basic Inequality 7.6 to these $q_n(z) = \Sigma c_k z^k$ to obtain

$$\|q_n\|_\Delta = \sup_\Delta \Big| \sum_{|k|=n} c_k z^k \Big| \leq \sum_{|k|=n} |c_k| \leq (n+1)^N \|q_n\|_E / \beta_n(E)^n \ .$$

Thus in view of (7.11),

$$1/\sigma(E) = \lim_{n\to\infty} \|q_n\|_\Delta{}^{1/n} \leq \liminf_{n\to\infty} 1/\beta_n(E),$$

whence (7.10).

Remark 7.12. The same method shows that the standard LH-capacity $\rho(E)$ is equal to the infimum and the limit of polynomial approximation numbers $\rho_n(E)$, defined by approximating (on E) not z^k as in Definition 7.5, but $z^k/\|z^k\|_B$. For the proof, inequality (7.2) is replaced by

$$|c_k| \cdot \|z^k\|_B \leq \|q_n\|_B$$

which is valid for each k. This inequality is a consequence of the fact that B contains the polydisc $\Delta(0,r) = \{|z_\nu| < r_\nu,\ \nu = 1,\ldots,N\}$, where $r = (r_1,\ldots,r_N) \in \partial B$ is chosen such that, for the given k, one has

$$r_1^{k_1}\ldots r_N^{k_N} = \|z^k\|_B .$$

8. The K-W estimate for mixed derivatives and applications.

Inequality 7.6 and Theorem 7.9 imply a useful estimate for mixed derivatives in terms of directional derivatives in $\mathbb{R}^N$. Let f be a C^∞ function on a neighborhood of a in $\mathbb{R}^N$. The Taylor expansion for $f(a + x)$ may (formally) be written as

$$f(a + x) = \sum_0^\infty q_n(x) \quad \text{where} \quad q_n(x) = \sum_{|k|=n} c_k x^k,$$

$$c_k = D^k f(a)/k! = D_1^{k_1}\ldots D_N^{k_N} f(a)/k_1!\ldots k_N!.$$

On the other hand, setting $x = ty$, $y \in S^{N-1} \subset \mathbb{R}^N$, $t \in \mathbb{R}$, one has

$$f(a + ty) = \sum_0^\infty q_n(y)t^n,\ q_n(y) = \frac{1}{n!}\left(\frac{d}{dt}\right)^n f(a + ty)|_{t=0}$$

Now let E be any nonempty subset of the unit sphere S^{N-1} in $\mathbb{R}^N$. then the Basic Inequality 7.6 gives the estimate

$$(8.1)\qquad \begin{aligned} \max_{|k|=n} \frac{1}{k!}|D^k f(a)| &= \max |c_k| \le \|q_n\|_E/\beta_n(E)^n \\ &= \frac{1}{n!} \sup_{y\in E} \left|\left(\frac{d}{dt}\right)^n f(a + ty)|_{t=0}\right| / \beta_n(E)^n . \end{aligned}$$

We put this in words, combining (8.1) and Theorem 7.9:

Theorem 8.2. The mixed derivatives $D^k f(a)$ of order $|k| = n \ge 1$ may be estimated by (8.1) in terms of the supremum of the absolute values of the nth order directional derivatives of f at a, corresponding to the directions given by E. For fixed n, the uniform bound in (8.1) is best possible. One may finally replace the polynomial approximation numbers $\beta_n(E)$ in (8.1) by their infimum or limit $\beta(E) = \sigma(\overline{E})$. The latter is strictly positivewhenever $\overline{E}_c$ is non-pluripolar as a subset of $\mathbb{C}^N$.

This Theorem gives a precise form to the "Main Lemma" of Wiegerinck and the author [11] on the estimation of mixed derivatives in terms of directional derivatives. The present form owes much to a letter by Siciak in which he pointed out the relevance of inequality (7.4).

Applications [11,22,24].

(i) Real-analyticity of functions in $\mathbb{R}^N$. Let f be of class C^{∞} on a (convex) domain D in $\mathbb{R}^N$. Suppose there are a subset E of S^{N-1} with $\beta(E) = \beta > 0$ and positive constants M and C such that

$$\text{(8.3)} \qquad \frac{1}{n!}\left|\left(\frac{d}{dt}\right)^n f(a + ty)\Big|_{t=0}\right| \le MC^n, \; a \in D, \; y \in E, \; n = 0,1,2,\dots .$$

then f is real-analytic on D. Indeed, by Theorem 8.2 the Taylor series for f around a will converge throughout the polydisc $\Delta_a = \Delta(a,\beta/C)$. The sum will agree with f on $\Delta_a \cap D$ since the remainder in the Taylor series will tend to zero.

It is not necessary to suppose that f by C^{∞} on D: under the conditions (8.3), assuming the existence of the derivatives involved, continuity of f on D will suffice. (Convolve f with an approximate identity).

(ii) A simple edge-of-the-wedge theorem. Let D be a domain in $\mathbb{R}^N = \mathbb{R}^N + i0 \subset \mathbb{C}^N$ and let E be a connected open subset of S^{N-1} in (another) $\mathbb{R}^N$. To D and E we associate two domains in $\mathbb{C}^N$ as follows:

$$W^+ = D + iJ\cdot E, \; W^- = D - iJ\cdot E, \quad J : \text{interval } 0 < \tau < r$$

("wedges" with common "edge" D). We also set

$$W = W^+ \cup D \cup W^-.$$

Observe that in general, W will contain no $\mathbb{C}^N$ neighborhood of any point of D. The result in (i) implies the following simple edge-of-the-wedge theorem:

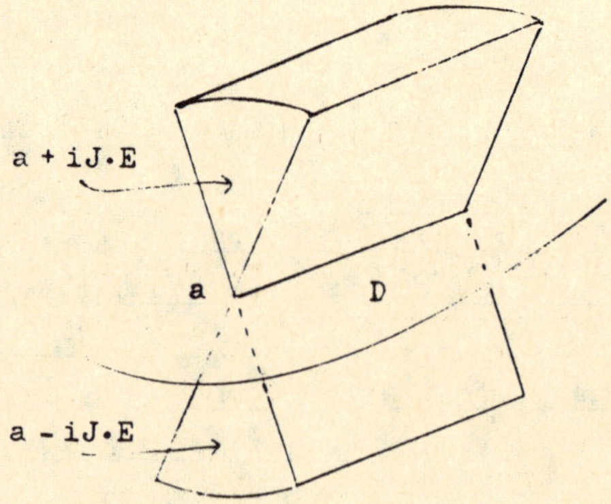

There exists an open neighborhood Ω of W in $\mathbb{C}^N$ such that every continuous function f on W which is holomorphic on W^+ and on W^- has a holomorphic extension to Ω.

For the proof, take a subdomain D_0 with $\bar{D}_0 \subset D$ compact and take $0 < \rho < \min\{d(D_0,\partial D),r\}$. Then the family of one-dimensional discs

$$\Delta_y(a,\rho) = \{z \in \mathbb{C}^N: \quad z = a + wy, \ |w = t + i\tau| < \rho\}, \ a \in D_0, \ y \in E$$

is contained in a compact subset W_0 of W. The restrictions $f(a+wy)$ of f to these discs are continuous and they are holomorphic for $\operatorname{Im} w = \tau \neq 0$, hence by Morera's theorem, they will be (one-dimensionally) holomorphic. The Cauchy inequalities now give (8.3) with $M = \sup |f|$ on W_0, $C = 1/\rho$, $a \in D_0$ and $y \in E$; observe that $\beta(E) > 0$.

More precise information on Ω could be obtained from the results in Section 10.

(iii) <u>A support theorem for the Radon transformation</u>. Let g be a rapidly decreasing continuous function on $\mathbb{R}^N$ and let $\hat{g}(y,\lambda)$ be its Radon transform, obtained by integration over the hyperplanes $x \cdot y = \lambda$:

$$\hat{g}(y,\lambda) = \int_{x\cdot y=\lambda} g(x)dm(x), \quad (y,\lambda) \in S^{N-1} \times \mathbb{R}.$$

Helgason's support theorem asserts that g <u>must have bounded support whenever</u> $\hat{g}$ <u>does</u>, cf. [7]. Wiegerinck [22] has given a new proof and has established bounded support of g under weaker hypotheses, namely, the vanishing of $\hat{g}$ on a set $E \times \{|\lambda| > C\}$ with $\beta(E) > 0$, coupled with a suitable smallness condition on $\hat{g}$ for y outside E and $|\lambda| \to \infty$.

His proof depends on the observation that the Fourier transform $f = \mathcal{F}g$ will satisfy a condition of the form (8.3) for $a = 0$, but <u>without</u> the $n!$. Indeed,

$$f(ty) = \int_{\mathbb{R}^N} g(x)e^{-ity\cdot x}dx = \int_{-C}^{C} \hat{g}(y,\lambda)e^{-it\lambda}d\lambda, \quad y \in E .$$

Applying Theorem 8.2, the conclusion is that f can be extended to an entire function of exponential type on $\mathbb{C}^N$, hence by the Paley-Wiener/Plancherel-Pólya theorem, g must have bounded support.

9. <u>Dependence of Green functions and capacities on the defining set</u>.

Homogeneous capacities and Green functions are important tools in the derivation of convergence theorems for series of homogeneous polynomials $\Sigma q_n(z)$, cf. Siciak [21]. One has to know then how the capacities and Green functions vary with the defining set. To that end, it is convenient to introduce Green functions for arbitrary bounded sets E in $\mathbb{C}^N$. The polynomial definition is not fine enough here: it does not distinguish between E and its closure on E. One therefore works with the class LH (studied by Lelong [12]) of logarithmically homogeneous plurisubharmonic functions v: $v(wb) = v(b) + \log|w|$.

Definition 9.1 (pre-Green function h_E). For any bounded set E in $\mathbb{C}^N$,

$$h_E(z) = \sup\{v(z):\ v \in LH,\ v(\zeta) \le 0 \text{ on } E\},\ z \in \mathbb{C}^N.$$

Cf. Siciak [21] who used $\exp h_E$. With the aid of Proposition 5.7 one readily shows that Definition 9.1 is consistent with Definition 5.1 in the case of compact sets E (see [21], and cf. [10] for a short proof based on the fact that for compact E, the polynomially convex hull $\hat{E}_c$ is also the PSH convex hull of E_c).

The Green function h_E^* is again defined by upper regularization of h_E, cf. 5.8. The Principal Fact 5.9 still obtains: either $h_E^* \equiv +\infty$ (for pluripolar E), or $h_E^* < +\infty$ everywhere and then h_E^* belongs to the class LH.

The work of Bedford and Taylor [3] and Siciak [21] shows that h_E, h_E^* and $\rho(E)$ (Definition 6.1, first two lines) have the following regularity properties as set functions.

Regularity Properties 9.2:

$$h_E = \sup\{h_\Omega:\ \Omega \supset E,\ \Omega \text{ bounded, open}\},$$
$$h_\Omega = \inf\{h_K:\ K \subset \Omega,\ K \text{ compact}\},\ h_\Omega^* = h_\Omega,$$
$$h_{K_n} \uparrow h_K,\ \rho(K_n) \downarrow \rho(K) \text{ when } K_n \downarrow K \text{ (compact sets)},$$
$$h_{E_n}^* \downarrow h_E^*,\ \rho(E_n) \uparrow \rho(E) \text{ when } E_n \uparrow E \text{ (bounded sets)}.$$

Also,

$$(9.3)\qquad \rho(E) = \rho(E_c) = 0 \rightleftarrows E \text{ is LH-polar} \rightleftarrows E_c \text{ is pluripolar}.$$

The above properties justify the name generalized capacity for $\rho(E)$. The function $c(E)$ introduced in (3.7) has similar properties, cf. also Korevaar [10] for proofs in the case of $\mathbb{C}$ (which extend to $\mathbb{C}^N$).

10. Series of homogeneous polynomials and holomorphic extension.

For any logarithmically homogeneous PSH function v, the set

$$\Omega_v = \{z \in \mathbb{C}^N:\ \ v(z) < 0\}$$

is a starlike circular domain of holomorphy (maximal domain of existence for some holomorphic function). It is also the maximal domain of (locally uniform) convergence for some series of homogeneous polynomials, cf. Lelong [12]. Siciak has proved the following sharp convergence theorem [21]:

Theorem 10.1. Let $\sum_0^\infty q_n(z)$ be a series of homogeneous polynomials (4.1) which converges pointwise on a bounded set E in $\mathbb{C}^N$. If $\rho(E) > 0$ or $h_E^* < +\infty$, the series will converge (and converge locally uniformly) throughout the domain

$$\Omega = \{z \in \mathbb{C}^N:\ \ h_E^*(z) < 0\}. \tag{10.2}$$

Proof. The proof uses the most difficult of the Regularity Properties 9.2 Set

$$E_m = \{z \in E:\ \ \log|q_n(z)| \le m,\ n = 0,1,\dots\},\ m = 1,2,\dots\ .$$

Then $\{E_m\}$ is an increasing sequence of bounded sets with union E and by the definition of h_{E_m}:

$$\frac{1}{n}\log|q_n(z)| \le \frac{m}{n} + h_{E_m}(z),\ z \in \mathbb{C}^N\ . \tag{10.3}$$

Thus

$$\varphi(z) \overset{\text{def}}{=} \limsup_{n\to\infty} \frac{1}{n}\log|q_n(z)| \le h_{E_m}(z) \quad \text{on} \quad \mathbb{C}^N\ .$$

Hence by 9.2,

$$\varphi(z) \le \varphi^*(z) \le \lim h^*_{E_m}(z) = h^*_E(z) \quad \text{on} \quad \mathbb{C}^N\ . \tag{10.4}$$

It follows that $\Sigma q_n(z)$ converges at every point $z \in \Omega$. By (10.3), the convergence is uniform on every bounded set given by $\{h^*_{E_m}(z) \le -\varepsilon\}$.
For $m = 1,2,\dots$ these sets form an increasing sequence whose union contains the set $\{h^*_E(z) < -\varepsilon\}$.

Remark. For any F_σ set E with $\rho(E) = 0$ there are series $\Sigma q_n(z)$ which converge on E, but which have no open set of convergence in $\mathbb{C}^N$ (Siciak [21]).

Siciak's sophisticated Theorem 10.1 may be used to give an alternate proof of Wiegerinck's recent general result [23] on N-dimensional holomorphic extension of functions, defined and one-dimensionally holomorphic on discs in a nonpluripolar pencil of complex lines:

Theorem 10.5. Let Ω_0 be a bounded starlike circular domain in $\mathbb{C}^N$ and let K be its Shilov boundary relative to polynomials. Let $f(z) = \sum_0^\infty q_n(z)$ be a formal power or polynomial series with the following property: for each point b in K, the restriction $f(wb) = \Sigma q_n(b)w^n$ represents a holomorphic function on the disc $|w| < 1$. Then these holomorphic functions extend to asingle N-dimensional holomorphic function on the interior $\Omega = \{h_K^*(z) < 0\}$ of the polynomially convex hull $\hat{K}$ of K (which contains Ω_0).

Proof. Apply Theorem 10.1 to the series $\Sigma q_n(z)$ and the sets $E = \lambda K$, $0 < \lambda < 1$; one has $\rho(E) > 0$ since $\hat{E}$ contains $\lambda\Omega_0$. The conclusion is that Σq_n converges locally uniformly on the domains $\{h_E^*(z) < 0\} = \{h_K^*(z) < \log \lambda\}$.

Remarks. The case $\Omega_0 = B(0,1)$ of Theorem 10.5 is related to a theorem of Forelli [6], cf. Rudin [16]. Other special cases have been considered in [24]. Another application of Theorem 10.1 to formal power series has been made by Levenberg and Molzon [13].

References

1. H. Alexander, Projective capacity. In: Recent developments in several complex variables, pp. 3-27. J.E. Fornaess, ed., Princeton Univ. Press, 1981.

2. E. Bedford, Extremal plurisubharmonic functions and pluripolar sets in $\mathbb{C}^2$. Math. Ann. 249 (1980) 205-223.

3. E. Bedford and B.A. Taylor, A new capacity for plurisubharmonic functions. Acta Math. 149 (1982) 1-40.

4. U. Cegrell, Capacities and extremal plurisubharmonic functions on subsets of $\mathbb{C}^n$. Ark. Mat. 18 (1980) 199-206.

5. J.-P. Demailly, Monge-Ampère measures and plurisubharmonic measures. Preprint, Univ. de Grenoble, 1986.

7a. L. Hörmander, An introduction to complex analysis in several variables. North-Holland Publ. Col, Amsterdam, 1973.

8. B. Josefson, On the equivalence between locally polar and globally polar sets for plurisubharmonic functions on $\mathbb{C}^n$. Ark. Mat. 16 (1978) 109-115.

9. C.O. Kiselman, Sur la définition de l'opérateur de Monge-Ampère complexe. Analyse complexe. Proc. Toulouse 1983, pp. 139-150. Lecture Notes in Math. 1094, Springer, Berlin, 1984.

10. J. Korevaar, Polynomial approximation numbers, capacities and extended Green functions for $\mathbb{C}$ and $\mathbb{C}^N$. Math. Dept., Univ. of Amsterdam, Report 85-25. To appear in Proc. Fifth Texas Symposium on Approximation Theory, Acad. Press, New York, 1986.

11. J. Korevaar and J. Wiegerinck, A representation of mixed derivatives with an application to the edge-of-the-wedge theorem. Nederl. Akad. Wetensch. Proc. Ser. A 88 (1985) 77-86.

12. P. Lelong, Fonctions entières de type exponential dans $\mathbb{C}^N$. Ann. Inst. Fourier 16 (1966) 269-318.

13. N. Levenberg and R.E. Molzon, Convergence sets of formal power series. Preprint, Univ. of Kentucky, 1985.

14. R.E. Molzon, B. Shiffman and N. Sibony, Average growth estimates for hyperplane sections of entire analytic sets. Math. Ann. 257 (1981) 43-59.

15. L. Ronkin, Introduction to the theory of entire functions of several variables. Amer. Math. Soc. Transl. of Math. Monographs 44, Providence, 1974.

16. W. Rudin, Function theory in the unit ball of $\mathbb{C}^n$. Springer, Berlin, 1980.

17a. A. Sadullaev, Plurisubharmonic measures and capacities on complex manifolds. Uspekhi Mat. Nauk 36 (1981) no. 4, 53-105.

17b. A. Sadullaev, Plurisubharmonic functions. In: Contemporary problems of mathematics vol. 8 (Complex analysis of several variables part 2), pp. 65-113. A.G. Vitushkin and G.M. Khenkin, eds., Acad. of Sciences USSR, Moscow, 1985.

18. N. Sibony and P.-M. Wong, Some results on global analytic sets. In: Sém. Lelong-Skoda 1978/79, pp. 221-237. Lecture Notes in Math. 822, Springer, Berlin, 1980.

19. J. Siciak, On some extremal functions and their applications in the theory of analytic functions of several variables. Trans. Amer. Math. Soc. 105 (1962) 322-357.

20. J. Siciak, Extremal plurisubharmonic functions in $\mathbb{C}^n$. Proc. First Finnish-Polish Summer School in Complex Analysis, 1977, pp. 115-152; cf. Ann. Polon. Math. 39 (1981) 175-211.

21. J. Siciak, Extremal plurisubharmonic functions and capacities in $\mathbb{C}^n$. Sophia Kokyoroku in Math. 14, Sophia Univ., Tokyo, 1982.

22. J. Wiegerinck, A support theorem for Radon transforms on $\mathbb{R}^n$. Nederl. Akad. Wetensch. Proc. Ser. A 88 (1985) 87-93.

23. J. Wiegerinck, Convergence of formal power series and analytic extension. To appear in Proc. Special Year in Complex Analysis, Univ. of Maryland 1985/86-Springer Lecture Notes in Math.

24. J. Wiegerinck and J. Korevaar, A lemma on mixed derivatives and results on holomorphic extension. Nederl. Akad. Wetensch. Proc. Ser. A 88 (1985) 351-362.

25. V.P. Zakharyuta, Extremal plurisubharmonic functions, orthogonal polynomials and the Bernstein-Walsh theorem for analytic functions of several complex variables. Ann. Polon. Math. 33 (1976) 137-148.

26. A. Zériahi, Capacité, constante de Tchebysheff et polynomes orthogonaux associés à un compact de $\mathbb{C}^n$. To appear in Bull. Sc. Math.

Math. Institute
Univ. of Amsterdam
Roetersstraat 15
1018 WB Amsterdam

SPLITTING OF SLOWLY DECREASING IDEALS IN WEIGHTED ALGEBRAS OF ENTIRE FUNCTIONS

Reinhold Meise, Siegfried Momm, B. Alan Taylor

Mathematisches Institut der Universität
Universitätsstraße 1
D-4000 Düsseldorf 1
Federal Republic of Germany

Department of Mathematics
University of Michigan
347 West Engineering Building
Ann Arbor, Michigan 48109
United States of America

For p a nonnegative plurisubharmonic function on $\mathbb{C}^n$, let $A_p(\mathbb{C}^n)$ denote the algebra of all entire functions f such that $|f(z)| \leq Ae^{Bp(z)}$ for constants $A,B > 0$ depending on f. Algebras of this type arise at various places in complex analysis and functional analysis, e.g. as Fourier transforms of certain convolution algebras. The structure of their closed ideals has been studied for a long time, primarily in the work of Schwartz [22], Ehrenpreis [9], Malgrange [13], and Palamodov [20] in connection with the existence and approximation of solutions of (systems of) convolution equations. The question whether a certain parameter dependence of the right hand side of such an equation is shared also by its solutions, is closely related with the question of the existence of a continuous linear right inverse. The existence of such a right inverse is equivalent to the splitting of the closed ideal I associated to the corresponding equation. Also, since the quotient space $A_p(\mathbb{C}^n)/I$ is quite often identified with the space $A_p(V)$ of holomorphic functions on the variety V of I which satisfy the restricted growth conditions, the latter question is equivalent to the existence of a linear extension operator from $A_p(V)$ to $A_p(\mathbb{C}^n)$.

Recently, Meise and Taylor [15] have shown that the splitting of all closed ideals I in $A_p(\mathbb{C}^n)$, which have a discrete zero-variety and which are generated by n slowly decreasing functions, is equivalent to the fact that the strong dual $A_p(\mathbb{C}^n)_b'$ of $A_p(\mathbb{C}^n)$ has the linear topological invariant (DN). This invariant was introduced by Vogt [25] to characterize the linear topological subspaces of the space s of all rapidly decreasing sequences. In [15] it was also shown that the invariant (DN) for $A_p(\mathbb{C}^n)'_b$ is equivalent to other interesting properties, like the splitting of a certain weighted $\bar{\partial}$-complex and the existence of continuous linear extension operators for all strongly p-interpolating submanifolds of $\mathbb{C}^n$. For radial weight functions p (i.e. $p(z) = p(|z|)$), the space $A_p(\mathbb{C}^n)'_b$ has (DN) iff p^{-1} has a certain convexity property. For such radial weights p, which also satisfy $p(2z) = O(p(z))$, it was shown in Meise and Taylor [16] that each principal ideal in $A_p(\mathbb{C}^n)$ splits.

The aim of the present paper is to give a brief survey of [15] and [16] and to prove the following complementary results: First we characterize the componentwise radial weight functions p, for which $A_p(\mathbb{C}^n)'_b$ has (DN), which also gives a new direct proof in the radial case. Then we look at the localized ideals $I_{loc}(F_1,\ldots,F_N)$ of N slowly decreasing generators in $A_p(\mathbb{C})$ for $p(z) = |\mathrm{Im}\, z| + r(z)$, where r is a radial weight function on $\mathbb{C}$ with $r(z) = o(|z|)$ which satisfies some technical conditions. We show that $I_{loc}(F_1,\ldots,F_N)$ splits in $A_p(\mathbb{C})$ if and only if

$$\sup\left\{ \frac{|\mathrm{Im}\, a|}{1+r(a)} \;\middle|\; a \in \mathbb{C},\ F_j(a) = 0,\ 1 \leq j \leq N \right\} < \infty.$$

This extends previous results of Taylor [23] and Meise and Vogt [19] and implies that for certain systems $(T_{\mu_1},\ldots,T_{\mu_N})$ of convolution operators on the space $\mathcal{E}_\omega(\mathbb{R})$ of ω-ultradifferentiable functions on $\mathbb{R}$, the kernel K of the system is a complemented subspace of $\mathcal{E}_\omega(\mathbb{R})$, i.e. there exists a continuous linear projection P on $\mathcal{E}_\omega(\mathbb{R})$ with range K. The proof of this characterization is based on the idea of proof of Taylor [23], Thm. 5.2, and on a variation of some arguments out of Meise, Taylor and Vogt [18]. By a remark of Ehrenpreis, this class of ideals is strictly smaller than the class of all closed ideals in $A_p(\mathbb{C})$, though each closed ideal in $A_p(\mathbb{C})$ is of the form $I_{loc}(F_1,F_2)$ for suitable functions $F_1,F_2 \in A_p(\mathbb{C})$.

The authors thank Professor L. Ehrenpreis for the communication of this result.

1. Preliminaries

In this section we recall some definitions and results which are needed to formulate or to prove the statements of the following sections. Without further reference we use the standard notation from complex analysis and from functional analysis.

<u>1.1 Definition.</u> A function $p : \mathbb{C}^n \to [0,\infty[$ is called a weight function if it has the following properties:

(1) p is continuous and plurisubharmonic

(2) $\log(1+|z|^2) = O(p(z))$

(3) There exists $C \geq 1$ such that for all $w \in \mathbb{C}^n$ we have

$$\sup_{|z-w|\leq 1} p(z) \leq C(1 + \inf_{|z-w|\leq 1} p(z)).$$

A weight function p is called radial (resp. componentwise radial) if $p(z) = p(|z|)$ (resp. $p(z_1,\ldots,z_n) = p(|z_1|,\ldots,|z_n|)$) for all $z \in \mathbb{C}^n$.

<u>1.2 Examples.</u> The following functions p are typical examples of weight functions on $\mathbb{C}^n$:

(1) $p(z) := |z|^\rho$, $\rho > 0$

(2) $p(z) := (\log(1+|z|^2))^s$, $s > 1$

(3) $p(z) := |\mathrm{Im}\, z|^\alpha + |z|^\beta$, $0 < \beta < \alpha$ and $\alpha \geq 1$

(4) $p(z) := |\mathrm{Im}\, z| + \log(1+|z|^2)$

(5) $p(z) := |\mathrm{Im}\, z| + r(z)$, where r is a radial weight function with $r(z) = o(|z|)$.

For an open set Ω in $\mathbb{C}^n$ we denote by $A(\Omega)$ the algebra of all holomorphic functions on Ω. For each weight function p on $\mathbb{C}^n$ we define a subalgebra $A_p(\mathbb{C}^n)$ of $A(\mathbb{C}^n)$ in the following way:

<u>1.3 Definition.</u> For a weight function p on $\mathbb{C}^n$ we put

$$A_p(\mathbb{C}^n) := \{f \in A(\mathbb{C}^n) \mid \text{there exists } k \in \mathbb{N} : \sup_{z\in\mathbb{C}^n} |f(z)|\exp(-kp(z)) < \infty\},$$

and endow $A_p(\mathbb{C}^n)$ with its natural inductive limit topology. Then $A_p(\mathbb{C}^n)$ is a locally convex algebra and (DFN)-space, i.e. $A_p(\mathbb{C}^n)$ is the strong dual of a nuclear Fréchet space (see e.g. Meise [14],2.4).

The algebras of type $A_p(\mathbb{C}^n)$ arise at various places in complex analysis and functional analysis. We are particularly interested in certain closed ideals in $A_p(\mathbb{C}^n)$. From Berenstein and Taylor [2] we recall:

<u>1.4 Definition.</u> Let p be a weight function on $\mathbb{C}^n$ and let $F = (F_1,\ldots,F_N) \in (A_p(\mathbb{C}^n))^N$. F is called slowly decreasing if

$$V(F) := \{z \in \mathbb{C}^n \mid F_j(z) = 0 \text{ for } 1 \leq j \leq N\}$$

is discrete (which implies $N \geq n$) and if there are $\varepsilon > 0$, $C > 0$ and $D > 0$ such that for each component S of the set

$$S(F,\varepsilon,C) := \{z \in \mathbb{C}^n \mid (\sum_{j=1}^{N} |F_j(z)|^2)^{1/2} < \varepsilon \exp(-Cp(z))\}$$

we have

$$\sup_{z \in S} p(z) \leq D(1 + \inf_{z \in S} p(z)).$$

1.5 Definition. Let p be a weight function on $\mathbb{C}^n$, let I be an ideal in $A_p(\mathbb{C}^n)$ and let F denote an N-tuple of functions $F_1,\ldots,F_N$ in $A_p(\mathbb{C}^n)$.

(a) The localization I_{loc} of I is defined by

$$I_{loc} := \{f \in A_p(\mathbb{C}^n) \mid [f]_a \in I_a \text{ for all } a \in \mathbb{C}^n\},$$

where I_a denotes the ideal in the local ring $\mathcal{O}_a$ which is generated by the germs $[g]_a$ of all $g \in I$.

(b) By $I(F) = I(F_1,\ldots,F_N)$ we denote the ideal in $A_p(\mathbb{C}^n)$ which is algebraically generated by $F_1,\ldots,F_N$.

(c) $I_{loc}(F) := (I(F))_{loc}$.

It is easy to check that I_{loc} is a closed ideal in $A_p(\mathbb{C}^n)$ which contains I. Hence we have $I(F) \subset \overline{I(F)} \subset I_{loc}(F)$ for each $F \in (A_p(\mathbb{C}^n))^N$. By Kelleher and Taylor [11], Thm. 4.6, resp. Berenstein and Taylor [2], Thm. 4.2, the following holds:

1.6 Theorem. Let p be a weight function on $\mathbb{C}^n$.

(a) If $F \in (A_p(\mathbb{C}^n))^N$ is slowly decreasing, then $I_{loc}(F) = \overline{I(F)}$.

(b) If $F \in (A_p(\mathbb{C}^n))^n$ is slowly decreasing, then $I(F) = \overline{I(F)} = I_{loc}(F)$.

1.7 Definition. Let V be a complex submanifold of $\mathbb{C}^n$ of complex dimension k and let p be a weight function on $\mathbb{C}^n$.

(a) V is said to be strongly interpolating for p if there exist $F_1,\ldots,F_m$ in $A_p(\mathbb{C}^n)$ and positive numbers ε and C with

(1) $V = \{z \in \mathbb{C}^n \mid F_j(z) = 0 \text{ for } 1 \leq j \leq m\}$ and

(2) $\sum |\Delta_{I,J}(z)| \geq \varepsilon \exp(-Cp(z))$ for all $z \in \mathbb{C}^n$,
where the sum is taken over all the determinants $\Delta_{I,J}$ of the $(n-k) \times (n-k)$ submatrices of the matrix $\left(\frac{\partial F_i}{\partial z_j}\right)_{1\leq i\leq m, 1\leq j\leq n}$.

(b) We put

$$A_p(V) := \{f \in A(V) \mid \text{there is } B > 0 \text{ with } \sup_{z\in V} |f(z)| \exp(-Bp(z)) < \infty\}$$

and endow $A_p(V)$ with its natural inductive limit topology. Then we define the restriction map $\rho : A_p(\mathbb{C}^n) \to A_p(V)$ by $\rho(f) := f|V$.

By Berenstein and Taylor [3], Thm. 1, we have:

1.8 Theorem. Let p be a weight function on $\mathbb{C}^n$ and let V be a complex submanifold of $\mathbb{C}^n$ which is strongly interpolating for p. Then the restriction map $\rho : A_p(\mathbb{C}^n) \to A_p(V)$ is continuous, linear and surjective.

1.9 Definition. For a weight function p on $\mathbb{C}^n$ we define

$$K(p,n) := \{f \in C^\infty(\mathbb{C}^n) \mid \sup_{|\alpha|\le k} \sup_{z\in\mathbb{C}^n} |f^{(\alpha)}(z)| \exp(kp(z)) < \infty \text{ for all } k \in \mathbb{N}\}$$

and endow K(p,n) with its natural Fréchet space topology. The properties of p imply that K(p,n) is a nuclear Fréchet space. By $K(p,n)_b'$ we denote the strong dual of K(p,n), and by $K'_{(r,s)}(p,n)$ we denote the locally convex space of all distributional differential forms of bidegree (r,s) with coefficients in $K(p,n)_b'$.

By Meise and Taylor [15], 1.9, we have:

1.10 Proposition. For each weight function p on $\mathbb{C}^n$

$$0 \to A_p(\mathbb{C}^n) \to K'_{(0,0)}(p,n) \xrightarrow{\bar\partial} K'_{(0,1)}(p,n) \xrightarrow{\bar\partial} \dots \xrightarrow{\bar\partial} K'_{(0,n)}(p,n) \to 0$$

is a topologically exact sequence.

2. Splitting of closed ideals in $A_p(\mathbb{C}^n)$ and the property (DN)

Let p be a weight function on $\mathbb{C}^n$ and let I be a closed ideal in $A_p(\mathbb{C}^n)$. We shall say that I splits or that I is complemented in $A_p(\mathbb{C}^n)$ if there exists a continuous linear projection P on $A_p(\mathbb{C}^n)$ with range I. It is easy to check that I splits if and only if the quotient map $q : A_p(\mathbb{C}^n) \to A_p(\mathbb{C}^n)/I$ admits a continuous linear right inverse.
The fact that there are weight functions p on $\mathbb{C}$ for which each closed ideal I in $A_p(\mathbb{C})$ splits, was first observed by Taylor [23]. For $p(z) = |z|^s$, $s \geq 1$, he showed the existence of a continuous linear right inverse R for $q : A_p(\mathbb{C}) \to A_p(\mathbb{C})/I$ in the following way: First he constructed a continuous linear right inverse $\tilde{R}$ of q with values in a

suitable space $C_p^\infty(\mathbb{C})$ of weighted C^∞-functions, using interpolation formulas and appropriate cut-off functions. Then he showed that the $\bar\partial$-operator on $C_p^\infty(\mathbb{C})$ admits a continuous linear right inverse, so that he could obtain R from $\tilde{R}$ by solving a certain $\bar\partial$-equation in a continuous linear way.
Later, Meise [14] showed that Taylor's result holds for a larger class of weight functions. To get this, he used the observation that for slowly decreasing $F \in (A_p(\mathbb{C}^n))^n$, the quotient space $A_p(\mathbb{C}^n)/I_{loc}(F)$ has a rather particular locally convex structure. Because of this structure, the splitting of $I_{loc}(F)$ follows from the splitting theorem of Vogt and Wagner [28], provided that the nuclear Fréchet space $A_p(\mathbb{C}^n)_b'$ has the property (DN).
Finally, the work of Meise and Taylor [15], [16] showed that the property (DN) of $A_p(\mathbb{C}^n)_b'$ is in fact crucial for the existence of many splitting ideals in $A_p(\mathbb{C}^n)$ and that the approaches of Taylor [23] and Meise [14] are essentially equivalent. To state the main results of [15] and [16] and to explain the functional analytic tools a bit, we first introduce some more notation.

<u>2.1 Definition.</u> Let $\alpha = (\alpha_j)_{j\in\mathbb{N}_o}$ be an increasing unbounded sequence of positive real numbers. Then the power series space (of infinite type) $\Lambda_\infty(\alpha)$ is defined by

$$\Lambda_\infty(\alpha) := \{x \in \mathbb{C}^{\mathbb{N}_o} \mid \|x\|_k := \sum_{j\in\mathbb{N}_o} |x_j| \exp(k\alpha_j) < \infty \text{ for each } k \in \mathbb{N}\}.$$

We endow $\Lambda_\infty(\alpha)$ with the l.c. topology which is induced by the system $(\| \ \|_k)_{k\in\mathbb{N}}$ of norms. Then $\Lambda_\infty(\alpha)$ is a Fréchet space.

Power series spaces of infinite type come up at various places in analysis, quite often as sequence spaces of coefficients of certain classical series expansions. For example, the Taylor expansion at the origin induces an isomorphism between $A(\mathbb{C})$ and $\Lambda_\infty(j)$, while the Fourier expansion of 2π-periodic C^∞-functions induces an isomorphism between $C_{2\pi}^\infty(\mathbb{R})$ and $\Lambda_\infty(\log(1+j)) =: s$, the space of all rapidly decreasing sequences.

Note that a power series space $\Lambda_\infty(\alpha)$ is nuclear if and only if

$$\sup_{j\in\mathbb{N}} \frac{\log j}{\alpha_j} < \infty.$$

For our purposes, the following result of Meise [14],3.8, will be useful:

2.2 Proposition. Let p be a weight function on $\mathbb{C}^n$ and let $F = (F_1,\ldots,F_N)$ be a slowly decreasing N-tuple in $A_p(\mathbb{C}^n)$. Then $(A_p(\mathbb{C}^n)/I_{loc}(F))_b'$ is either finite dimensional or isomorphic to a nuclear power series space of infinite type.

The following properties (DN) and (Ω) were introduced by Vogt [25] and Vogt and Wagner [28]:

2.3 Definition. Let E be a metrizable locally convex space with a fundamental system $(\| \ \|_k)_{k\in\mathbb{N}}$ of semi-norms and put

$$U_k := \{x \in E \mid \|x\|_k < 1\}.$$

(a) E has property (DN) if there exists $m \in \mathbb{N}$ such that for each $k \in \mathbb{N}$ there exist $n \in \mathbb{N}$ and $C > 0$ with $\| \ \|_k^2 \leq C\| \ \|_m \| \ \|_n$.

(b) E has property (Ω) if for each $p \in \mathbb{N}$ there exists $q \in \mathbb{N}$ such that for each $k \in \mathbb{N}$ there exist $d > 0$ and $C > 0$ such that for all $r > 0$

$$U_q \subset Cr^d U_k + \frac{1}{r} U_p.$$

2.4 Remark. (a) It is easy to check that (DN) is a linear topological invariant which is inherited by linear topological subspaces. By Vogt [25],1.3, a nuclear metrizable locally convex space E has (DN) iff E is isomorphic to a linear topological subspace of s.

(b) It is easy to check that (Ω) is a linear topological invariant which is inherited by quotient spaces. By Vogt and Wagner [28],1.8, a nuclear Fréchet space E has (Ω) iff E is isomorphic to a quotient space of s.

(c) Every power series space $\Lambda_\infty(\alpha)$ has (DN) and (Ω).

The significance of the properties (DN) and (Ω) is underlined by the splitting theorem of Vogt and Wagner [28] (see also Vogt [26]):

2.5 Theorem. Let E,F,G be nuclear Fréchet spaces and let

$$0 \longrightarrow E \xrightarrow{i} F \xrightarrow{q} G \longrightarrow 0$$

be an exact sequence with continuous linear maps. If E has (Ω) and G has (DN) then the exact sequence splits, i.e. there exists a continuous linear map $R : G \to F$ with $q \circ R = id_G$.

From 2.2 and 2.5 we get (see Meise [14],4.4):

2.6 Proposition. Let p be a weight function on $\mathbb{C}^n$ and assume that $A_p(\mathbb{C}^n)_b'$ (the strong dual of $A_p(\mathbb{C}^n)$) has (DN). Then, for each slowly decreasing n-tuple $F = (F_1,\dots,F_n)$ in $A_p(\mathbb{C}^n)$, the closed ideal I(F) in $A_p(\mathbb{C}^n)$ splits.

Proof. Since F is slowly decreasing, I(F) is closed by 1.6. Since $A_p(\mathbb{C}^n)$ is a (DFN)-space, we can dualize the exact sequence

(1) $$0 \longrightarrow I(F) \longrightarrow A_p(\mathbb{C}^n) \longrightarrow A_p(\mathbb{C}^n)/I(F) \longrightarrow 0$$

to get an exact sequence

(2) $$0 \longrightarrow (A_p(\mathbb{C}^n)/I(F))_b' \longrightarrow A_p(\mathbb{C}^n)_b' \longrightarrow I(F)_b' \longrightarrow 0$$

of nuclear Fréchet spaces. Without loss of generality we can assume that $A_p(\mathbb{C}^n)/I(F)$ is infinite dimensional. Then 1.6, 2.2 and 2.4(c) imply that $(A_p(\mathbb{C}^n)/I(F))_b'$ has property (Ω). Since

$$I(F) = \sum_{j=1}^{n} F_j A_p(\mathbb{C}^n),$$

it follows easily from the hypothesis, that $I(F)_b'$ has property (DN). Thus, the exact sequence (2) splits by Theorem 2.5 and hence the exact sequence (1) splits, too.

The converse of Proposition 2.6 also holds, as Meise and Taylor [15], 2.17, have shown. The proof of this result is by far more technical and is obtained by showing that several properties are equivalent, which so far had not been known to be related.

2.7 Theorem. Let p be a weight function on $\mathbb{C}^n$ satisfying $\log(1+|z|^2) = o(p(z))$. Then the following assertions are equivalent:

(1) for each complex submanifold V of $\mathbb{C}^n$ which is strongly interpolating for p, there exists a continuous linear extension operator $E : A_p(V) \to A_p(\mathbb{C}^n)$ (i.e. $\rho \circ E = id_{A_p(V)}$)

(2) for each slowly decreasing n-tuple $F = (F_1,\dots,F_n)$ in $A_p(\mathbb{C}^n)$, the ideal I(F) in $A_p(\mathbb{C}^n)$ splits

(3) the exact sequence

$$0 \to A_p(\mathbb{C}^n) \to K'_{(o,o)}(p,n) \xrightarrow{\bar\partial} K'_{(o,1)}(p,n) \xrightarrow{\bar\partial} \dots \xrightarrow{\bar\partial} K'_{(o,n)}(p,n) \to 0$$

splits

(4) $A_p(\mathbb{C}^n)_b'$ has the property (DN)

(5) for each $k \in \mathbb{N}$ there exist $0 < \varepsilon < 1$, $A_o > 0$ and $m \in \mathbb{N}$ such that for all $A \geq A_o$ the greatest plurisubharmonic minorant of $\min(p+A, mp-A)$ is at least as big as kp at all points $z \in \mathbb{C}^n$ with

$p(z) = \varepsilon A.$

If the weight function p is radial (i.e. $p(z) = q(|z|)$, where $q : [0,\infty[\to [0,\infty[$ is eventually strictly increasing) then (1)-(5) are also equivalent to

(6) for each $C > 1$ there exist $R_o > 0$ and $0 < \delta < 1$ with

$$q^{-1}(CR)\cdot q^{-1}(\delta R) \leq (q^{-1}(R))^2 \text{ for all } R \geq R_o.$$

If we restrict our attention to the case of one variable and of radial weight functions then we get the following extension of Taylor [23], Thm. 5.1 (for the proof see Meise and Taylor [15],3.4).

2.8 Theorem. Let p be a radial weight function on $\mathbb{C}$ with $p(2z) = O(p(z))$ and $\log(1+|z|^2) = o(p(z))$. Then the following conditions are equivalent:

(1) each closed ideal in $A_p(\mathbb{C})$ splits

(2) $A_p(\mathbb{C})$ is a complemented subspace of $K(p,1)'_b$

(3) $A_p(\mathbb{C})'_b$ has (DN)

(4) $A_p(\mathbb{C})'_b$ is isomorphic to a power series space of infinite type

(5) for each $C > 1$ there exist $R_o > 0$ and $0 < \delta < 1$ such that for all $R \geq R_o$ we have $p^{-1}(CR)p^{-1}(\delta R) \leq (p^{-1}(R))^2$.

2.9 Examples. (a) It is easy to check that the following radial weight functions satisfy condition 2.8(5) resp. 2.7(6):

$p(z) = |z|^\rho(\log(1+|z|^2))^\sigma$, $\rho > 0$, $\sigma \geq 0$

$p(z) = \exp(|z|^\alpha)$, $0 < \alpha \leq 1$

$p(z) = \exp((\log(1+|z|^2))^\alpha)$, $0 < \alpha < 1$

p satisfies $p(2z) = O(p(z))$ and $2p(z) \leq p(Az) + A$ for some $A \geq 1$ and all $z \in \mathbb{C}^n$.

(b) For $s > 1$ the radial weight function $p(z) = (\log(1+|z|^2))^s$ does not satisfy condition 2.7(6). By Meise [14],2.13(2) and 4.12(1), no infinite codimensional non-trivial ideal in $A_p(\mathbb{C})$ is complemented.

Assume that p is a weight function on $\mathbb{C}^n$ for which $A_p(\mathbb{C}^n)'_b$ has (DN) and let I be a closed ideal in $A_p(\mathbb{C}^n)$ which is finitely generated. Then the arguments out of the proof of Proposition 2.6 imply that I splits, as soon as we know that $(A_p(\mathbb{C}^n)/I)'_b$ has the property (Ω). Under the hypotheses of Proposition 2.2 we have a nice sequence space representation of $(A_p(\mathbb{C}^n)/I)'_b$. In more general cases it is not known whether $(A_p(\mathbb{C}^n)/I)'_b$

has such a representation. Nevertheless, Meise and Taylor [16] used arguments of Berenstein and Taylor [2] together with some easy functional analysis, to show that $(A_p(\mathbb{C}^n)/I)_b'$ has (Ω) for many closed ideals in $A_p(\mathbb{C}^n)$ (see [16], Thm. 12). As a consequence of their result we only mention the following special case which is particularly interesting because of its application to convolution operators (see Meise and Taylor [16], Thm. 17. and Cor. 18) and which extends a previous result of Djakov and Mityagin [8].

2.10 Theorem. Let p be a radial weight function on $\mathbb{C}^n$ satisfying $p(2z) = O(p(z))$ and $\log(1+|z|^2) = o(p(z))$. If $A_p(\mathbb{C}^n)_b'$ has (DN) (which is equivalent to p satisfying condition 2.7(6)) then every principal ideal I in $A_p(\mathbb{C}^n)$ splits.

In concluding this section we prove a necessary and sufficient condition for $A_p(\mathbb{C}^n)_b'$ having (DN) for componentwise radial weight functions, which extends condition 2.7(6). To formulate this condition it is appropriate to introduce the following notation:

Notation: We put $\mathbb{R}_+ := [0,\infty[$ and for subsets A,B of $\mathbb{R}_+^n$ we define

$$A\cdot B := \{(a_1b_1,\ldots,a_nb_n) \in \mathbb{R}_+^n \mid a = (a_1,\ldots,a_n) \in A,\ b = (b_1,\ldots,b_n) \in B\}.$$

2.11 Proposition. For $p : \mathbb{R}_+^n \to \mathbb{R}_+$ assume that $\tilde{p} : z \to p(|z_1|,\ldots,|z_n|)$ is a weight function on $\mathbb{C}^n$ satisfying $\log(1+|z|^2) = o(\tilde{p}(z))$. Then the following assertions are equivalent:

(1) $A_{\tilde{p}}(\mathbb{C}^n)_b'$ has (DN)

(2) for each $C > 1$ there exist $R_0 > 0$ and $0 < \delta < 1$ such that for each $R \geq R_0$ we have $p^{-1}([0,CR])\cdot p^{-1}([0,\delta R]) \subset p^{-1}([0,R])\cdot p^{-1}([0,R])$

(3) for each $C > 1$ there exist $R_0 > 0$ and $0 < \delta < 1$ such that for each $R \geq R_0$ we have $p^{-1}(CR)\cdot p^{-1}(\delta R) \subset p^{-1}([0,R])\cdot p^{-1}([0,R])$.

Proof. Without loss of generality we shall assume that the weight function $\tilde{p}$ is infinitely differentiable (see Meise and Taylor [15],2.9 and 2.4). Then p is a C^∞-function on $\mathbb{R}_+^n$ for which

$$\varphi : (t_1,\ldots,t_n) \to p(e^{t_1},\ldots,e^{t_n})$$

is convex. Consequently, each component of grad φ is non-negative. W.l.o.g. we can even assume that each component of grad φ is positive. For $k \in \mathbb{N}$ we define the functions $b_k, c_k : \mathbb{R}_+^n \to \mathbb{R}_+$ by

$b_k(\alpha) := \inf\{r^{-\alpha}\exp(kp(r)) \mid r \in \mathbb{R}_+^n,\ r_j > 0 \text{ for } 1 \le j \le n\}$

$c_k(\alpha) := \inf\{r^{-\alpha}\exp(kp(r)) \mid r \in \mathbb{R}_+^n,\ r_j \ge 1 \text{ for } 1 \le j \le n\}$,

where we apply the usual multi-index notation. Since $\tilde{p}$ has the property 1.1(3), it is easily checked that the following holds:

(4) for each $k \in \mathbb{N}$ there exist $l \in \mathbb{N}$ and $D > 0$ with $b_k(\alpha) \le c_k(\alpha) \le Db_l(\alpha)$ for all $\alpha \in \mathbb{R}_+^n$.

Since $\tilde{p}$ has the property 1.1(2), we get

(5) for each $k \in \mathbb{N}$ there exist $l \in \mathbb{N}$ and $D > 0$ such that for each $\alpha, \beta \in \mathbb{R}_+^n$ with $\alpha \ge \beta$ we have $c_k(\alpha) \le c_k(\beta) \le Dc_l(\beta+(1,\ldots,1))$.

Now note that by Meise and Taylor [15],3.2, $A_{\tilde{p}}(\mathbb{C}^n)_b'$ is isomorphic to the sequence space

$$\lambda(B,\mathbb{N}_o^n) := \{(x_\alpha)_{\alpha\in\mathbb{N}_o^n} \in \mathbb{C}^{\mathbb{N}_o^n} \mid \|x\|_k := \sum_{\alpha\in\mathbb{N}_o^n} |x_\alpha| b_k(\alpha) < \infty\ \forall k \in \mathbb{N}\}.$$

Hence it follows from Vogt [25],2.3,(4) and (5) that $A_{\tilde{p}}(\mathbb{C}^n)_b'$ has the property (DN) iff the following holds:

(6) there exists $m \in \mathbb{N}$ such that for each $k \in \mathbb{N}$ there exist $l \in \mathbb{N}$ and $L > 0$ with $b_k^2(\alpha) \le Lb_m(\alpha)b_l(\alpha)$ for all $\alpha \in \mathbb{R}_+^n$.

Now we shall use the equivalence of (1) and (6) to prove the proposition.

(1) ⇒ (2): To show that (6) implies (2) let $C > 1$ be given. Obviously, we can assume $C \in \mathbb{N}$. Then put $k := 2Cm$, where $m \in \mathbb{N}$ exists by (6); choose $l \in \mathbb{N}$ according to (6) and put $\delta := \frac{k}{2l}$. Next, let $r, s \in \mathbb{R}_+^n$ with $r > 0$, $s > 0$ be given and put $\sqrt{rs} := (\sqrt{r_1s_1},\ldots,\sqrt{r_ns_n})$. By elementary calculus and by the convexity of φ, it follows, that for $\alpha := k \operatorname{grad}\varphi(\log\sqrt{rs})$ we have $b_k(\alpha) = \sqrt{rs}^{-\alpha}\exp(2kp(\sqrt{rs}))$. Hence (6) implies

$$\sqrt{rs}^{-2\alpha}\exp(2kp(\sqrt{rs})) \le Lr^{-\alpha}\exp(mp(r))s^{-\alpha}\exp(lp(s))$$

and consequently

$$2kp(\sqrt{rs}) \le mp(r) + lp(s) + \log L. \tag{7}$$

Now put $R_o := \frac{\log L}{k}$ and fix $R \ge R_o$. Then for $r \in p^{-1}([0,CR])$ and $s \in p^{-1}([0,\delta R])$ with $r > 0$, $s > 0$, we have by (7) and our choice of k, δ and R

$$2kp(\sqrt{rs}) \le mCR + l\delta R + \log L \le 2kR,$$

which proves $\sqrt{rs} \in p^{-1}([0,R])$. Hence (2) holds.

(2) ⇒ (3): This holds trivially.

(3) ⇒ (1): For $C > 1$, choose R_o and δ according to (3) and fix $R \geq R_o$. For $r \in p^{-1}([0,CR])$ and $s \in p^{-1}([0,\delta R])$ we can choose $\rho, \sigma \in \mathbb{R}^n$ satisfying $\exp(\rho) := (\exp(\rho_1),\dots,\exp(\rho_n)) \geq r$, $\exp(\sigma) \geq s$, $p(\exp(\rho)) = CR$, $p(\exp(\sigma)) = \delta R$. Then we note that the convexity of φ and (3) imply

$$p(\sqrt{rs}) \leq \varphi(\tfrac{\rho+\sigma}{2}) = p(\sqrt{\exp(\rho)\exp(\sigma)}) \leq R.$$

Hence (3) implies (2). From (2) we get easily

(8) for each $C > 1$ there exist $R_o > 0$ and $0 < \delta < 1$ such that for all $R,S \geq R_o$ we have

$$p^{-1}([0,CR])\cdot p^{-1}([0,\delta S]) \subset p^{-1}([0,R+S])\cdot p^{-1}([0,R+S]).$$

To complete the proof, we show that (8) implies (6). To do this, we put $m := 1$ and fix $k \in \mathbb{N}$ arbitrarily. Then we put $C := 2k$, choose R_o and δ according to (8) and put $l := \frac{2k}{\delta}$. Obviously, we can assume $l \in \mathbb{N}$. Then (8) implies that for each $r,s \in \mathbb{R}^n_+$ with $r,s > 0$ and $p(r) \geq CR_o$, and $p(s) \geq \delta R_o$, we have

$$p(\sqrt{rs}) \leq \frac{p(r)}{C} + \frac{p(s)}{\delta}.$$

By our choice of C and l this implies the existence of $L \geq 1$ with

(9) $\quad 2kp(\sqrt{rs}) \leq p(r) + lp(s) + \log L$ for all $r,s \in \mathbb{R}^n_+, r,s > 0$.

It is easy to check that (9) implies (6).

<u>2.12 Remark.</u> The equivalence of 2.7(4) and 2.7(6) for radial weight functions p on $\mathbb{C}^n$ with $p(z) = q(|z|)$ follows easily from Proposition 2.11, since $p(z) = q_o(|z_1|,\dots,|z_n|)$, where

$$q_o(x_1,\dots,x_n) = q((\sum_{j=1}^{n} x_j^2)^{1/2}).$$

3. Splitting ideals in $A_p(\mathbb{C}^n)$ for non-radial weight functions p

For non-radial weight functions p like $|\text{Im } z| + \log(1+|z|^2)$ or $|\text{Im } z| + |z|^\alpha$, $0 < \alpha < 1$, the algebras $A_p(\mathbb{C}^n)$ come up as Fourier-Laplace transforms of distributions or ultradistributions with compact support. In these cases $A_p(\mathbb{C}^n)'_b$ is isomorphic to a Fréchet space of C^∞-functions which does not admit a continuous norm and hence fails the property (DN). More generally, we have by Meise and Taylor [15],4.1:

<u>3.1 Proposition.</u> Let r and q be nonnegative continuous functions on

$[0,\infty[$ which have the following properties:

(i) q is convex, strictly increasing and satisfies $q(2t) = O(q(t))$

(ii) $z \to r(|z|)$ is a weight function on $\mathbb{C}^n$

(iii) $\lim_{t\to\infty} \frac{r(t)}{q(t)} = 0$.

Then $p : z \to q(|\mathrm{Im}\, z|) + r(|z|)$ is a weight function on $\mathbb{C}^n$, for which $A_p(\mathbb{C}^n)_b'$ fails the property (DN).

For p as in 3.1 it follows from Theorem 2.7 and Proposition 3.1 that there exists a slowly decreasing n-tuple $F \in (A_p(\mathbb{C}^n))^n$ for which the ideal $I(F)$ does not split. On the other hand one can construct slowly decreasing n-tuples $G \in (A_p(\mathbb{C}^n))^n$ for which $I(G)$ does split. Hence $A_p(\mathbb{C}^n)$ contains splitting as well as non-splitting closed ideals. Therefore it is reasonable to ask for a characterization of the splitting closed ideals. So far, only partial results are known, see e.g. Treves [24] for a result of Grothendieck, Cohoon [7], Vogt [27] and Taylor [23].

The first general result concerning ideals which are not necessarily generated by polynomials was proved by Taylor [23], Thm. 5.2. For $p(z) = |\mathrm{Im}\, z| + \log(1+|z|^2)$ and discrete sequences $(a_j)_{j\in\mathbb{N}}$ in $\mathbb{C}^n$ he looks at ideals $I = \{f \in A_p(\mathbb{C}^n) \mid f(a_j) = 0 \text{ for all } j \in \mathbb{N}\}$, assuming that for each complex sequence $(x_j)_{j\in\mathbb{N}}$ with $\sup_{j\in\mathbb{N}} |x_j|\exp(-kp(a_j)) < \infty$ for some $k \in \mathbb{N}$, there exists $g \in A_p(\mathbb{C}^n)$ with $g(a_j) = x_j$ for all $j \in \mathbb{N}$. For ideals I of this form he showed that I splits iff

$$\text{(V)} \qquad \sup \left\{ \frac{|\mathrm{Im}\, a_j|}{1+\log(1+|a_j|^2)} \;\middle|\; j \in \mathbb{N} \right\} < \infty.$$

To prove this, one identifies $A_p(\mathbb{C}^n)/I$ in an obvious way with the sequence space $\Lambda_\infty((p(a_j))_{j\in\mathbb{N}})_b'$. If one denotes by $(e_j)_{j\in\mathbb{N}}$ the canonical basis vectors of this sequence space then it is easy to see that the quotient map $q : A_p(\mathbb{C}^n) \to A_p(\mathbb{C}^n)/I$ admits a continuous linear right inverse R iff there exist functions $f_j \in A_p(\mathbb{C}^n)$ with certain growth properties, for which $q(f_j) = e_j$ for all $j \in \mathbb{N}$. By the Phragmen-Lindelöf principle, such functions can only exist, if (V) holds. To prove that (V) in fact implies the existence of such functions one first chooses $g_j \in A_p(\mathbb{C}^n)$ with $q(g_j) = e_j$ for $j \in \mathbb{N}$ such that $\{g_j \mid j\in\mathbb{N}\}$ is a bounded set in $A_p(\mathbb{C}^n)$. Since $A_p(\mathbb{C}^n)$ contains functions which fall down very rapidly on $\mathbb{R}^n$, one can use condition (V) to obtain the functions f_j by multiplying the functions g_j with suitable correction functions.

For the weight functions $p(z) = |\mathrm{Im}\, z| + \omega(|z|)$, which are interesting for the convolution operators on the ω-ultradifferentiable functions $\mathcal{E}_\omega(\mathbb{R})$ of Beurling [4] and Björck [5], Meise and Vogt [19] have used the sequence space representation of Meise [14],3.7, and the Phragmen-Lindelöf principle to show that the corresponding condition (V) is necessary for the splitting of $I_{loc}(F)$ in $A_p(\mathbb{C})$ for each slowly decreasing N-tuple F. Moreover, they characterized the surjective convolution operators on $\mathcal{E}_\omega(\mathbb{R})$ which admit a continuous linear right inverse by various properties. This characterization implies also a characterization of the splitting principal ideals in $A_p(\mathbb{C})$ which have a slowly decreasing generator (see [19],3.9).

In this section we look at the class of all ideals $I_{loc}(F)$ in $A_p(\mathbb{C})$, where F is a slowly decreasing N-tuple and where $p(z) = |\mathrm{Im}\, z| + \omega(|z|)$. We show that Taylor's idea of proof can be used to characterize the splitting ideals in this class by the corresponding condition (V). This class of ideals in $A_p(\mathbb{C})$ is strictly smaller than the class of all closed ideals, as L. Ehrenpreis has pointed out to the authors.

Our main lemma, which provides a sufficient condition for the splitting of slowly decreasing ideals $I_{loc}(F)$ in $A_p(\mathbb{C}^n)$ will be formulated for weight functions p of the following type:

<u>3.2 Notation.</u> Let r be a radial weight function on $\mathbb{C}^n$ and let $q : \mathbb{R}_+ \to \mathbb{R}_+$ be convex and strictly increasing. Furthermore we assume that there exist $K \geq 1$ and $L \geq 1$ such that $r(2t) \leq Kr(t)$ and $q(2t) \leq Lq(t)$ for all $t \geq 0$ and that $\lim_{t\to\infty} \frac{r(t)}{q(t)} = 0$. Then

$$p : \mathbb{C}^n \to \mathbb{R}_+ ,\ p(z) := q(|\mathrm{Im}\, z|) + r(z)$$

is a weight function on $\mathbb{C}^n$.

<u>3.3 Lemma.</u> For p as in 3.2 let $F = (F_1,\ldots,F_N) \in (A_p(\mathbb{C}))^N$ be slowly decreasing. Assume that the following conditions are satisfied:

(1) $\sup \{ \frac{q(|\mathrm{Im}\, a|)}{1+r(a)} \mid a \in \mathbb{C},\ F_j(a) = 0 \text{ for } 1 \leq j \leq N\} = Q < \infty$

(2) there exist positive numbers ε,δ and C such that for each component S of the set $S(F,\varepsilon,C)$ with $S \cap V(F) \neq \emptyset$ we have for each $a \in S \cap V(F)$

$$\delta r(a) \leq \inf_{z\in S} r(z) + \frac{1}{\delta}$$

(3) there exists $v \in A(\mathbb{C})$ with $v(o) = 1$ which satisfies the following estimates:

(3) there exists $B > 0$ such that for each $k \in \mathbb{N}$ there exist $C_k > 0$ with $\log|v(z)| \leq Bq(|\mathrm{Im}\, z|) - kr(z) + C_k$ for all $z \in \mathbb{C}$.

Then the ideal $I_{loc}(F)$ in $A_p(\mathbb{C})$ splits.

Proof. Without loss of generality we can assume that

$$V = V(F) = \{a \in \mathbb{C} \mid F_j(a) = 0 \text{ for } 1 \leq j \leq N\}$$

is an infinite set. If we choose ε, δ, C and D so that (2) and the slowly decreasing condition 1.4 hold, then $S(F,\varepsilon,C)$ has infinitely many components S with $S \cap V \neq \emptyset$. For each of these components S we fix a point $a \in S \cap V$ and label all these components (resp. points) by $(S_j)_{j\in\mathbb{N}}$ (resp. $(a_j)_{j\in\mathbb{N}}$) in such a way that the sequence $\alpha := (p(a_j))_{j\in\mathbb{N}}$ is increasing.
Then we denote by I_a the ideal generated by the germs $[F_1]_a, \ldots, [F_N]_a$ in $\mathcal{O}_a$ and we define

$$E_j := \prod_{a\in V\cap S_j} \mathcal{O}_a/I_a \text{ and } n_j := \dim_{\mathbb{C}} E_j,\ j \in \mathbb{N}.$$

By the proof of Meise [14],3.7, there exists a norm $\|\ \|_j$ on E_j for which the map

$$\rho_j : A^\infty(S_j) \to E_j,\ \rho_j(g) := ([g]_a + I_a)_{a\in V\cap S_j},\ j \in \mathbb{N}$$

is continuous and open. Moreover, $A_p(\mathbb{C})/I_{loc}(F)$ can be identified with

$$k^\infty(\mathbb{E},\alpha) := \{x = (x_j)_{j\in\mathbb{N}} \in \prod_{j\in\mathbb{N}} E_j \mid \exists m \in \mathbb{N} :$$
$$|||x|||_m := \sup_{j\in\mathbb{N}} \|x_j\|_j \exp(-m\alpha_j) < \infty\},$$

which is endowed with its natural inductive limit topology. This identification is induced by the surjective continuous linear map

$$\rho : A_p(\mathbb{C}) \to k^\infty(\mathbb{E},\alpha),\ \rho(f) = (\rho_j(f|S_j))_{j\in\mathbb{N}},$$

the kernel of which is $I_{loc}(F)$.

To show that ρ admits a continuous linear right inverse, we choose for each $j \in \mathbb{N}$ an Auerbach basis $\{e_{j,l}\}_{l=1}^{n_j}$ of $(E_j, \|\ \|_j)$ with coefficient functionals $\{f_{j,l}\}_{l=1}^{n_j}$ (see e.g. Jarchow [10],p. 291). Since $\{e_{j,l} \mid 1 \leq l \leq n_j,\ j \in \mathbb{N}\}$ is a bounded subset of $k^\infty(\mathbb{E},\alpha)$ and since ρ is an open map, we can find $g_{j,l} \in A_p(\mathbb{C})$ with $\rho(g_{j,l}) = e_{j,l}$ for $1 \leq l \leq n_j$, $j \in \mathbb{N}$, such that for suitable positive numbers A' and B' we have

(4) $|g_{j,l}(z)| \leq A' \exp(B'p(z))$ for all $z \in \mathbb{C}$, $1 \leq l \leq n_j$ and $j \in \mathbb{N}$.

Next we choose v as in (3) and define $u_j : \mathbb{C} \to \mathbb{R}$ by

$$u_j(z) := \sup_{a\in S_j} \log|v(z-a)|.$$

Then u_j is a continuous subharmonic function. By our assumptions on r and q stated in 3.2 and by (3) it follows from Meise, Taylor and Vogt [18],3.3, that

(5) for each $n \in \mathbb{N}$ and each $j \in \mathbb{N}$ we have for all $z \in \mathbb{C}$

$$u_j(z) \le LBq(|\mathrm{Im}\, z|) + nKr(z) + \sup_{a\in S_j} LBq(|\mathrm{Im}\, a|) - \frac{n}{K} \inf_{a\in S_j} r(a) + C_n.$$

To estimate this, note that (1) implies for each $j \in \mathbb{N}$

$$(6) \qquad \alpha_j = p(a_j) \le Q(1+r(a_j)) + r(a_j) = Q + (Q+1)r(a_j).$$

Hence the slowly decreasing condition 1.4 implies

$$\sup_{a\in S_j} q(|\mathrm{Im}\, a|) \le \sup_{a\in S_j} p(a) \le D(1+\inf_{a\in S_j} p(a)) \le D(1+p(a_j))$$

$$\le D(Q+1)(1+r(a_j)).$$

From this we get by (3) and (5)

$$u_j(z) \le LBq(|\mathrm{Im}\, z|) + nKr(z) + LBD(Q+1)r(a_j) - \frac{n\delta}{K}r(a_j) + C_n + LBD(Q+1) + \frac{n}{\delta K}.$$

Since L,B,K,D,Q and δ do not depend on n and j, this implies:

For each $k \in \mathbb{N}$ there exist positive numbers A_k' and B_k' such that
(7) for all $j \in \mathbb{N}$ and all $z \in \mathbb{C}$ we have

$$u_j(z) \le A_k' p(z) - kr(a_j) + B_k'.$$

It is easy to check that the proof of Meise, Taylor and Vogt [18],3.2, also applies to the present weight function p. Hence there exist positive numbers A,E and T such that for each $j \in \mathbb{N}$ there exist $h_j \in A_p(\mathbb{C})$ with

$$(8) \qquad \rho_n(h_j|S_n) = 0 \text{ for } j \neq n \text{ and } \rho_j(h_j|S_j) = \rho_j(1|S_j)$$

$$(9) \qquad |h_j(z)| \le A \exp(EDp(a_j)) \sup_{|z-w|\le 1} \exp(u_j(w)+Tp(w)), \ z \in \mathbb{C}.$$

From (7),(9) and (6) and the fact that p is a weight function, we get the existence of $M \ge 1$ such that for each $k,j \in \mathbb{N}$ and all $z \in \mathbb{C}$ we have

$$(10) \qquad |h_j(z)| \le A \exp(EDp(a_j) + (A_k'+T)M(p(z)+1) + B_k' - kr(a_j))$$
$$= A_k'' \exp(B_k'' p(z) - (k-(Q+1)ED)r(a_j)).$$

Now we define $E_{j,l} := h_j g_{j,l}$, $1 \le l \le n_j$, $j \in \mathbb{N}$, and note that (4) and (10) imply:

(11) for each $k \in \mathbb{N}$ there exist positive numbers A_k and B_k such that for each $j \in \mathbb{N}$ and $1 \le l \le n_j$ we have for all $z \in \mathbb{C}$

(11) $$|E_{j,l}(z)| \leq A_k \exp(B_k p(z) - kr(a_j)).$$

Moreover, (8) and the properties of the functions $g_{j,l}$ imply

(12) $$\rho_n(E_{j,l}|S_n) = 0 \text{ for } j \neq n \text{ and } \rho_j(E_{j,l}|S_j) = e_{j,l}.$$

Furthermore note that by Berenstein and Taylor [1], Lemma 4.(f) or Meise [14], p. 80, there exists $d > 0$ with $\sum_{j=1}^{\infty} n_j \exp(-d\alpha_j) < \infty$. Next fix an arbitrary number $m \in \mathbb{N}$ and choose $k = k(m) \in \mathbb{N}$ so large that $m - \frac{k}{Q+1} \leq -d$. Then it follows from (6),(11) and our choice of the Auerbach bases that for each $x \in k^{\infty}(\mathbb{E},\alpha)$ with $|||x|||_m < \infty$ we have the following estimate for each $z \in \mathbb{C}$:

(13) $$\begin{aligned} &\sum_{j=1}^{\infty} \sum_{l=1}^{n_j} |< f_{j,l}, x_j > E_{j,l}(z)| \exp(-B_k p(z)) \\ &\leq \sum_{j=1}^{\infty} \sum_{l=1}^{n_j} A_k |||x|||_m \exp(m\alpha_j - kr(a_j)) \\ &\leq A_k e^k |||x|||_m \sum_{j=1}^{\infty} n_j \exp((m - \frac{k}{Q+1})\alpha_j) \\ &\leq A_k e^k |||x|||_m \sum_{j=1}^{\infty} n_j \exp(-d\alpha_j) \leq \tilde{A}_k |||x|||_m . \end{aligned}$$

From (13) it follows easily that the map $R : k^{\infty}(\mathbb{E},\alpha) \to A_p(\mathbb{C})$ can be defined by

$$R((x_j)_{j\in\mathbb{N}}) := \sum_{j=1}^{\infty} \sum_{l=1}^{n_j} < f_{j,l}, x_j >_j E_{j,l}$$

and that R is continuous and linear. From (12) we get

$$\begin{aligned} \rho \circ R((x_j)_{j\in\mathbb{N}}) &= (\sum_{l=1}^{n_j} < f_{j,l}, x_j >_j \rho_j(E_{j,l}|S_j))_{j\in\mathbb{N}} \\ &= (\sum_{l=1}^{n_j} < f_{j,l}, x_j >_j e_{j,l})_{j\in\mathbb{N}} = (x_j)_{j\in\mathbb{N}} . \end{aligned}$$

Hence R is a continuous linear right inverse of the quotient map ρ. Since $\ker \rho = I_{loc}(F)$, this implies that the ideal $I_{loc}(F)$ in $A_p(\mathbb{C})$ splits.

3.4 Remark. Lemma 3.3 extends literally to functions in several complex variables by the same proof. One only has to note that 3.2 and 3.3 out of Meise, Taylor and Vogt [18] extend to several variables. For [18], 3.3,this is obvious, for [18],3.2,this follows from the method of proof Berenstein and Taylor [2],2.2.

For certain weight functions $p(z) = |\text{Im } z| + r(z)$ satisfying the conditions of 3.2 it turns out that condition 3.3(2) is a consequence of 3.3(1) and the slowly decreasing condition 1.4. To obtain this, we use a variation of the arguments out of section 2 of Meise, Taylor and Vogt [18], to prove the following proposition, which extends [18],2.6.

3.5 Proposition. Let $p(z) = |\text{Im } z| + r(z)$ satisfy the conditions of 3.2 and assume moreover, that $\int_{-\infty}^{+\infty} \frac{r(t)}{1+t^2} dt < \infty$.

Then $F = (f_1,\ldots,f_N) \in (A_p(\mathbb{C}))^N$ is slowly decreasing if and only if the following condition (E) holds:

(E) there exists $A \geq 1$ such that for each $x \in \mathbb{R}$ there exists $t \in \mathbb{R}$ with $|t - x| \leq Ar(x)$ with $|F(t)| := (\sum_{j=1}^{N} |f_j(t)|^2)^{1/2} \geq \exp(-Ar(t))$.

Proof. (a) Arguing by contraposition, we assume that F does not satisfy condition (E) and show that then F is not slowly decreasing. If F does not satisfy condition (E) then we can find a sequence $(x_j)_{j\in\mathbb{N}}$ in $\mathbb{R}$ with $\lim_{j\to\infty} |x_j| = \infty$ such that for each $j \in \mathbb{N}$ we have

(1) $|F(t)| \leq \exp(-jr(t))$ for all $t \in \mathbb{R}$ with $|t - x_j| \leq jr(x_j)$.

Since p satisfies 3.2, the proof of Meise, Taylor and Vogt [18],2.5 implies the existence of functions $F_j \in A(\mathbb{C})$, $j \in \mathbb{N}$, which have the properties [18],2.5(3)-(6). In particular, they satisfy the following estimates for all $j \in \mathbb{N}$:

(2) $\sup_{z\in\mathbb{C}} |F_j(z)|\exp(-\pi|z|) < \infty$

(3) $|F_j(x_j)| \geq \frac{1}{e} \exp(jr(x_j))$.

Next we put $G_j := |F|\cdot|F_j|$ for $j \in \mathbb{N}$ and note that the estimate

(4) $G_j(x) \leq |F(x)| + \exp(2r(x))$ for all $x \in \mathbb{R}$ and all $j \in \mathbb{N}$, $j \geq 2$

is derived in the same way as [18],2.5(7). Moreover, we get from (2) and (4)

(5) there exists $R > 0$ such that for $1 \leq k \leq N$ and each $j \in \mathbb{N}$

$$\sup_{z\in\mathbb{C}} |f_k(z)F_j(z)|\exp(-(R+\pi)|z|) < \infty$$

(6) there exists $B > 0$ with $\sup_{1\leq k\leq N} \sup_{j>1} \sup_{x\in\mathbb{R}} |f_k(x)F_j(x)|e^{-Br(x)} < \infty$.

By the Phragmen-Lindelöf principle (see Boas [6],6.5.6) it follows from (5) and (6) that there exist positive numbers S and T with

(7) $$\sup_{j>1} \sup_{z\in\mathbb{C}} G_j(z)\exp(-Sp(z)) \le T.$$

Now assume that F is slowly decreasing and choose ε and C according to 1.4. Then (7) implies that for all $z \in \complement S(F,\varepsilon,C)$ we have for all $j > 1$

(8) $$|F_j(z)| = \frac{G_j(z)}{|F(z)|} \le \frac{T}{\varepsilon} \exp((S+C)p(z)).$$

By the maximum principle and the slowly decreasing condition, (8) implies the existence of positive numbers m and M with

(9) $$\sup_{j>1} \sup_{z\in\mathbb{C}} |F_j(z)|\exp(-mp(z)) \le M,$$

which contradicts (3). This contradiction shows that F is not slowly decreasing.

(b) If E satisfies condition (F) then an obvious modification of the proof of Meise, Taylor and Vogt [18],2.3, shows that F is slowly decreasing.

<u>3.6 Theorem.</u> Let r be a radial weight function on $\mathbb{C}$ which has the following properties:

(α) $$0 = r(0) \le r(s+t) \le r(s) + r(t) \text{ for all } s,t \in \mathbb{R}$$

(β) $$\int_{-\infty}^{+\infty} \frac{r(t)}{1+t^2}dt < \infty$$

and define the weight function p on $\mathbb{C}$ by $p(z) = |\mathrm{Im}\, z| + r(z)$.
Then, for each slowly decreasing N-tuple $F = (F_1,\ldots,F_N) \in (A_p(\mathbb{C}))^N$ the following conditions are equivalent:

(1) the ideal $I_{loc}(F)$ in $A_p(\mathbb{C})$ splits

(2) $\sup\{\frac{|\mathrm{Im}\, a|}{1+r(a)} \mid a \in \mathbb{C},\ F_j(a) = 0 \text{ for } 1 \le j \le N\} < \infty$.

<u>Proof.</u> (1) ⇒ (2): From Björck [5],1.2.8, it follows that $\lim_{t\to\infty} \frac{r(t)}{t} = 0$.

Because of this, condition (β) and Björck [5],1.3.11, condition (2) follows from Meise and Vogt [19],3.6.

(2) ⇒ (1): By our hypotheses, this will follow from Lemma 3.3 if we show that the conditions 3.3(2) and (3) are satisfied.
To show that condition 3.3(2) holds, note that by Proposition 3.5 the slowly decreasing condition for F is equivalent to F satisfying condition 3.5(E). For the proof of this we have refered to Meise, Taylor

and Vogt [18],2.3. There one defines positive numbers ε and C so that the components S of $S(F,\varepsilon,C)$ satisfy the diameter estimate [18],2.3(23). From this estimate and the properties of r it follows easily that condition 3.3(2) is satisfied.

The existence of functions v having all the properties which are required in condition 3.3(1) is a consequence of Björck [5],1.3.7,1.4.1 and 1.3.11 together with the remark that r(z) is always estimated from above by the Poisson integral of $r|\mathbb{R}$.

3.7 Remark. Let $\omega : \mathbb{R} \to [0,\infty[$ be an even continuous function which satisfies the conditions (α), (β) and (γ) of Björck [5],1.3.22 and for which $t \to \omega(e^t)$ is convex. Then $p : z \to |\mathrm{Im}\, z| + \omega(|z|)$ is a weight function on $\mathbb{C}$ and $A_p(\mathbb{C})$ is isomorphic to the space $\mathcal{E}_\omega(\mathbb{R})_b'$ by the Fourier-Laplace transform $\wedge$ (see Björck [5],1.8.14), where the space $\mathcal{E}_\omega(\mathbb{R})$ of ω-ultradifferentiable functions is defined in Björck [5], 1.5.1. For $\mu \in \mathcal{E}_\omega(\mathbb{R})'$ one defines the convolution operator $T_\mu : \mathcal{E}_\omega(\mathbb{R}) \to \mathcal{E}_\omega(\mathbb{R})$ by $T_\mu(\varphi) := \mu*\varphi$.

Let $\mu_1,\ldots,\mu_N \in \mathcal{E}_\omega(\mathbb{R})'$ be given and assume that $(\hat{\mu}_1,\ldots,\hat{\mu}_N)$ is a slowly decreasing N-tuple in $A_p(\mathbb{C})$. Then it follows from Theorem 3.6 by Meise, Schwerdtfeger and Taylor [17],2.4, and Kelleher and Taylor [11],6.14, that $\bigcap_{j=1}^{N} \ker T_{\mu_j}$ is a complemented linear subspace of $\mathcal{E}_\omega(\mathbb{R})$ if and only if we have

$$\text{(V)} \qquad \sup\left\{ \frac{|\mathrm{Im}\, a|}{1+\omega(|a|)} \;\middle|\; a \in \mathbb{C},\ \hat{\mu}_j(a) = 0 \text{ for } 1 \le j \le N\right\} < \infty.$$

The corresponding result also holds for slowly decreasing systems $(T_{\mu_1},\ldots,T_{\mu_N})$ of convolution operators on the spaces $\mathcal{E}^{(M_j)}(\mathbb{R})$ of ultradifferentiable functions in the sense of Komatsu [12], provided that the sequence $(M_j)_{j\in\mathbb{N}_0}$ satisfies the conditions (M1),(M2),(M3)' and (M4) stated in Meise and Vogt [19],1.4. By Meise, Taylor and Vogt [18],2.7 and 2.8 these two results extend Meise and Vogt [19],3.9.

For p as in 3.2 it follows from Kelleher and Taylor [11],6.14, and an argument indicated in Berenstein and Taylor [1],p. 120, that each closed ideal I in $A_p(\mathbb{C})$ is of the form $I = I_{loc}(F_1,F_2)$, for suitable functions $F_1,F_2 \in A_p(\mathbb{C})$. Therefore Theorem 3.6 suggests the question, whether every closed ideal I in $A_p(\mathbb{C})$ is of the form $I_{loc}(F)$ for some slowly decreasing N-tuple $F = (F_1,\ldots,F_N) \in (A_p(\mathbb{C}))^N$. The answer to this question is negative, as L. Ehrenpreis has pointed out to the authors during the April-session of the Special Year in Complex Analysis.

Ehrenpreis' idea is to show that the multiplicities of the common zeros of a slowly decreasing N-tuple in $A_p(\mathbb{C})$ satisfy a certain growth condition on the real axis and that there are functions f in $A_p(\mathbb{C})$ having a zero-variety which does not satisfy this condition. The precise growth condition is described in the following lemma.

3.8 Lemma. Let $p(z) = |\mathrm{Im}\, z| + r(z)$ be as in 3.5 and let $F = (F_1,\dots,F_N) \in (A_p(\mathbb{C}))^N$ be slowly decreasing. For $x \in \mathbb{R}$ put $m_F(x) := \min_{1\le j\le N} m_j(x)$, where $m_j(x)$ denotes the zero-multiplicity of F_j at x. Then we have $m_F(x) = O(r(x))$ as $|x|$ tends to infinity.

Proof. Let $0 \neq x \in \mathbb{R}$ be a common zero of $F_1,\dots,F_N$. Then note that by Proposition 3.5, F satisfies condition 3.5(E). Hence we can choose $t \in \mathbb{R}$ with $|t - x| \le Ar(x)$ and $|F(t)| \ge \exp(-Ar(t))$, where A is the constant from 3.5(E). Then Jensen's formula implies that the number of zeros of F_j, $1 \le j \le N$, in the closed disk of radius $R = Ar(x)$ around t is estimated from above by

$$\sup_{|z-t|=eR} \log|F_j(z)| - \log|F_j(t)|.$$

Since there exists k with $|F_k(t)| \ge \frac{1}{\sqrt{N}}|F(t)| \ge \frac{1}{\sqrt{N}} \exp(-Ar(t))$, this implies

$$\begin{aligned} m_F(x) &\le \sup_{|z-t|=eR} \log|F(z)| + Ar(t) + \log\sqrt{N} \\ &\le B(eAr(x) + r(t + eAr(x))) + Ar(t) + \log\sqrt{N} + C, \end{aligned}$$

where B and C are suitable constants which are determined by the growth estimates of $F_1,\dots,F_N$. Since $|t - x| \le Ar(x)$, the properties of r imply by standard arguments the existence of a constant $D \ge 1$ with $m_F(x) \le D(1+r(x))$.

3.9 Lemma. Let r and s be radial weight functions on $\mathbb{C}$ and put $p(z) := |\mathrm{Im}\, z| + r(z)$. Assume that s satisfies 3.6(α) and (β) and that the conditions (1) and (2) are satisfied:

(1) $\lim_{t\to\infty} \frac{r(t)}{s(t)} = 0$

(2) there exists $A > 1$ with $As(t) \le s(2t)$ for all $t \ge 0$.

Then there exists a non-zero function $g \in A_p(\mathbb{C})$ having infinitely many zeros such that $I_{loc}(g) \neq I_{loc}(F)$ for each slowly decreasing N-tuple

$F \in (A_p(\mathbb{C}))^N$.

Proof. For $n \in \mathbb{N}$ put $x_n := 2^n$ and $m_n := [s(x_n)]$, where $[a] \in \mathbb{Z}$ satisfies $[a] \leq a < [a] + 1$. Then denote by $(a_j)_{j\in\mathbb{N}}$ the sequence which is obtained from the sequence $(x_n)_{n\in\mathbb{N}}$ by repeating each x_n m_n times. Then put

$$n(t) := \#\{j \in \mathbb{N} \mid a_j \leq t\},\ t \geq 0$$

and note that (2) implies for each $k \in \mathbb{N}$

$$n(2^k) \leq \sum_{l=1}^{k} s(2^l) = \sum_{l=1}^{k} s(2^k 2^{l-k}) \leq \Big(\sum_{j=o}^{\infty} A^{-j}\Big) s(2^k).$$

Hence we have $n(t) = O(s(t))$. Using (2) and 3.6(α) we get for $2^k \leq \rho < 2^{k+1}$, $k \in \mathbb{N}$

$$\int_1^{\rho} \frac{s(t)}{t}dt \leq \sum_{l=o}^{k} \int_{2^l}^{2^{l+1}} \frac{s(t)}{t}dt \leq \sum_{l=o}^{k} s(2^{l+1}) = O(s(\rho)).$$

Hence we have

$$\int_o^{\rho} \frac{n(t)}{t}dt = O(s(\rho)).$$

By Rubel and Taylor [21],3.5 and 5.2, this implies the existence of $f \in A_s(\mathbb{C})$, $f \not\equiv 0$, such that f vanishes at x_n at least of order m_n. From the proof of Theorem 3.6 we get the existence of an entire function v satisfying condition 3.3(3) with r replaced by s. Hence $g := v\cdot f$ is in $A_p(\mathbb{C})$.

Assume now that there exists a slowly decreasing N-tuple $F \in (A_p(\mathbb{C}))^N$ with $I_{loc}(g) = I_{loc}(F)$. Then Lemma 3.8 implies

$$[s(x_n)] = m_n \leq m_F(x_n) = O(r(x_n))$$

for all $n \in \mathbb{N}$. Hence we derive a contradiction to (1).

3.10. Example. It is easy to check that for $\alpha > 1$ there exists a radial weight function s on $\mathbb{C}$ satisfying 3.6(α), 3.6(β) and 3.9(2) with $s(z) = |z|(\log|z|)^{-\alpha}$ for large $|z|$. Hence the conclusion of Lemma 3.9 holds in particular for the following weight functions

$$p(z) = |\mathrm{Im}\, z| + (\log(1+|z|^2))^a,\ a \geq 1$$

$$p(z) = |\mathrm{Im}\, z| + |z|^b,\ 0 < b < 1$$

$$p(z) = |\mathrm{Im}\, z| + |z|(\log(2+|z|^2))^{-c},\ c > 1.$$

References

[1] Berenstein, C. A.; Taylor, B. A.: A new look at interpolation theory for entire functions of one variable, Adv. Math. 33, (1979), 109-143.

[2] Berenstein, C. A.; Taylor, B. A.: Interpolation problems in $\mathbb{C}^n$ with applications to harmonic analysis, J. Anal. Math. 38, (1980), 188-254.

[3] Berenstein, C. A.; Taylor, B. A.: On the geometry of interpolating varieties, pp. 1-25 in Seminaire Lelong-Skoda, Springer LNM 919 (1982).

[4] Beurling, A.: Quasi-analyticity and general distributions, Lectures 4. and 5. AMS Summer Institute, Stanford (1961).

[5] Björck, G.: Linear partial differential operators and generalized distributions, Ark. Mat. 6, (1965), 351-407.

[6] Boas, R. P.: Entire Functions, Academic Press (1954).

[7] Cohoon, D. K.: Nonexistence of a continuous right inverse for linear partial differential operators with constant coefficients, Math. Scand. 29, (1971), 337-342.

[8] Djakov, P. B.; Mityagin, B. S.: The structure of polynomial ideals in the algebra of entire functions, Stud. Math. 68, (1980), 85-104.

[9] Ehrenpreis, L.: Fourier Analysis in Several Complex Variables, New York: Wiley-Interscience Publ. (1976).

[10] Jarchow, H.: Locally Convex Spaces, Stuttgart: Teubner (1981).

[11] Kelleher, J. J.; Taylor, B. A.: Closed ideals in locally convex algebras of entire functions, J. Reine Angew. Math. 255, (1972), 190-209.

[12] Komatsu, H.: Ultradistributions I, Structure theorems and a characterization, J. Fac. Sci. Tokyo Sec. IA, 20, (1973), 25-105.

[13] Malgrange, B.: Existence et approximation des solutions des équations aux dérivées partielles et des équations de convolution, Ann. Inst. Fourier (Grenoble) 6,(1955/56), 271-355.

[14] Meise, R.: Sequence space representations for (DFN)-algebras of entire functions modulo closed ideals, J. Reine Angew. Math. 363, (1985), 59-95.

[15] Meise, R.; Taylor, B. A.: Splitting of closed ideals in (DFN)-algebras of entire functions and the property (DN), preprint.

[16] Meise, R.; Taylor, B. A.: Each non-zero convolution operator on the entire functions admits a continuous linear right inverse, preprint.

[17] Meise, R.; Schwerdtfeger, K.; Taylor, B. A.: Kernels of slowly decreasing convolution operators, Doga, Tr. J. Math. 10, (1986), 176-197.

[18] Meise, R.; Taylor, B. A.; Vogt, D.: Equivalence of slowly decreasing conditions and local Fourier expansions, preprint.

[19] Meise, R.; Vogt, D.: Characterization of convolution operators on spaces of C^∞-functions admitting a continuous linear right inverse, preprint.

[20] Palamodov, V. P.: Linear Differential Operators with Constant Coefficients, Springer 1970.

[21] Rubel, L. A.; Taylor, B. A.: A Fourier series method for meromorphic and entire functions, Bull. Soc. Math. Fr. 96, (1968), 53-96.

[22] Schwartz, L.: Théorie générale des fonctions moyenne-périodiques, Ann. Math, II. Ser. 48, (1947), 857-929.

[23] Taylor, B. A.: Linear extension operators for entire functions, Mich. Math. J. 29, (1982), 185-197

[24] Treves, F.: Locally Convex Spaces and Linear Partial Differential Equations, Springer (1967).

[25] Vogt, D.: Characterisierung der Unterräume von s, Math. Z. 155, (1977), 109-117.

[26] Vogt, D.: Subspaces and quotient spaces of (s), pp. 167-187 "Functional Analysis: Surveys and Recent Results", K.-D. Bierstedt, B. Fuchssteiner (Eds.), North-Holland Mathematics Studies 27, (1977).

[27] Vogt, D.: On the solvability of P(D)f = g for vector valued functions, RIMS Kokyuroku 508, (1983), 168-182.

[28] Vogt, D.; Wagner, M. J.: Charakterisierung der Quotientenräume von s und eine Vermutung von Martineau, Stud. Math. 67, (1980), 225-240.

CONVOLUTORS IN SPACES OF HOLOMORPHIC FUNCTIONS

A. Meril and D.C. Struppa

Abstract. Let Ω be an open convex set in $\mathbb{C}^n$ and K a convex compact subset of $\mathbb{C}^n$: we provide sufficient, as well as necessary, conditions for the surjectivity of convolution operators between the spaces $\mathcal{H}(\Omega+K)$ and $\mathcal{H}(\Omega)$. Under natural hypotheses on the convolutors, we prove integral representation theorems for solutions $f\in\mathcal{H}(\Omega+K)$ of systems of homogeneous convolution equations. We apply this analysis to provide necessary and sufficient conditions for the hyperbolicity and the ellipticity of given systems of convolution equations; we also study the extension of solutions of homogeneous convolution equations to some sets which can be defined in terms of Ω, K and the convolutor.

1. Introduction.

This paper is devoted to the study of several problems in the theory of convolution equations for analytic functions in convex domains of $\mathbb{C}^n$; the results which we obtain can be applied, in particular, to infinite order differential equations and to difference-differential equations with constant coefficients.

Due to the interest that convolution equations have for both theoretical questions and for more applied problems, there is now a wide literature on this subject, e.g. [3], [5], [14], [17], [23], etc. Most results, however, deal with convolution equations in the whole space $\mathbb{C}^n$ or $\mathbb{R}^n$, even though the space of solutions is often allowed to be a space of generalized functions (an AU-space in the sense of Ehrenpreis, for example).

---

*Partially supported by the G.N.S.A.G.A. of the Italian C.N.R.

Key words: Convolution equations, holomorphic functions, difference-differential equations.

AMS Subject Classification: 32A15, 42B99, 43A45.

Only recently some results have been obtained for convolution operators acting on the space $\mathcal{H}(\Omega$ of holomorphic functions on a convex domain Ω of $\mathbb{C}^n$ (see [19], [20] [21]); the purpose of this paper is to provide some answers to basic question which are still open for this kind of equations.

Let Ω be a convex domain of $\mathbb{C}^n$ and let μ be an analytic functional carried by a convex compact set $K\subseteq\mathbb{C}^n$; then, in a standard way the functional μ defines a convolution operator

$$\mu^* : \mathcal{H}(\Omega+K) \to \mathcal{H}(\Omega).$$

The first problem we consider, in section 2, is to find sufficient conditions on the functional μ which make μ^* a surjection. Some results in this direction had already been obtained by Napalkov, [20], [21], and Morzhakov, [19], but the conditions which we provide in theorem 1 are of a completely different nature. Apart from the simple case in which n=1, no good necessary conditions are known, but recently Gruman and Lelong, [8], have shown that one of the sufficient conditions of Morzhakov (the other is taken as hypothesis) is also necessary, when the boundary of Ω is sufficiently regular; in theorem 2 we show that, when the boundary of K is sufficiently regular, also the second condition of Morzhakov is necessary: in particular one deduces that, when μ^* is surjective, then K is a support of μ.

Of great interest in the theory of convolution equations is the so called "approximation problem", which, in the case $\Omega=\mathbb{C}^n$, has been solved by Malgrange [14]; more precisely, if W is the set of all solutions, in $\mathcal{H}(\Omega+K)$, of $\mu^*f=0$, and if E is the linear hull of the set of exponential polynomial solutions, one might ask whether $W=\bar{E}$ or not. A very thorough study of this question has been carried out by Malgrange, [14], Martineau, [16] and, more recently, Napalkov, [21]; in section 3, however, we deal with a more sophisticated question in this direction: namely we prove that under quite natural hypotheses, not only $W=\bar{E}$, but every solution of a convolution equation as above (and even every solution of a system of such equations) can be represented as a convergent series of integrals of exponential polynomial solutions (theorems 7 and 11); this result amounts to restoring, for this kind of convolution equations, the well known Fundamental Principal proved by Ehrenpreis, for linear constant coefficients partial differential equations [7], by Berenstein and Taylor, for convolution equations in $\mathcal{H}(\mathbb{C}^n)$ and $C^\infty(\mathbb{R}^n)$ [3] and, quite recently, by Berenstein and Struppa, for convolution equations in $C^\infty(\Omega)$, Ω being a convex subset of $\mathbb{R}^n$ [2]. In this section our methods follow very closely those of [2], with the modifications which are necessary to take into account the different topological structures of $\mathcal{H}(\Omega)$ and $C^\infty(\Omega)$; for this reason most proofs are only sketched, and the reader is often referred to [2] and [3].

In section 4 we consider some hyperbolicity problems which can be solved as a consequence of the analysis carried through in section 3. To be precise, we show in Theorem 12 that a solution $f \in \mathcal{H}(\Omega+K)$ of a slowly decreasing system of convolution equations $\mu_1 * f = \ldots = \mu_r * f = 0$ (see §3 for definitions and examples) extends to a solution $\tilde{f} \in \mathcal{H}(\mathbb{C}^n)$ of the same system if and only if it is r=n and the characteristic variety $V=\{z \in \mathbb{C}^n: \hat{\mu}_1(z)=0\}=\ldots=\hat{\mu}_r(z)=0\}$ consists only of a finite number of points. In [2], the same problem is considered, for C^∞-solutions, and a sufficient condition is provided to ensure the possibility of extending every solution C^∞ on a convex set to a solution C^∞ everywhere; our method, essentially due to Meril, [17], besides proving theorem 12, can also be used to prove that this is also a necessary condition, so that corollary 19 in [2] actually provides a necessary and sufficient condition for the hyperbolicity of systems of convolution equation in $C^\infty(\Omega)$. Finally we employ the same ideas to discuss an ellipticity question, to give necessary and sufficient conditions which ensure that a distribution solution, on a convex set, of a system of convolution equations is, actually, a C^∞ solution of the same system, on the same convex set.

In the last section we also study, in some special cases, the extension of solutions of $\mu * f = 0$ from $\Omega+K$ to a set $\Gamma_{\alpha(W_{\hat{\mu}})}(\Omega)+K$, where $\alpha(W_{\hat{\mu}})$ is the asymptotic cone of the zero set $W_{\hat{\mu}}$ of $\hat{\mu}$. This is done in the spirit of [12] and [22]. We notice, in this respect, that $\alpha(W_{\hat{\mu}})$ is related to the "principal part" of μ; more precisely, when μ is a partial differential operator, then $\alpha(W_{\hat{\mu}})$ is the zero set of the homogeneous polynomial of highest degree in the polynomial $\hat{\mu}$.

The authors would like to thank Professors R. Gay and A. Yger for their many useful suggestions and ideas, and for the time they spent with them during the preparation of this paper. The second author gratefully acknowledges the hospitality of the University of Bordeaux, where this paper was written. Finally the authors would like to thank the Organizers of the Special Year in Complex Analysis at the University of Maryland for their kind invitation to participate in the events.

2. Surjectivity conditions.

In this section we consider the problem of surjectivity of convolutors acting on spaces of holomorphic functions. In particular, in theorem 1, we give two sufficient conditions for the surjectivity of such operators, while in theorem 2 we give a necessary condition for it.

Let Ω_2 be an open convex set of $\mathbb{C}^n$ and let K be a convex compact set in $\mathbb{C}^n$; define $\Omega_1=\Omega_2+K$ and consider a non-zero analytic functional $\mu\in\mathcal{H}'(\mathbb{C}^n)$, carried by K. Then μ acts on $\mathcal{H}(\Omega_1)$ by

$$\mu^* : \mathcal{H}(\Omega_1) \longrightarrow \mathcal{H}(\Omega_2)$$

$$f \longrightarrow \mu^*f(\zeta)=<\mu, z \longrightarrow f(z+\zeta)>;$$

indeed, for $\zeta\in\Omega_2$, the function $z \to f(z+\zeta)$ belongs to $\mathcal{H}(K)$, the space of germs of holomorphic functions on K, and representing μ by a measure one can easily see that μ^*f belongs to $\mathcal{H}(\Omega_2)$.

We recall that the Fourier-Borel transform $\hat{\mu}$ of μ is the entire function $\hat{\mu}(\zeta)=<\mu, z \to \exp<z, \zeta>>$, where $<z,\zeta>=z_1\zeta_1+\ldots+z_n\zeta_n$; as it is well known the space of Fourier-Borel transforms of elements of $\mathcal{H}'(K)$ is the space Exp $(K)=\{f\in\mathcal{H}(\mathbb{C}^n):\ \forall\ \varepsilon>0\ \exists A_\varepsilon>0$, such that $|f(z)|\le A_\varepsilon \exp(H_K(z)+\varepsilon|z|)\}$, where $H_K(z)=\sup_{\zeta\in K} \mathrm{Re}<z, \zeta>$ is the supporting function of the convex compact set K (this is a convex and positively homo- geneous (of degree one) function), [15], [24].

Let $\Delta(\zeta_0,\varepsilon)$ denote the closed polydisk of center ζ_0 and radius ε, $\varepsilon>0$. Consider the following conditions:

(C_1) for any $\varepsilon>0$, there exists $C(\varepsilon)>0$ such that for any $\zeta_0\in\mathbb{C}^n$ and any function $f\in\mathcal{H}(\Delta(\zeta_0,\varepsilon))$ we have

$$|f(\zeta_0)|\exp(H_K(\zeta_0))\le C(\varepsilon)) \sup_{\zeta\in\Delta(\zeta_0,\varepsilon)} |f(\zeta)\hat{\mu}(\zeta)|;$$

(C_2) there exists a family $\mathcal{L}=\{L\}$ of complex lines, covering the variety $V=\{z\in\mathbb{C}^n: \hat{\mu}(z)=0\}$, and a positive constant A such that, for every L in $\mathcal{L}$, the set

$$\mathcal{O}_L=\{z\in L : |\hat{\mu}(z)| < A \exp(H_K(z))\}$$

has relatively compact connected components; also, we require that for z_1, z_2 in the same connected component of $\mathcal{O}_L$, and for any compact subset K_2 of Ω_2 one has $H_{K_2}(z_1) \leq$ $\leq H_{K_2}(z_2) + B$. where the constant B may depend only on K_2, but does not depend on L or on the connected component of $\mathcal{O}_L$ (Note that this last condition is always satisfied when all components of $\mathcal{O}_L$ are uniformly bounded).

We will show that each of these conditions is sufficient to ensure the surjectivity of μ^*, and, later on, we will also discuss a necessary condition for it. In the case of n=1, necessary and sufficient conditions for the surjectivity of $\mu^*:\mathcal{H}(\Omega_2+K) \longrightarrow H(\Omega_2)$ have been given by Napalkov, [20], [21]; μ^* is surjective, for all convex domains $\Omega_2\subseteq\mathbb{C}$, if and only if $\hat{\mu}(z)$ is a function of completely regular growth, [11]. We notice that for $\Omega_1=\Omega_2=\mathbb{C}^n$, any convolutor from $\mathcal{H}(\mathbb{C}^n)$ into itself is always surjective, as it follows from the arguments in theorem 1, by applying Lindelöf theorem (this fact was first proved by Ehrenpreis [6]). Finally we should remark that, for n>1, a different sufficient condition for the surjectivity is provided in [19]. Here Morzhakov proves that, for μ^* to be surjective from $\mathcal{H}(\Omega_2+K)$ onto $\mathcal{H}(\Omega_2)$, it is sufficient that:

i) the regularized radial indicator of $\hat{\mu}$ coincides with H_K (this is trivially satisfied for n=1).

ii) $\hat{\mu}$ is of completely regular growth.

Notice that condition ii) is, in spirit, very similar to (C_2) since both require to consider the restriction of $\hat{\mu}$ to a family of complex lines. In [8], Gruman and Lelong show that if the boundary of Ω_2 is sufficiently regular, and if i) is assumed, then condition ii) is necessary for surjectivity of μ^*; in theorem 2, on the other hand, we provide a proof (based on an idea due to R. Gay) of the necessity of i), at least when the boundary of K is sufficiently regular.

The situation is completely different in the C^∞ case; here a necessary and sufficient condition for a distribution of compact support $\mu\in E'$ to act surjectively from $C^\infty(\Omega_1)$ onto $C^\infty(\Omega_2)$ is that μ is invertible and that (Ω_1, Ω_2) is μ-convex (see [10]).

Finally, in the case of partial differential operators acting on $\mathcal{H}(\Omega)$, for Ω a non convex subset of $\mathbb{C}^n$, no necessary and sufficient conditions for the surjectivity are known. We can only say that, for $n=1$, and connected domains Ω, a differential operator $P(d/dz):\mathcal{H}(\Omega) \longrightarrow \mathcal{H}(\Omega)$ is surjective if and only if Ω is simply connected.

We can now prove:

<u>THEOREM 1</u>. *If Ω_1, Ω_2 and K are as above, and if $\mu\in\mathcal{H}'(K)$ satisfies either (C_1) or (C_2), then $\mu*:\mathcal{H}(\Omega_1) \longrightarrow \mathcal{H}(\Omega_2)$ is surjective.*

Proof. By using standard arguments we know that $\mu*$ is surjective if and only if the adjoint map

$$\mu*:\mathcal{H}'(\Omega_2) \rightarrow \mathcal{H}'(\Omega_1)$$
$$\nu \rightarrow \mu*\nu$$

is injective and has closed range: in fact it is only needed that it has weakly closed range, but since the space $\mathcal{H}(\Omega)$ is a Frèchet nuclear space for any open $\Omega\subseteq\mathbb{C}^n$, the notion of weak closure and of strong closure coincide for subspaces of $\mathcal{H}'(\Omega)$. Take now the Fourier-Borel transform and recall, [15], that for $\nu\in\mathcal{H}'(\Omega)$, Ω any convex open set in $\mathbb{C}^n$, $\hat{\nu}(z)=\langle\nu,\exp\langle z,\xi\rangle\rangle$ is an entire function, and the space $\mathcal{H}'(\Omega)$ is topologically isomorphic, via the Fourier-Borel transform, to $\mathrm{Exp}(\Omega)=\{f\in\mathcal{H}(\mathbb{C}^n)$: there exists $K\subset\Omega$, compact, such that $|f(z)|\leq A\exp(H_K(z))$, for all z in $\mathbb{C}^n\}$; it is then clear that the injectivity of $\mu*:\mathcal{H}'(\Omega_2) \rightarrow \mathcal{H}'(\Omega_1)$ follows. If (C_1) holds, then the proof of the closure of $\mu*\mathcal{H}'(\Omega_2)$ in $\mathcal{H}'(\Omega_1)$ follows as in [24], page 477 (where a similar estimate is used to prove the surjectivity of constant coefficients partial differential operators on $\mathcal{H}(\Omega)$). If (C_2) holds, we simply apply a standard maximum modulus argument, as in [18]. □

REMARK. When $\hat{\mu}$ is an exponential polynomial, i.e. there exist polynomials $(P_j)_{j=1,\dots,N}$ and points $(\lambda_j)_{j=1,\dots,N}$ in $\mathbb{C}^n$ such that

$\hat{\mu}(z) = \sum_{j=1}^{N} P_j(z) \exp \langle\lambda_j, z\rangle$ (in particular, when $\hat{\mu}$ is a polynomial), then (C_1) is satisfied, see [1], [24]; in this case, in fact μ^* is a difference-differential operator. We do not know, at the moment, whether (C_2) implies (C_1).

Another example in which the surjectivity of μ^* follows immediately is when $K=\{0\}$, i.e. when μ^* is a differential operator of infinite order. In this case $\hat{\mu}$ is of minimal type (of exponential type zero) and the closure of $\hat{\mu}\cdot\mathrm{Exp}(\Omega)$ follows from Lindelöf theorem; this example, as well as the example of exponential polynomials, are also a consequence of Morzhakov's conditions, [19], since it is possible to verify that an exponential polynomial is always of completely regular growth. More generally, if $\hat{\mu}$ is the product of an exponential polynomial and of a function of minimal type, when μ^* is surjective.

We now conclude this section by proving the necessity of condition i) for the surjectivity of μ^*; more precisely we have:

THEOREM 2. *Let Ω be a bounded convex set in $\mathbb{C}^n$ and K a (geometrically) strictly convex compact subset of $\mathbb{C}^n$, with C^1 boundary ∂K. Let μ be an analytic functional carried by K, whose Fourier-Borel transform we denote by F, and let Λ_F^* be the radial regularized indicator of F. If $\mu^*: \mathcal{H}(\Omega+K)\to\mathcal{H}(\Omega)$ is surjective, then $\Lambda_F^*\equiv H_K$; this implies, in particular, that μ has a unique support which is equal to K.*

Proof. Since μ is carried by K, then $\Lambda_F^*\le H_K$. Suppose that there exists a point z_0, which we can take on the unit sphere, i.e. with $|z_0|=1$, for which $\Lambda_F^*(z_0) < H_K(z_0)$; by semicontinuity we have that for every small $\varepsilon>0$, there exists a conic neighborhood $V_\varepsilon(z_0)$ of z_0 such that for any $z\in V_\varepsilon(z_0)$, it is

$$\Lambda_F^*(z)<H_K(z)-\varepsilon|z|.$$

Since $H_K(z)=\max_{\zeta\in K} \mathrm{Re}\langle z,\zeta\rangle=\max_{\zeta\in K} (\bar{z}.\zeta)$, we can take ζ in ∂K such that $H_K z_0)=\bar{z}_0.\zeta_0$. Let now β be such that $\Lambda_F^*(z_0)<\beta<H_K(z_0)$ and, for α sufficiently close to $H_K(z_0)$, with $\beta<\alpha<H_K(z_0)$, consider the hyperplane E_α defined by the equation $\bar{z}_0.\zeta=\alpha$ (this hyperplane is parallel to $\bar{z}_0.\zeta=H_K(z_0)$ and is such that $E_\alpha\cap K$ is non-empty). Take now $K_\alpha=K\cap E_\alpha^+$,

where E_α^+ is the half space defined by E_α, and which does not contain ζ_0. Note that for $\zeta \in E_\alpha \cap K$ the tangent hyperplanes are such that the directions defining them are contained in $V_\varepsilon(z_0)$; therefore, for $z \in V_\varepsilon(z)$, we have $H_{K_\alpha}(z)=H_K(z)$. On the other hand, we know that $H_{K_\alpha}(z_0)>\beta>\Lambda_F^*(z_0)$; hence, if we take a small conic neighborhood $V_{\varepsilon'}(z_0)$ (ε', a priori, smaller than ε), we have that

$$z \in V_{\varepsilon'}(z_0) \cap S^{2n-1} \qquad \text{implies} \qquad H_{K_\alpha}(z)>\beta-\varepsilon'/2>\Lambda_F^*(z).$$

If we now make V_ε smaller, and take α sufficiently close to $H_K(z_0)$, we still get

$$H_{K_\alpha}(z)=H_K(z), \text{ for } z \in V_\varepsilon(z_0),$$

and if ε is small enough so that $V_\varepsilon(z_0) \subset V_{\varepsilon'}(z_0)$,

$$H_K(z)>H_{K_\alpha}(z)>\Lambda_F^*(z), \qquad \text{for all } z \in V_\varepsilon(z_0).$$

Thus, by only using the regularity conditions on K, we have shown that, if $\Lambda_F^*(z_0)<H_K(z_0)$, there exists a compact K_1, strictly contained in K, such that

$$\Lambda_F^*(z) \le H_{K_1}(z) \le H_K(z), \qquad \text{for all } z.$$

Consider now a sequence of points z_n in Ω which converge to a point $\tilde{z}_0$ in $\partial\Omega$, and lie on the direction along which K has been cut, to provide K_1. Then $\{\mu*\delta_{z_n}\}_{n\in N}$ is a normal family in $\mathcal{H}'(\Omega+K)$, for δ_{z_n} the Dirac measure in z_n. Indeed

$$|\mu*\delta_{z_n}(z)| = |\hat{\mu}(z)|\exp(\mathrm{Re}<z_n,z>) \le A\exp(\Lambda_F^*(z)+\varepsilon|z|+\mathrm{Re}<z_n,z>) \le$$

$$\le A\exp(H_{K_1}(z)+\varepsilon|z|+\mathrm{Re}<z_n,z>) \le A\exp(H_{K_2}(z)+\varepsilon|z|),$$

where K_2 is a compact convex subset of $\Omega+K$ which, for any n, contains the compact K_1+z_n (such a K_2 certainly exists). Since $\{\mu*\delta_{z_n}\}_{n\in N}$ is a normal family, and since $\mathcal{H}'(\Omega+K)$ is a Montel space, we can extract a convergent subsequence $\{\mu*\delta_{z_{n_p}}\}_{p\in N}$ converging to some $\nu \in \mathcal{H}'(\Omega+K)$. Now $\nu=\mu*S$, because the map $\mu*:\mathcal{H}'(\Omega) \to \mathcal{H}'(\Omega+K)$ has closed range. This forces that $\delta_{z_{n_p}} \to S$ in $\mathcal{H}''(\Omega)$, but then, necessarily, $S=\delta_{\tilde{z}_0}$, which does not belong to $\mathcal{H}''(\Omega)$; this contradiction proves the result. □

REMARK. Notice that the argument of normal families immediately implies that, when $\mu^*:\mathcal{H}(\Omega) \to \mathcal{H}(\Omega)$ is surjective, then K is in fact a support, since one cannot take a smaller carrier, for μ, contained strictly in K.

3. Representation theorem.

This section is devoted to the proof of a representation theorem for holomorphic solutions of suitable systems of convolution equations: as pointed out in the introduction, this strengthens considerably some results obtained by Napalkov, [21], and Morzhakov, [19].

Representation theorems of this kind are well known in the literature (see e.g. Ehrenpreis, [7], for the case of solutions of systems of partial differential equations, and Berenstein and Taylor, [3], for solutions - defined on all of $\mathbb{R}^n$ or $\mathbb{C}^n$ - of suitable systems of convolution equations). Quite recently, [2], a representation theorem has been obtained for solutions, in the space of C^∞ functions on a convex subset of $\mathbb{R}^n$, of systems of convolution equations; this section mainly follows the ideas of this last paper, with the modifications which are necessary since we are dealing with holomorphic functions instead of C^∞ functions; we will often refer the reader to [2], when only trivial changes are needed.

Let Ω_1, Ω_2 and K be as in section 2; consider r convolutors ($1\le r \le n$) $\mu_1,\ldots,\mu_r$ belonging to $\mathcal{H}'(K)$, such that the variety $V=\{z\in\mathbb{C}^n: \hat{\mu}_1(z)=\ldots=\hat{\mu}_r(z)=0\}$ is a complete intersection; we are going to give conditions on the μ_j in order to be able to extend analytic functions with growth conditions from the variety V to all of $\mathbb{C}^n$, with a good control on the bounds. Our first step is the following semilocal to global interpolation theorem.

THEOREM 3. *Let* $\varepsilon>0$ *be sufficiently small,* $A>0$, *and let* λ *be a function analytic on*

$$\tilde{S}(\hat{\mu};\varepsilon,A) = \{z\in\mathbb{C}^n: |\hat{\mu}(z)|=(\sum_{j=1}^{r}|\hat{\mu}_j(z)|^2)^{1/2}\le A\exp(-\varepsilon|z|+H_K(z))\}$$

such that, for some constant A_1 *and some compact* $K_1\subset\Omega_1$, *it is*

$$|\lambda(z)|\le A_1\exp(H_{K_1}(z)), \qquad \textit{for every } z\in\tilde{S}(\hat{\mu};\varepsilon,A)$$

Then there exists an entire function $\tilde{\lambda}$ *such that, for some* $A_2>0$,

$$|\tilde{\lambda}(z)|\le A_2\exp(H_{K_1}(z)), \qquad \textit{for every } z\in\mathbb{C}^n.$$

Moreover, there are analytic functions $\alpha_1,\ldots,\alpha_r$ *such that*

$$\tilde{\lambda}(z)=\lambda(z)+\sum_{j=1}^{r}\alpha_j(z)\hat{\mu}_j(z), \qquad \textit{for every } z\in\tilde{S}(\hat{\mu};2\varepsilon,A),$$

and such that, for suitable $A_3>0$ *and* K_2, *convex compact in* Ω_2,

$$|\alpha_i(z)|\le A_3\exp(H_{K_2}(z)) \qquad \text{for } z\in\tilde{S}(\hat{\mu};2\varepsilon,A) \textit{ and } i=1,\ldots,r.$$

Proof. We only sketch the proof, since it follows the well known pattern of [2], [3]. Since for every (i,j) and every $\varepsilon>0$ there exists $A_\varepsilon>o$ such that

$$|\partial\hat{\mu}_j(z)/\partial z_i|\le A_\varepsilon\exp(H_K(z)+\varepsilon|z|),$$

by using the mean value theorem and the argument in [3], one shows that there exist positive constants A', A" such that the distance from $\tilde{S}(\hat{\mu};2\varepsilon,A)$ to $\tilde{S}(\hat{\mu};\in,A')$ is at least $A"\exp(-\varepsilon|z|)$. We can then construct a cut-off function $\chi\in C^\infty$, $0\le\chi\le1$, $\chi\equiv1$ on $S(\mu;2\varepsilon, A)$, $\chi\equiv0$ on a neighborhood of the complement of $\tilde{S}(\hat{\mu}_i;\varepsilon,A')$ and

$$|\bar{\partial}\chi(z)|\le A'''\exp(-\varepsilon|z|) \qquad (\text{for some } A'''>0).$$

Then $\chi\lambda\in C^\infty(\mathbb{C}^n)$ and $w=\lambda\bar{\partial}\chi\in C^\infty_{(0,1)}(\mathbb{C}^n)$ is $\bar{\partial}$-closed. Using Hörmander's results [9], as in [3], one can find $\bar{\partial}$-closed forms $w_1,\ldots,w_r$ such that $w=w_1\hat{\mu}_1+\ldots+w_r\hat{\mu}_r$, with good L^2-growth control on the w_i; the usual $\bar{\partial}$-technique finally shows that we can solve $\bar{\partial}\alpha_i=w_i$ with bounds, and we take $\lambda=\lambda+\Sigma\alpha_i\hat{\mu}_i$. This is the desired function. □

To obtain a representation theorem, we first start considering the discrete case, i.e. r=n. We need the following

DEFINTION 4. An n-tuple $\mu=(\mu_1,\ldots,\mu_n)$ of analytic functionals carried by K is said to be slowly decreasing with respect to $(\Omega_1=\Omega_2+K,\Omega_2)$ if there exists a constant C>0 and $m\in N$ such that, for all $z\in\mathbb{C}^n$,

$$|\hat{\mu}_i(z)|\ge c(\tilde{d}(z,V_i))^m\exp(H_K(z))(1+|z|)^{-m}$$

for i=1,...,m, where $V_i=\{z\in\mathbb{C}^n\colon \hat{\mu}_i(z)=0\}$, and $\tilde{d}(z,V_i)=$

min(1, distance from z to V_i),

$$|\hat{\mu}(z)| \geq c(\tilde{d}(z,V))^m \exp(H_K(z))(1+|z|)^m,$$

and for every $\varepsilon>0$ there are constants A, B>0 such that the set

$$S(\hat{\mu};\varepsilon,A)=\{z\in\mathbb{C}^n:\tilde{d}(z,V_i)\leq A\exp(-\varepsilon|z|), \qquad 1\leq i\leq m\}$$

has relatively compact connected components such that for any z_1 and z_2 belonging to the same component of $S(\hat{\mu};\varepsilon,A)$ we have

$$H_{K_2}(z_1)\leq H_{K_2}(z_2)+B,$$

for all compact convex $K_2\subset\Omega_2$ (this is obviously satisfied when the components have uniformly bounded diameters).

The reader interested in the different role played by the families of sets S and $\tilde{S}$ is referred to [2] where all the details are given.

REMARK. Clearly any slowly decreasing n-tuple will be showly decreasing for <u>all</u> pairs (Ω_1, Ω_2) with $\Omega_1=\Omega_2+K$ when the components of $S(\hat{\mu};\varepsilon,A)$ have uniformly bounded diameters. This observation also applies to the definition in [2] as well as to the notion of slowly decreasing which will be given for r-tuples, $1\leq r\leq n$.

Using the Jacobi interpolation formulas from [3] and the hypotheses above, one can prove, in the usual fashion, the following

<u>PROPOSITION 5</u>. *Let $\mu=(\mu_1,\ldots,\mu_n)$ be slowly decreasing with respect to (Ω_2+K,Ω_2), and let λ be holomorphic in $S(\hat{\mu};\varepsilon,A)$; if $\lambda(z)=f_1(z)\hat{\mu}_1(z)+\ldots+f_n(z)\hat{\mu}_n(z)$ for $f_1,\ldots,f_n$ analytic in $S(\hat{\mu};\varepsilon.A)$, and if there exist $K_1\subset\Omega_1$, convex and compact, and $A_1>0$ such that $|\lambda(z)|\leq A_1\exp(H_{K_1}(z))$ on $S(\hat{\mu};\varepsilon,A)$, then, on a smaller set $S(\hat{\mu};\varepsilon',A')$, we can find holomorphic functions a_i such that*

$$\lambda(z)=\sum_{i=1}^{n}a_i(z)\hat{\mu}_i(z) \text{ on } S(\hat{\mu};\varepsilon',A')$$

and

$$|a_i(z)|\leq A''\exp(H_{K_2}(z)), \qquad \text{for } z\in S(\hat{\mu};\in',A')$$

where K_2 is some convex compact in Ω_2.

The representation theorem we are interested in is (see [2], [3], [7], [17], [23]), a consequence of a division and an extension theorem with bounds; as far as the extension is concerned, this is an inmediate consequence of the Jacobi interpolation formula and of the semilocal to global interpolation theorem (theorem 3). The division theorem, on the other hand, is essentially a consequence of proposition 5 (which is a semilocal division result), and the usual $\bar{\partial}$-techniques. Therefore we we get:

<u>THEOREM 6</u>. *Suppose that* $\lambda \in \mathrm{Exp}(\Omega_1)$ *belongs to the ideal generated in* $\mathcal{H}(\mathbb{C}^n)$ *by the slowly decreasing* n*-tuple* $(\hat{\mu}_1, \ldots, \hat{\mu}_n)$*; then there are analytic functions* $\gamma_1, \ldots \gamma_n$ *in* $\mathrm{Exp}(\Omega_2)$ *such that* $\lambda = \gamma_1 \hat{\mu}_1 + \ldots + \gamma_n \hat{\mu}_n$.

We are now able to state the main representation theorem for functions which are solutions of a system of convolution equations, in the case in which the number of equations is equal to n.

<u>THEOREM 7.</u> *Let* $\mu = (\mu_1, \ldots, \mu_n)$ *be an* n*-tuple of analytic functionals carried by a convex compact set* K*. If* μ *is slowly decreasing for the pair* $(\Omega_1 = \Omega_2 + K, \Omega_2)$*, then every function of* $\mathcal{H}(\Omega_1)$*, solution of the system*

$$\mu_1 * f = \ldots = \mu_n * f = 0$$

can be represented in the form

$$f(z) = \sum_{k=1}^{+\infty} \sum_{j=1}^{J_k} P_{k,j}(z) \exp\langle \alpha_{k,j}, z\rangle$$

where $(\alpha_{k,j})_{k,j}$ *are the common zeroes of the* $\hat{\mu}_j$*, i.e.* $V = \{z \in \mathbb{C}^n : \hat{\mu}_1(z) = \ldots = \hat{\mu}_n(z) = 0\} = \bigcup_{k=1}^{+\infty} \{\alpha_{k,j} : j = 1, \ldots, J_k < \infty\}$ *and where the polynomials* $P_{k,j}$ *are such that the functions* $P_{k,j}(z)\exp\langle \alpha_{k,j}, z\rangle$ *are solutions of the given system. The series converges in* $\mathcal{H}(\Omega_1)$.

Proof. The same as in [3] or [2]. □

In order to extend theorem 7 to the general case, i.e. to the case in which the variety V is non-discrete, one needs to be more careful ; indeed the Jacobi interpolation formulas on which proposition 5 and theorem 6 are based, only hold on discrete varieties. To overcome this difficulty, Berenstein and Taylor, in [3], introduced a notion of slow decrease for r-tuples $\mu=(\mu_1,\ldots, \mu_r)$ which requires the cutting of the variety V with a suitable family of r-dimensional complex planes (such families are called analytic almost parallel). These complex planes are used to reduce the variety V to discrete portions, to which one can apply the division and extension theorems mentioned before; of course one then needs to be able to paste together these local results, and this is achieved with the use of the Koszul complex associated to the map induced by multiplication by $(\hat{\mu}_1,\ldots,\hat{\mu}_r)$. We will not enter the technical details of these notions, since they have now become standard (see, e.g., [2], [3]), and the reader is referred to [2] for the precise description which is needed in our situation (with obvious modifications). For the sake of brevity, and since no new ideas need to be introduced, we will only give the main definition and its most relevant consequences:

DEFINITION 8. An r-tuple $\mu=(\mu_1,\ldots,\mu_r)$ of analytic functionals carried by a compact $K\subset\mathbb{C}^n$ is said to be slowly decreasing (with respect to the pair (Ω_2,Ω_2) if there exists an analytic almost parallel family $\mathcal{L}=\{L\}$ of r-dimentional complex planes covering $\mathbb{C}^n$ such that, for some constants C, m>0, it is, for all $L\in L$,

$$|\hat{\mu}_i(z)|\geq C(\tilde{d}(z,V_i\cap L))^m\exp(H_K(z))(1+|z|)^{-m}, \text{ for } z\in L,$$

as well as

$$|\hat{\mu}(z)|\geq C(\tilde{d}(z,V\cap L)^m\exp(H_K(z))(1+|z|)^{-m}, \text{ for } z\in L.$$

Moreover, for every $\varepsilon>0$, there are constants A, B>0 such that, for all $L\in\mathcal{L}$, the set

$$\mathcal{O}_L(\hat{\mu};\varepsilon,A) = \{z\in L:\tilde{d}(z,V_i\cap L)\leq A\exp(-\varepsilon|z|) \quad \text{for all } i=1,\ldots,r\}$$

has relatively compact connected components, and for any z_1 and z_2 belonging to the same component of $\mathcal{O}_L(\hat{\mu};\varepsilon,A)$, we have

$$H_{K_2}(z_1)\leq H_{K_2}(z_2)+B.$$

By arguing as in [2], one can show that the space $\hat{\mathcal{H}}'(\Omega_1)/(\hat{\mu}).\hat{\mathcal{H}}'(\Omega_2)$ (whose dual is topologically isomorphic to the space of solutions, in $\mathcal{H}(\Omega_1)$, to the system $\mu_1*f=..=\mu_r*f=0$) can be identified with a suitable space of cochains with good bounds and coefficients in "good" open sets (see [2], Theorems 10 and 11). In particular we have:

THEOREM 9. (Division theorem) *Suppose that* $\lambda\in\text{Exp}(\Omega_1)$ *belongs to the ideal generated in* $\mathcal{H}(\mathbb{C}^n)$ *by the slowly decreasing r-tuple* μ. *Then there are analytic functions* $\gamma_1,\ldots,\gamma_r$ *in* $\text{Exp}(\Omega_2)$ *such that* $\lambda=\sum_{i=1}^{r}\gamma_i\hat{\mu}_i$.

THEOREM 10. *Let* $\mu=(\mu_1,\ldots,\mu_r)$ *be slowly decreasing with respect to the analytic almost parallel family* $\mathcal{L}$. *Let* λ *be analytic on* V, *the multiplicity variety associated to* μ. *Assume that, for some constant* $A>0$, *some compact* $K_1\subset\Omega_1$ and for every $L\in\mathcal{L}$, *there exists a function* $\tilde{\lambda}_L$, *analytic on* L, *such that*

i) $|\tilde{\lambda}_L(z)|\leq A\exp(H_{K_1}(z))$ *for all* $z\in L$

ii) the restriction (with multiplicities) of $\tilde{\lambda}_L$ *to* $V\cap L$ *is equal to the restriction of* λ *to* $V\cap L$.

Then there exists $\tilde{\lambda}$, *entire, such that its restriction to* V *coincides with* λ, *and which satisfies, for some other constant* $A_1>0$, *and some other compact* $K_1'\subset\Omega_1$,

$$|\tilde{\lambda}(z)|\leq A_1\exp(H_{K_1'}(z)) \qquad \textit{for all } z\in\mathbb{C}^n.$$

These interpolation results, together with the identification of $\hat{\mathcal{H}}'(\Omega_1)/(\hat{\mu}).\hat{\mathcal{H}}'(\Omega_2)$, are sufficient to recover a representation theorem for slowly decreasing systems:

THEOREM 11. *Suppose* $\mu=(\mu,\ldots,\mu_r)$ *is a slowly decreasing r-tuple for the pair* (Ω_1,Ω_2). *Then there exists a locally finite family of closed sets* $\{V_j : j\in J\}$, *a partition of* J *into finite subsets* J_k, *and partial differential operators* ∂_j *on* $\mathbb{C}^n$, *such that:*

i) $\bigcup_{j\in J} V_j\subset V=\{z\in\mathbb{C}^n:\ \hat{\mu}_1(z)=\ldots=\hat{\mu}_r(z)=0\}$

ii) each function $x\to\partial_j(e^{ix.z})$, *with* $z\in V_j$, *is a solution of* $\mu_1*f=\ldots=\mu_r*f=0$

iii) to each solution $f\in\mathcal{H}(\Omega_1)$ *of* $\mu_1*f=\ldots=\mu_r*f=0$. *there corresponds a family* (ν_j) *of complex valued Borel measures, with* ν_j *supported in* V_j, *such that, with convergence in the topology of* $\mathcal{H}(\Omega_1)$,

One might notice that condition (1) actually implies the discreteness of the variety V (see [4] for a proof of this result) and hence one might suspect that this requirement could be much stronger than necessary; this is not the case: indeed we can prove, with the use of an argument already employed in [17], that the extension property mentioned before actually implies (1), and hence the discreteness of V. We have:

THEOREM 12. *Let* $U \subset \mathbb{R}^n$ *be a convex set and let* $\mu=(\mu_1,\ldots,\mu_r)$, $1 \le r \le n$, *be a slowly decreasing* r*-tuple of compactly supported distributions with* $supp(\mu_i)=T$, $i=1,\ldots,r$, *for* T *a compact subset of* $\mathbb{R}^n$. *Then every solution* $f \in C^\infty(U+T)$ *of* $\mu_1 * f = \ldots = \mu_r * f = 0$ *extends to a solution* $\tilde{f} \in C^\infty(\mathbb{R}^n)$ *of the same system, if and only if there is a constant* $A>0$ *such that, for all* $z \in V$, *(1) holds. In particular, the extension property may hold only if* $r=n$ *and* V *is discrete.*

Proof. The sufficiency of (1) (remember that slowly decreasing r-tuples define complete intersection varieties) is proved in [2]; we only have to prove necessity. Denote by $C^\infty_\mu(\mathbb{R}^n)$ and $C^\infty_\mu(U+T)$ the space of C^∞ solutions (in $\mathbb{R}^n$ and in U+T) to $\mu_1 * f = \ldots = \mu_r * f = 0$, and consider the map (restriction)

$$R: C^\infty_\mu(\mathbb{R}^n) \to C^\infty_\mu(U+T).$$

This map is surjective if and only if its transpose

$${}^tR: (C^\infty_\mu(U+T))' \to C^\infty_\mu(\mathbb{R}^n))'$$

is injective and of closed range. But, taking the Fourier transform, this is the same as

$${}^t\hat{R}: (\widehat{C^\infty(U+T)'})/(\hat{\mu}).(\widehat{C^\infty(U)'}) \to (\widehat{C^\infty(\mathbb{R}^n)'})/(\hat{\mu}).(\widehat{C^\infty(\mathbb{R}^n)'}).$$

It is obvious (using the density of polynomials) that ${}^t\hat{R}$ has dense range, while the injectivity follows from the division theorems of [2], [3], i.e. the fact that the local ideal general by $(\hat{\mu})$ is (in both spaces) equal to the global ideal. Therefore R is surjective (i.e. the extension property holds) if and only if $\varphi={}^t\hat{R}$ is surjective, that is (since injectivity has been proved) R is surjective if and only if φ is a continuous bijection. By Banach theorem (which can be applied since we are dealing with Fréchet spaces), also φ^{-1} is

$$f(z) = \sum_{k=1}^{+\infty} \Big(\sum_{j\in J_k} \int_{V_j} \partial_j (e^{iz.\zeta}) d\nu_j(\zeta) \Big).$$

REMARK. We conclude this section by pointing out that the same arguments used in [3], section 7, and [2], section 4, show that any single exponential polynomial satisfies the hypotheses of theorem 11 (thus providing a representation theorem for holomorphic - non entire- functions solutions of difference-differential equations), and that most n-tuples of exponential sums satisfy the hypotheses of theorem 7.

4. Hyperbolicity.

In this section we deal with the following natural extension problem:

Let $f\in\mathcal{H}(\Omega+K)$ and suppose that, for some slowly decreasing r-tuple $\mu=(\mu_1,\ldots,\mu_r)$ of analytic functionals carried by K, it is $\mu_1*f=\ldots=\mu_r*f=0$. Which geometrical conditions must be asked on the characteristic variety $V=\{z\in\mathbb{C}^n:\mu_1(z)=..=\mu_r(z)=0\}$ in order that f extends to an entire function $\tilde{f}$ satisfying the same system of convolution equations?

As pointed out by Ehrenpreis in [7], such a question, which can also be asked for spaces other than $\mathcal{H}$, amounts to hyperbolicity requirement on the system defined by μ, and its solution depends on the possibility of proving a representation theorem for the solutions of such a system (see chapter VIII of [7], where Ehrenpreis deals with the case of linear constant coefficients differential operators).

In [2], a similar problem, for the space of C^∞ functions on subsets of $\mathbb{R}^n$, is dealt with; since we have something to add to what is proved in [2], we recall their result: let $U\subseteq\mathbb{R}^n$ be a convex set and $\mu=(\mu_1,\ldots,\mu_n)$ a slowly decreasing n-tuple of compactly supported distributions, with $supp(\mu_i)=T$, $i=1,\ldots,n$, where T is a compact subset of $\mathbb{R}^n$; if there exists a constant A>0 such that

(1) $$|\operatorname{Im} z| \le A(1+\log(1+|z|))$$

for all z in V, then every solutions $f\in C^\infty(U+T)$ of $\mu_1*f=\ldots=\mu_n*f=0$ extends to a solution $\tilde{f}\in C^\infty(\mathbb{R}^n)$ of the same system.

continuous, and takes bounded sets into bounded sets. Consider now a compact $S\subset\mathbb{R}^n$; then the family of exponentials $\{e_x\}_{x\in S}$, with $e_x=\hat{\delta}_x$ is a bounded set in $(\widehat{C^\infty(\mathbb{R}^n)'})$, and hence (if $\dot{e}_x$ denotes the equivalence class of e_x in the quotient space) also $\{\varphi^{-1}(\dot{e}_x)\}_{x\in S}$ is bounded in $(\widehat{C^\infty(U+T)'})\ (\hat{\mu}).(\widehat{C^\infty(U)'})$. This shows that, for every $x\in S$, there exists $g_x\in(\widehat{C^\infty(U+T)'})$ such that g_x-e_x belongs to $(\hat{\mu}).(\widehat{C^\infty(\mathbb{R}^n)'})$, and such that $\{g_x\}_{x\in S}$ is bounded in $(\widehat{C^\infty(U+T)'})$. This means that, for any compact subset T_1 of U, there exists m such that

$$|g_x(z)|\le A(1+|z|)^m\exp(H_{T_1}(\mathrm{Im}\ z)+H_T(\mathrm{Im}\ z)).$$

But, for z in V, one has

$$|e^{ix.z}|\le A(1+|z|)^m\ \exp(H_{T_1+T}(\mathrm{Im}\ z)).$$

This shows that, for $z\in V$,

$$H_S(\mathrm{Im}\ z)\le C+H_{T_1+T}(\mathrm{Im}\ z)+m\ \log(1+|z|).$$

By taking S suitably large, for example $S=B(0,R+T_1+T)$, one then obtains the first part of the theorem. Finally, the fact that (1) implies r=n and V discrete is proved in [4]. □

The same technique can be employed to treat the corresponding problem in the case of holomorphic solutions of systems of convolution equations. In this case we obtain:

THEOREM 13. *Let* Ω *be a convex subset of* $\mathbb{C}^n$ *and let* $\mu=(\mu_1,\ldots,\mu_r)$, $1\le r\le n$, *be a slowly decreasing* r-*tuple of analytic functionals carried by the same compact* $K\subset\mathbb{C}^n$. *Then every solution* $f\in\mathcal{H}(\Omega+K)$ *of* $\mu_1*f=\ldots=\mu_r*f=0$ *extends to a solution* $\tilde{f}\in\mathcal{H}(\mathbb{C}^n)$ *of the same system, if and only if the characteristic variety* V *has only a finite number of points. In particular the extension property may hold only if* r=n.

Proof. The sufficiency of the condition is a trivial consequence of the representation theorem (theorem 7), which shows that every solution is, if V is a finite set, a finite sum of exponential polynomials, and hence entire. The proof of the necessity of the condition, on the other hand, is similar to the proof of theorem 12, keeping into account the different topological structure of $\mathcal{H}(\Omega+K)$; more precisely, if we follow the proof of theorem 12 we obtain, in our case, that condition (1) has to be substituted by $|z|\le C$ for all $z\in V$, which clearly concludes the proof. □

REMARK. We wish to conclude this section by noticing that the arguments used in theorem 12 and 13 are also applicable to the study of other, similar, situations. For example one might study, with the tools developed in [2], the space of solutions, in $\mathcal{D}'(U+T)$, of a given system of convolution equations ($\mathcal{D}'$denotes, as usual, the space of Schwartz distributions), and it would be natural to ask for conditions on the system under which any distribution solution would actually become a C^∞ solution. It is completely clear that the methods described before, would enable to write explicit conditions for this to happen.

5. <u>Extension of solutions</u>.

In this last section we deal with more sophisticated extension problems. Let Ω be a convex open set of $\mathbb{C}^n$ and let K_0 be a convex compact set $\mathbb{C}^n$. We want to find, as in [12], [22] good conditions in order that any solution of the equation $\mu*f=0$ defined in the set $\Omega+K_0$ can be extended to a larger set, this set being defined by the asymptotic cone of the zero set of $\hat{\mu}$. We follow the main ideas of [12],[22] and we adopt the same notations, with the exception of Int A, which will denote the interior of the set A. Let us briefly recall the enveloping map defined by a set: let $U\subseteq\mathbb{C}^n$ be open and convex, denote by $\mathcal{K}(U)$ the family of non-empty compact convex sets contained in U; if $A\subseteq\mathbb{C}^n$ is non-empty, and $K\in\mathcal{K}(\mathbb{C}^n)$, we define

$$\gamma_A(K) = \{z\in\mathbb{C}^n\colon \operatorname{Re}\langle z,\zeta\rangle\le H_K(\zeta), \text{ for all } \zeta\in A\}$$

and

$$\Gamma_A(U) = \bigcup_{K\in\mathcal{K}(U)} \operatorname{Int}\gamma_A(K).$$

In [12] it is shown that if A is a closed cone, then, for every open convex set U, we have

$$\Gamma_A(U) = \operatorname{Int}\{z\in\mathbb{C}^n\colon \operatorname{Re}\langle z,\zeta\rangle\le H_U(\zeta)\}.$$

Another ingredient we need is the θ-map; for A as before and $r\in\mathbb{R}^+$, define

$$\theta_{A,r}(K) = \{z\in\mathbb{C}^n: \operatorname{Re}\langle z,\zeta\rangle\le H_K(\zeta)+r\},$$

$$\theta_A(K) = \overline{\bigcup_{r\ge 0}\theta_{A,r}(K)},$$

and

$$\theta_A(U) = \bigcup_{K\in\mathcal{K}(U)} \operatorname{Int}\theta_A(K).$$

We recall that if A is an unbounded set of $\mathbb{C}^n$, then the asymptotic cone $\alpha(A)$ of A is the real cone generated by zero and the adherence of the set of sequences $z_j/|z_j|$, with $z_j\in A$ and $\lim_{j\to+\infty}|z_j|=+\infty$. If A is unbounded in $\mathbb{C}^n$ then, for every convex set $U\subseteq\mathbb{C}^n$, we have, [12],

$$\Gamma_{\alpha(A)}(U) = \theta_A(U).$$

REMARK. In the extension theorem which will follow (theorem 15), we will use the set $\Gamma_A(\Omega)+K_0$; notice that, even if A is a closed cone, it is not true, in general, that $\Gamma_A(\Omega+K_0)=\Gamma_A(\Omega)+K_0$. In fact we only have the inclusion $\Gamma_A(\Omega)+K_0\subseteq\Gamma_A(\Omega+K_0)$. To prove it, let $z\in\Gamma_A(\Omega)+K_0$; then $z=z_1+z_2$ with $z_2\in K_0$, and there exists $\varepsilon>0$ such that the open ball $B(z_1,\varepsilon)$ is contained in $\{s\in\mathbb{C}^n: \operatorname{Re}\langle s,\zeta\rangle\le H_\Omega(\zeta)$, for all $\zeta\in A\}$. Let now $\xi\in B(z,\varepsilon)$; then $\xi=t+z_2$, with $t\in B(z_1,\varepsilon)$. Hence, for $\zeta\in A$, we have $\operatorname{Re}\langle\xi,\zeta\rangle=\operatorname{Re}\langle z_2,\zeta\rangle+\operatorname{Re}\langle t,\zeta\rangle\le H_{K_0}(\zeta)=H_{\Omega+K_0}(\zeta)$, i.e. $z\in\Gamma_A(\Omega+K_0)$. The converse inclusion is not true in general; indeed this depends strongly on the relation between A and K_0: so, for example, one can show that $\Gamma_A(\Omega)+K_0=\Gamma_A(\Omega+K_0)$ when A contains only one or two directions, or when $A=\mathbb{R}\cup i\mathbb{R}^+$, $K_0=[0,1]\subset\mathbb{R}$ and Ω is the open unit disk in $\mathbb{C}$. That the equality fails in general is shown by the following counterexample: take $A=\mathbb{R}\cup i\mathbb{R}$, Ω the open unit disk in $\mathbb{C}$ and $K_0=\bar\Omega$; then $\Gamma_A(\Omega)=\{z\in\mathbb{C}: |\operatorname{Re} z|<1, |\operatorname{Im} z|<1\}$, while $\Gamma_A(\Omega+K_0)=\{z\in\mathbb{C}: |\operatorname{Re} z|<2, |\operatorname{Im} z|<2\}$, therefore $\Gamma_A(\Omega+K_0)\ne\Gamma_A(\Omega)+K_0$.

Before proving the extension theorem we need a division result for which, as it happened in [22] with $K_0=\{0\}$, we need some technical conditions.

Let $K\subset L$ be two compact convex sets of $\mathbb{C}^n$ and let $f\in \operatorname{Exp}(K_0)$; we say that the function f satisfies condition (P) if for every $\varepsilon>0$, there exists $R_\varepsilon>0$ and $C_\varepsilon>0$ such that for every $z\in\mathbb{C}^n$ such that $|f(z)|\le 1$ and $z\notin B(0,R_\varepsilon)$ then $d(z,W_f)\le C_\varepsilon+(\varepsilon|z|-H_{K_0}(z))/(\alpha_L+\alpha_K+4\alpha_{K_0})$, where W_f is the zero set of f, and $\alpha_A=\sup|\zeta|$.

We will also need another condition, on f, namely

(*) $z\in\mathbb{C}^n$ and $|f(z)|\geq 1/2$ imply $|f(z)|\exp(-H_{K_0}(z))\geq \mathrm{Const}.$

REMARK. Condition (P) is the same as in [22], with $C_\varepsilon=0$ and condition (*) is not needed in case $K_0=\{0\}$; notice also that (P) implies that $0\in\partial K_0$.

We now have the following theorem:

THEOREM 14. *Let* f *be a function of* $\mathrm{Exp}(K_0)$. *Let* $K\subset L$ *be two convex compact sets in* $\mathbb{C}^n$ *such that for every* $z\in W_f$ *we have*

$$H_L(z)\leq H_K(z)+b\log(1+|z|).$$

If we suppose that the function f *satisfies condition (P) relatively to* K *and* L *and condition (*), then for every entire function* f_1 *satisfying*

$$\log|f_1(z)|\leq H_L(z)+H_{K_0}(z)+a_1\log(1+|z|) \qquad \text{for all } z\in\mathbb{C}^n$$

and for every $\varepsilon>0$, *we can find two entire functions* f_2 *and* f_3 *and two positive constants* $a_2(\varepsilon)$ *and* $a_3(\varepsilon)$ *(which do not depend on* f_2 *and* f_3*) such that*

$$f_1=ff_2+f_3.$$

and for every $z\in\mathbb{C}^n$ *we have:*

$$\log|f_2(z)|\leq H_{K_\varepsilon}(z)+a_2(\varepsilon)\log(1+|z|)$$

and

$$\log|f_3(z)|\leq H_{K_\varepsilon}(z)+H_{K_0}(z)+a_3(\varepsilon)\log(1+|z|)$$

(Here, and in the sequel, for $\varepsilon>0$, L_ε *and* K_ε *denote the* ε*-neighborhoods of* L *and* K *respectively.)*

Proof. We only sketch the proof, quoting the main differences from the proofs in [12] and [22]. Let ϕ be a standard function, i.e. $\phi\in\mathcal{D}(\mathbb{C})$, $0\leq\phi\leq 1$, and assume that $\phi(\tau)=1$ if $|\tau|\leq 1/2$, and $\phi(\tau)=0$ if $|\tau|\geq 1$.

We now look for a function v such that the required functions f_2 and f_3 can be written as

$$f_2=v+f_1(1-\phi of)/f$$

and

$$f_3=f_1(\phi of)-fv$$

The function f_2 is well defined and, in order for it to be entire, we see that $\bar{\partial}v=w$ must be a closed (0,1)-form. Using the fact that $\partial f=(\partial f/\partial z_1,\ldots,\partial f/\partial z_n)$ belongs to $(Exp(K_0))^n$, the hypotheses (P) and the inequality linking K to L on the zero set W_f, we can find a C^∞ function v with good L^2 growth conditions. As usual we can then replace the L^2 norms with sup norms in the inequality for v and this directly yields the desired inequality for f_3. The same procedure applies to f_2, where, now, we need to use (*). □

We are now ready to get the extension theorem: we will need to assume that the space of exponential polynomial solutions to $\mu*f=0$ is dense in the space $\mathcal{H}_\mu(\Omega+K_0)$ of all solutions, analytic in $\Omega+K_0$, to the same equation. This is always true when n=1, [13]; for $n\geq 2$ we have theconditions i), ii) of Morzhakov we quoted in section 2 (this contains the case $K_0=\{0\}$, which was already known to Martineau, [16]). Finally, this is always true for some special domains Ω, when $n\geq 2$, [21].

<u>THEOREM 15</u>. *Let K_0 be a compact convex set of $\mathbb{C}^n$, and Ω be an open set in $\mathbb{C}^n$. Let $\mu\in\mathcal{H}'(K_0)$, then $\mu*$ is a convolutor from $\mathcal{H}(\Omega+K_0)$ into $\mathcal{H}(\Omega)$. Suppose that μ satisfies conditions (P) and (*) for any compact convex sets K and L, $K\subset L\subset\Omega$, and that the space of exponential polynomial solutions to $\mu*f=0$ is dense in $\mathcal{H}_\mu(\Omega+K_0)$. Then every function $u\in\mathcal{H}_\mu(\Omega+K_0)$ can be extended to a function v holomorphic in $\Gamma_{\alpha(W_f)}(\Omega)+K_0\mu$ and solution of $\mu*v=0$.*

Proof. Consider, with the obvious notations, the restriction

$$R:\mathcal{H}_\mu(\Gamma_{\alpha(W_f)}(\Omega)+K_0)\rightarrow\mathcal{H}_\mu(\Omega+K_0);$$

we only need to prove that the transpose map

$${}^tR:\mathcal{H}'_\mu(\Omega+K_0)\rightarrow\mathcal{H}'_\mu(\Gamma_{\alpha(W_f)}(\Omega)+K_0)$$

is injective and with dense image. But the injectivity of ${}^tR$ is equivalent to the density of Im R in $\mathcal{H}_\mu(\Omega+K_0)$, which follows, as usual, from the density of exponential polynomial solutions in $\mathcal{H}_\mu(\Omega+K_0)$. Finally the density of Im ${}^tR$ can be proved by showing that ${}^tR$ is surjective, as in [22], with the use of the division theorem (theorem 14). □

REFERENCES

[1] Berenstein, C.A. and Dostal, M., A lower estimate for exponential polynomials, Bull. A.M.S. 80 (1974), 687-691.

[2] Berenstein, C.A. and Struppa, D.C., Solutions of convolution equations in convex sets, to appear in Am.J.Math.

[3] Berenstein, C.A. and Taylor, B.A., Interpolation problems in $\mathbb{C}^n$ with applications to harmonic analysis, Journ. An.Math. 38 (1980), 188-254.

[4] Berenstein, C.A. and Yger, A., Ideals generated by exponential polynomials, to appear in Adv. in Math.

[5] Ehrenpreis, L., Mean periodic functions I, Am. J. Math. 77 (1955), 293-328.

[6] Ehrenpreis, L., Solutions of some problems of division III, Am. J. Math. 78 (1956), 685-715.

[7] Ehrenpreis, L., Fourier Analysis in Several Complex Variables, Wiley-Interscience, New York, 1970.

[8] Gruman, L. and Lelong, P., Entire Functions and Several Complex Variables, Grundl. Math. Wiss. 282,Springer Verlag, New York-Heidelberg-Berlin, 1986.

[9] Hörmander, L., Complex Analysis in Several Variables, North Holland-American Elsevier, New York, 1973.

[10] Hörmander, L., The Analysis of Linear Partial Differential Operators II, Springer Verlag, Berlin, 1983.

[11] Levin, B., Zeroes of Entire Functions, Translations of the A.M.S., Providence, 1980.

[12] Kiselman, C.O., Prolongement des solutions d'une èquation aux dèrivèes partielles àcoefficients constants, Bull. Soc. Math. France 97 (1969), 329-354.

[13] Krasickov-Ternovskii, I.F., Invariant subspaces of analytic functions, II. Spectral synthesis on convex domains, Mat. Sb. 88 (130), no. 1 (1972), 3-30.

[14] Malgrange, B., Existence et approximation des solutionsdes équations aux derivées partielles et des équations de convolution, Ann. Inst. Fourier 6(1955), 271-354.

[15] Martineau, A., Sur les fonctionnelles analytiques et la transformation de Fourier-Borel, Journ. An. Math. 11 (1963),1-164.

[16] Martineau, A., Equations differentielles d'ordre infini, Bull. Soc. Math. France 95 (1967), 109-154.

[17] Meril, A., Problèmes d'interpolation dans quelques espaces de fonctions non entières, Bull. Soc. Math. France 111 (1983), 251-286.

[18] Meril, A. et Struppa, D.C., Phènomène de Hartogs et equations de convolution, preprint Université de Bordeaux I, 1984.

[19] Morzhakov, V.V., Convolution equations in spaces of functions holomorphic in convex domains and on convex compacta in $\mathbb{C}^n$, Matem. Zametki 16, no. 3 (1974), 431-440.

[20] Napalkov, V.V., On one class of inhomogeneous convolution type equations, Usp. Mat. Nauk, 29, no. 3 (1974), 217-218.

[21] Napalkov, V.V., Convolution equations in multidimensional spaces, Matem. Zametki, 25, no. 5 (1979), 761-774.

[22] Sebbar, A., Prolongement des solutions holomorphes de certains opèrateurs diffèrentiel d'ordre infini à coefficients constants. Seminaire Lelong-Skoda, Springer Lecture Notes in Mathematics 822 (1980), 199-220.

[23] Struppa, D.C., The fundamental principle for systems of convolution equations, Memoirs of the A.M.S., 41, no. 273 (1983).

[24] Treves, F., Linear Partial Differential Equations with Constant Coefficients, Gordon and Breach, New York-London-Paris, 1966.

Authors' addresses:

Alex Meril
Department de Mathematiques
Universitè de Bordeaux I
Lab.Associé an CNRS U.A.226
33405 TALENCE
FRANCE

Daniele C. Struppa
Scuola Normale Superiore
Piazza dei Cavalieri, 7
56100 PISA
ITALY

A REMARK ON "CONVOLUTORS IN SPACES OF HOLOMORPHIC FUNCTIONS"

by

Carlos A. Berenstein and Daniele C. Struppa

Theorem 12 in [8] can be somewhat refined if one considers a suitably weakened definition of hyperbolicity, which goes back to the classical work of Ehrenpreis [6]. For $t > 0$, let $B(t)$ denote the open ball centered at the origin and of radius t in R^n, and let E denote the topological vector space of C^∞ functions, endowed with the usual topology. We have the following

DEFINITION 1. Let $\mu_1,\ldots,\mu_r \in E'(R^n)$ be r compactly supported distributions with $\mathrm{supp}(\mu_i) \subset B(\alpha)$, for all j and some $\alpha > 0$. The system defined by convolution with $(\mu_1,\ldots,\mu_r)$ is called hyperbolic if the following holds: there exist $R > r > \rho > 0$ with $R > \alpha$ such that for each $f \in E(B(R))$, with $\vec{\mu}*f = (\mu_1*f,\ldots,\mu_r*f) = 0$ on $B(r)$, there exists a unique $g \in E(R^n)$, with $f|_{B(\rho)} = g|_{B(\rho)}$, and the map $f \to g$ is continuous with respect to the topologies of $E(B(R))$ and of $E(R^n)$.

REMARK 2. Note that this notion of hyperbolicity is weaker than the one used in [4], [8]; for this reason one expects to be able to prove an analogue of theorem 12 under somewhat weaker hypotheses. We will show that this is, indeed, the case.

REMARK 3. Before proceeding to our main theorem, we need to make a few simple observations on the kernel in $E(R^n)$ of an operator such as $\vec{\mu}*$. As it is known, the solutions of systems of (constant coefficients linear) partial differential equations do not satisfy any elementary uniqueness property: by this we mean that there may well be $f_1, f_2 \in E(R^n)$, both solutions of a same system of partial differential equations, for which $f_1 = f_2$ on some large open set in R^n, and yet $f_1 \not\equiv f_2$; in general one knows that uniqueness can however be restored by restricting the study to suitable subspaces of $E(R^n)$, defined in terms of particular growth conditions; in [1] it is shown (and examples are given) that the situation for systems of convolution equations is even more delicate, since there are concrete situations in which uniqueness cannot be achieved, even with the use of growth conditions.

REMARK 4. A quite simpler situation occurs, on the other hand, if one restricts the attention to the case of n slowly decreasing convolutors, in the sense of [3]. This, in particular, implies that the variety

$$V = \{z \in \mathbb{C}^n : \hat{\mu}_1(z) = \ldots = \hat{\mu}_n(z) = 0\}$$

is discrete, hence no integrals arise and the series expansion for mean-periodic functions is essentially unique, in the sense that if $\vec{\mu} * f = 0$, and $f \equiv 0$ on a sufficiently large open set, then $f \equiv 0$ on R^n. The reason for this (see [2], [3]) is that the coefficients of the exponentials in the series expansion for such an f are given by convolution with explicit kernels, and one can verify that a positive k exists such that $\vec{\mu} * f = 0$ and $f|_{B(k\alpha)} \equiv 0$ imply $f \equiv 0$.

REMARK 5. A final remark is in order on the meaning of the definition of slowly decreasing given in [4] (and in section 3 of [8]) as well as on the long computations carried over in there. Suppose $\vec{\mu}$ is an r-tuple in $E'(R^n)$ which is slowly decreasing in the classical sense of [3]: if one tries to directly prove a representation theorem for solutions $f \in E(\Omega_1)$ of $\vec{\mu} * f = 0$ (Ω_1 a suitable open set in R^n), one soon discovers that the task is not possible because the control on the weights may not be sufficient to yield a representation which is C^∞ in Ω_1: indeed one can keep track of all the bounds and the result is that every C^∞ solution of $\vec{\mu} * f = 0$ can be given an integral representation which converges to f on some smaller subset of Ω_1: if $\Omega_1 = B(R)$, then f admits an integral representation on $B(R-k\alpha)$, for some positive k which can be explicitly computed from the proof in [3]. This is the reason which lies behind the complicated definition of slowly decreasing which is given in [4], [8], and which ensures that the representation is actually convergent on $B(R)$.

We are now ready to state the following refinement of theorem 12 in [8]:

THEOREM 5. *Let* $\vec{\mu} = (\mu_1,\dots,\mu_r)$ *be slowly decreasing in the sense of* [3]. *Then* $\vec{\mu}$ *is hyperbolic if and only if there exists a positive constant* A *such that*

(1) $$|\mathrm{Im}\, t| \le A \log(1+|t|) \quad \text{on} \quad V.$$

In particular, V *has to be discrete, and* $r = n$.

Proof. Condition (1) is sufficient. Indeed let $R > 0$ be large enough, and let $f \in E_\mu(B(R))$ (the subspace of $E(B(R))$ whose elements are solutions to $\vec{\mu}*f = 0$). Then f has a series representation of the form $\sum_{\alpha\in V} c_\alpha(x)\, e^{i\alpha\cdot x}$, $c_\alpha(x)$ polynomials. Remember that, by [5], (1) implies that V is discrete. It converges in some sphere $B(\rho)$ in view of remark 4 (the actual value of ρ depends on α and on R, and it is therefore necessary that R be large enough to ensure $\rho > 0$). By (1) one immediately gets that $\sum_{\alpha\in V} c_\alpha(x)\, e^{i\alpha\cdot x}$ is actually convergent to a function $g \in E(R^n)$, which is the required extension. The continuity of the map $f \to g$ is an immediate consequence of its construction, while the unicity of g follows from remark 3.

To prove the necessity of (1) we consider $f \in E_\mu(B(R))$ and take $\varepsilon < R/2$. Denote by K a finite subset of $\partial B(R)$ such that, for all x,

(2) $$\sup_{y \in K}(x \cdot y) \geq H(x),$$

where H is the supporting function of the set $B(R-\varepsilon)$. For any $y \in K$ we define the functional

$$F_y\colon \mathcal{E}_\mu(B(R)) \to \mathbb{C}$$
$$f \to g(2y),$$

where g is the unique extension of f given by the hyperbolicity. By definition 1, F_y is continuous, and so we can find $C > 0$ and an integer m such that, for any $y \in K$,

(3) $$|F_y(f)| \leq C \sup_{\substack{x \in B(R) \\ |\alpha| \leq m}} |f^{(\alpha)}(x)|.$$

By uniqueness the extension g of $\exp(ix \cdot t)$ is still $\exp(ix \cdot t)$. Hence, applying inequality (3) to $f(x) = \exp(ix \cdot t)$, $t \in V$, and using (2), one immediately obtains the result.

REMARK 6. This result extends a theorem of Meise and Vogt [7] (corollary 2.3) to the case of several variables, and can be used to prove analogous statements concerning hyperbolicity in Beurling or Roumier spaces of C^∞ functions (proposition 3.4 of [7]) as well as to refine the hyperbolicity result for the space $H(C^n)$ given in [8].

REFERENCES

1. Berenstein, C. A. and Lesmes J., The Cauchy problem for convolution operators. Uniqueness. Mich. Math. J. 26 (1979), 333-349.

2. Berenstein, C. A. and Taylor, B. A., A new look at interpolation theory for entire functions of one variable. Adv. in Math. 33 (1979), 109-143.

3. ________________, Interpolation problems in $\mathbb{C}^n$ with applications to harmonic analysis. J. Anal. Math. 38 (1980), 188-254.

4. Berenstein, C. A. and Struppa, D. C., Solutions of convolution equations in convex sets. Amer. J. Math. (1986), to appear.

5. Berenstein, C. A. and Yger, A., Ideals generated by exponential-polynomials, Adv. in Math. (1986), to appear.

6. Ehrenpreis, L., Solutions of some problems of division. Part V, Hyperbolic operators, Amer. J. Math. 84 (1962), 324-348.

7. Meise, R. and Vogt, D., Characterization of convolution operators on spaces of C^∞ - functions admitting a continuous linear inverse, manuscript.

8. Meril, A. and Struppa, D. C., Convolutors in spaces of holomorphic functions, these proceedings.

Authors' addresses:

Department of Mathematics
University of Maryland
College Park, MD 20742.

and

Scuola Normale Superiore Piazza dei Cavalieri 7
56100 Pisa, Italy

The research of the first author was partly supported by a NSF grant.

INTEGRAL REPRESENTATIONS IN THE THEORY OF THE $\bar{\partial}$-NEUMANN PROBLEM

R. Michael Range*

1. Introduction.

In this report I will discuss some recent results which give a complete synthesis of concrete and explicit methods for solving the Cauchy-Riemann equations and abstract L^2 solution methods. Let us first briefly review the two methods.

Concrete integral solution operators for $\bar{\partial}u=f$ on a strictly pseudoconvex domain D in $\mathbb{C}^n$ were constructed around 1969 by Grauert and Lieb [GrLi], based on earlier work of Ramirez, and by Henkin [Hen], independently. These operators are generalizations of the classical Cauchy transform in one complex variable, and, together with appropriate estimates, they have found numerous applications. The original results have been extended in various ways by many authors. The principal ingredients of this method are:

(1) The Bochner-Martinelli-Koppelman formula.

(2) The calculus of Cauchy-Fantappiè forms introduced by Leray.

(3) The complex geometry of bD (in the form of the Levi polynomial of a defining function).

(4) Global methods from the classical theory of several complex variables.

(1) and (2) are closely connected with the Euclidean structure of $\mathbb{C}^n$. (4) is not really needed any longer; more direct construction are now known (see [Ran 1]). For a comprehensive discussion of these methods and for extensive references see [HeLe] or [Ran 3].

---

*Partially supported by NSF Grants MCS 83-00854 and DMS 85-01342.

A completely different, <u>abstract</u> solution operator for $\bar\partial$ arises in the theory of the $\bar\partial$-Neumann problem, as follows. Let $D\subset\subset X$ be a relatively compact strictly pseudoconvex domain in a Hermitian complex manifold (X,ds^2) of complex dimension n. For $0\le q\le n$ one considers the maximal closed operator

$$\bar\partial:\ L^2_{0,q}(D) \rightarrow L^2_{0,q+1}(D)$$

with Hilbert space adjoint $\bar\partial^*$, and one defines the self-adjoint complex Laplacian $\Box$ by

$$\Box = \bar\partial\,\bar\partial^* + \bar\partial^*\bar\partial:\ L^2_{0,q}(D) \rightarrow L^2_{0,q}(D),\ 0\le q\le n.$$

It is known that range $\Box$ is closed, with finite codimension if $q\ge 1$; this implies the existence and continuity of the $\bar\partial$-Neumann operator $N:\ L^2_{0,q}(D)\rightarrow L^2_{0,q}(D)$, which is the inverse of $\Box$, modulo the finite dimensional kernel of $\Box$. One easily shows that the operator

$$(5)\qquad \bar\partial^*N:\quad L^2_{0,q+1}(D) \rightarrow L^2_{0,q}(D),\quad 0\le q\le n-1,$$

satisfies $\bar\partial(\bar\partial^*Nf) = f$ if $\bar\partial f=0$ and $f\perp\ker\Box$, i.e., $u=\bar\partial^*Nf$ is the (unique) solution of $\bar\partial u=f$ of <u>minimal</u> L^2 norm. The principal results, including deep local regularity results and estimates in L^2 Sobolev norms for N, and hence also $\bar\partial^*N$, were obtained by J.J. Kohn in 1963 [Koh 1]. The proofs involve abstract Hilbert space theory, tangential pseudo-differential operators, and subelliptic a priori estimates. See [Koh 2] for a survey of this theory, including recent extensions to more general domains.

To combine the two methods is desirable and useful for at least two major reasons.

(I) To obtain a detailed and explicit understanding of the analytic and geometric structure of the kernels of fundamental abstract operators like N and $\bar\partial^*N$, including estimates in Hölder norms, L^p norms, C^k norms, etc. A detailed analysis of <u>estimates</u> for N and $\bar\partial^*N$ on (0,1)- forms in case the metric is of a special type known as Levi metric (see below for the definition of this concept) was carried out already in 1975 by Greiner and Stein [GrSt]. Their methods involve reduction to the boundary, osculation by the Heisenberg group, and a calculus of pseudodifferential operators on the Heisenberg group, leading to a complicated representation of N as a composition of several pseudodifferential operators, whose regularity properties are studied separately. Clearly there is need for a more direct and simpler approach. Moreover, the structure of the kernels of N and $\bar\partial^* N$ still remained obscure.

(II) To obtain C^k estimates for a solution operator

$$S_q: \ C_{0,q+1}(\bar D) \to C_{0,q}(\bar D)$$

for $\bar\partial$ defined by explicit kernels à la Grauert/Lieb and Henkin. More precisely, it was recognized in the early 1970's that such operators satisfy a Hölder 1/2-estimate

$$(6) \qquad |S_q f|_{\Lambda_{1/2}(D)} \lesssim |f|_{L^\infty(D)},$$

and it is natural to then look for corresponding higher order estimates

$$(7) \qquad |S_q f|_{C^{k+1/2}(\bar D)} \lesssim |f|_{C^k(\bar D)}, \quad k=1,2,\ldots$$

To reduce (7) to (6) one needs a calculus for the commutators of S_q with vector fields, but it turns out that such a calculus breaks down for vector fields which are not tangential to the boundary bD. In order to control the missing <u>normal</u> derivative of $S_q f$ one is led to express such a derivative in terms of <u>tangential</u> derivatives of $S_q f$, plus $\bar\partial(S_q f)$ and $\bar\partial^*(S_q f)$ (this is a well known consequence of the ellipticity of $\bar\partial+\bar\partial^*$). Now $\bar\partial(S_q f)=f$ is obviously controlled by f, but $\bar\partial^*(S_q f)$ is generally unknown, except when q=0, i.e., when $S_0 f$ is a function and $\bar\partial^* S_0 f=0$ by definition. (This explains - with hindsight - why the methods developed in [Alt] and [Siu] only allowed to prove (7) in case q=0, but not when $q\geq 1$.) On the other hand, if one had $S_q f=\bar\partial^* Nf$, then $\bar\partial^* S_q f=0$ also for $q\geq 1$, and the road would be clear to prove (7) in the general case.

2. <u>The Basic Integral Representation Formula</u>.

The first connection between concrete integral operators on strictly pseudoconvex domains and certain Hilbert space operators was discovered in 1977 by Kerzman and Stein, who found a representation of the (abstract) Szegö kernel in terms of the Henkin-Ramirez kernel [KeSt]. Their methods were extended in 1980 by Ligocka to treat the Bergman kernel [Lig]. The crucial ingredient in these results is an approximate symmetry in the explicit kernels.

In [LiRa 1], Lieb and I began a program to extend the above methods to more complicated matrix valued operators like $\bar\partial^* N$ and N, but it turned out that the necessary symmetry properties could only be established in dimension 2, or when the domain D is a ball. The difficulty was that the Euclidean metric, which was needed for (1) and (2), is in general not compatible with (3), i.e., with the complex geometry of bD. To overcome this obstruction we were forced to abandon the familiar setting of $D \subset \mathbb{C}^n$, and to formulate the $\bar\partial$-Neumann problem with respect to a Levi metric as in Greiner and Stein (op.cit.). Such metrics are defined as follows.

Suppose $D=\{x \in X: r(x)<0\}$, where r is smooth, $dr \neq 0$ on bD, and r is strictly plurisubharmonic near bD. The Hermitian metric on X is a *Levi metric for* D if on some neighborhood W of bD one has

$$(8) \qquad ds^2 = h \sum_{j,k=1}^{n} \frac{\partial^2 r}{\partial z_j \partial \bar z_k} dz_j \otimes d\bar z_k ,$$

for some positive smooth function h on W. The metric (8) is said to be *normalized* if $|\partial r|_{ds^2} \equiv 1$ near bD.

Normalized Levi metrics exist for every strictly pseudoconvex domain, and any two such metrics for D are conformally equivalent on the *complex* tangent bundle to bD.

I can now state our main result.

Main Theorem. (Lieb and Range [LiRa 2]). *Given* (X, ds^2) *and the strictly pseudoconvex domain* $D \subset\subset X$, *there exist bounded linear integral operators*

$$T_q: \ L^2_{0,q+1}(D) \to L^2_{0,q}(D)$$

with explicity known kernels, such that

a) *if* P_0: $L^2(D) \to \mathcal{O}(D) \cap L^2(D)$ *is the orthogonal projection, then*

(9) $f - P_0 f = T_0 \bar\partial f$ + errors *for* $f \in \operatorname{dom} \bar\partial \subset L^2(D)$;

(b) <u>if</u> $q \geq 1$ <u>and</u> ds^2 <u>is a normalized Levi metric for</u> D, <u>then</u>

(10) $$f = T_q \bar{\partial} f + T_{q-1}^* \bar{\partial}^* f + \text{errors}$$

for $f \in \text{dom}\,\bar{\partial} \cap \text{dom}\,\bar{\partial}^* \subset L^2_{0,q}(D)$.

<u>Remarks.</u> (i) The error terms depend on f, $\bar{\partial} f$, and -in (10)- on $\bar{\partial}^* f$, and they are negligible in comparison to the other terms.

(ii) The kernel of T_q if given explicitly in terms of the Levi polynomial of the defining function r for D, and the square of the geodesic distance function with respect to the metric ds^2. The precise formula is too complicated to be presented here, and the interested reader is referred to [LiRa 2].

(iii) The formula for T_q is natural in the following sense: if D is the unit ball in $\mathbb{C}^n$ with defining function $r=|z|^2-1$, and the metric is the Euclidean one (this is a Levi metric for the ball!), all error terms in (9) and (10) vanish identically (see [LiRa 1], §6).

<u>Ingredients of Proof</u>. (A) Generalize (1) and (2) -which depend on the Euclidean metric - to the case of an arbitrary Hermitian manifold. The relevant formulas will no longer hold exactly, but only modulo harmless error terms.

(B) Apply the new machinery developed in (A) to certain differential forms constructed from the Levi polynomial of r, and transform the resulting representation formula into the desired form by means of Stokes' theorem, integration by parts, *-operator, etc., so as to involve only inner products of forms over D with respect to ds^2.

(C) Analyse the error terms and prove the necessary smoothing properties by establishing a cancellation of singularities as a consequence of approximate symmetries. For (9), i.e. when q=0, this involves, essentially, the same sort of argument used by Ligocka (op. cit.) in the study of the Bergman projection on $D \subset\subset \mathbb{C}^n$. The case $q \geq 1$ involves long and delicate technical arguments; it is here that it is essential that the metric is Levi.

3. Applications

(A) Estimates

By well known methods one readily sees that the operators T_q are bounded form $L^{\infty}_{0,q+1}(D)$ into the space of (0,q)-forms with coefficients in the Hölder space $\Lambda_{1/2}(D)$ (compare with (6) above). One therefore obtains from (9) and (10) the following Lipschitz a priori estimates for $f \in \operatorname{dom} \bar\partial \cap \operatorname{dom} \bar\partial^* \subset L^2_{0,q}(D)$ (see [LiRa 2]):

(11) $$|f - P_0 f|_{\Lambda_{1/2}} \lesssim |\bar\partial f|_{L^\infty} + \|f\|_{L^2}, \quad \text{if } q=0 ,$$

(12) $$|f|_{\Lambda_{1/2}} \lesssim |\bar\partial f|_{L^\infty} + |\bar\partial^* f|_{L^\infty} + \|f\|_{L^2}, \quad \text{if } q\geq 1$$

and ds^2 is a (not necessarily normalized) Levi metric.

These estimates are analogous to the classical $1/2$-subelliptic estimate in Sobolev norms of Kohn [Koh 1].

By developing a calculus of commutators of the relevant class of "admissible kernels" with tangential vector fields, and by the ellipticity of $\bar\partial+\bar\partial^*$ we generalize (11) and (12) to

Theorem 1. For $k=0,1,2,...$ and $f \in \operatorname{dom} \bar\partial \cap \operatorname{dom} \bar\partial^* \subset L^2_{0,q}(D)$, one has

(13) $$|f - P_0 f|_{C^{k+1/2}(\bar D)} \lesssim |\bar\partial f|_{C^k(\bar D)} + \|f\|_{L^2} \text{ if } q=0 ,$$

(14) $$|f|_{C^{k+1/2}} \leq |\bar\partial f|_{C^k(\bar D)} + \|f\|_{L^2} \text{ if } q\geq 1 \text{ and } ds^2 \text{ is a Levi metric.}$$

By applying (13) or (14) to $\bar\partial^* Nf$ one obtains (notice that $\|\bar\partial^* Nf\|_{L^2} \lesssim \|f\|_{L^2} \lesssim |f|_{L^\infty}$):

Corollary. If $f \in L^2_{0.q+1}(D)$ satisfies $\bar\partial f=0$ and $f \perp \ker \square$, then the solution $\bar\partial^* Nf$ of $\bar\partial u=f$ of minimal L^2 norm satisfies

(15) $$|\bar\partial^* Nf|_{C^{k+1/2}(\bar D)} \leq |f|_{C^k(\bar D)}.$$

ds^2 needs to be a Levi metric only if $q\geq 1$.

Complete proofs of Theorem 1 and the Corollary are in [LiRa 3].

Remarks. (i) As mentioned earlier, Greiner and Stein (op.cit.) had proved (15) by different methods in the case q=0 and ds^2 a Levi metric. So even for q=0 the above result gives new information, as here there is no restriction on the metric if q=0.

(ii) In [LiRa 3] we also obtained estimates in non isotropic Hölder norms and in weighted L^P norms (both isotropic and non isotropic); the weighted L^P estimates are within $\epsilon>0$ of the optimal estimates. Because of the mixing of homogeneities in the relevant kernels, optimal L^p estimates present a new delicate analytic problem which is studied by Phong and Stein [PhSt] in a broader context.

B) The principal part of $\bar\partial^*N$

Suppose $f \in$ range $\bar\partial \subset L^2_{0,q+1}(D)$. Then $\bar\partial(\bar\partial^*Nf)=f$ and $\bar\partial^*(\bar\partial^*Nf)=0$; therefore, by applying the Main Theorem to $\bar\partial^*Nf$, one obtains

(16) $$\bar\partial^*Nf = T_q f + \text{errors}(f) \text{ if } f \in \text{range } \bar\partial.$$

One shows that the error terms in (16) are roughly twice as smoothing as the operator T_q. Hence (16) shows that T_q describes the principal part of $\bar\partial^*N$ restricted to the subspace range $\bar\partial$ of $L^2_{0,q+1}(D)$. Notice that for q=0 this holds for arbitrary metrics! This is the case of most interest in applications to complex analysis.

Since by Kohn's classical results one has the direct sum decomposition

(17) $$L^2_{0,q+1} = \text{range } \bar\partial \oplus \text{range } \bar\partial^* \oplus \ker \Box \quad ,$$

(16) is far from enough to describe $\bar\partial^*N$ on all of $L^2_{0,q+1}$. By using the results in [LiRa 1] it was proved in [Ran 2] that if D is the unit ball in $\mathbb{C}^n$, then one has $\bar\partial^*N \equiv T_q$. This representation for $\bar\partial^*N$ on the ball had been obtained earlier by Harvey and Polking [HaPo] by using the rotational symmetry of the ball. In contrast to the method of Harvey and Polking, the method developed in [Ran 2] generalizes to arbitrary strictly pseudoconvex domains, leading to the following result (see [LiRa 2]):

Theorem 2. If ds^2 is a normalized Levi metric, then the explicit integral operator T_q represents the principal part of $\bar\partial^*N$ on $L^2_{0,q+1}(D)$.

C) The kernel of N

Most recently we have been able to combine the Main Theorem and Theorem 2 with additional integrations by parts and with explicit integrations in order to obtain an explicit kernel for the principal part of the $\bar{\partial}$-Neumann operator on $L^2_{0,q}(D)$ in case of a normalized Levi metric, provided $1 \le q \le n-2$ (see [LiRa 4]). In case $q=n$, N is just the Green's operator for $\square$, which is well understood. The case $q=n-1$ needs further investigations. For $q=1$ and $n \ge 3$ related results have been announced by Phong [Pho]; his methods are based on an extremely delicate local osculation of D by the Siegel upper half space, the details of which have not yet been published.

REFERENCES

[Alt] Alt, W., Hölderabschätzungen für Ableitungen von Lösungen der Gleichung $\bar{\partial}u=f$ bei streng pseudokonvexem Rand, Man. Math. 13(1974), 381-414.

[GrLi] Grauert, H., and Lieb, I., Das Ramirezsche Integral und die Lösung der Gleichung $\bar{\partial}f=\alpha$ im Bereich der beschränkten Formen, Rice Univ. Studies 56(1970), 29-50.

[GrSt] Greiner, P., and Stein, E.M., Estimates for the $\bar{\partial}$-Neumann Problem, Princeton University Press, 1977.

[HaPo] Harvey, R., and Polking, J., The $\bar{\partial}$-Neumann solution to the inhomogeneous Cauchy-Riemann equations in the ball in $\mathbb{C}^n$, Trans. AMS 281(1984), 587-613.

[Hen] Henkin, G.M., Integral representations in strictly pseudoconvex domains and applications to the $\bar{\partial}$-problem, Mat. Sb. 82(1970), 300-308; Math. USSR Sb. 11(1970), 273-281.

[HeLe] Henkin, G.M., and Leiterer, J., Theory of Functions on Complex Manifolds. Birkhäuser, Boston, 1984

[KeSt] Kerzman, N., and Stein, E.M., The Szegö kernel in terms of Cauchy-Fantappiè kernels, Duke Math. J. 45 (1978), 197-224.

[Koh 1] Kohn, J.J., Harmonic integrals on strongly pseudoconvex manifolds, I, Ann. of Math. 78(1963), 112-148, II, ibid. 79(1964), 450-472.

[Koh 2] __________, A survey of the $\bar{\partial}$-Neumann Problem. Proc. Symp. Pure Math. 41, 137-145, Amer. Math. Soc., Providence, RI 1984.

[LiRa 1] Lieb, I., and Range, R.M., On integral representations and a priori Lipschitz estimates for the canonical solution of the $\bar{\partial}$-equation, Math. Ann. 265(1983), 221-251..

[LiRa 2] __________, Integral representations and estimates in the theory of the $\bar{\partial}$-Neumann problem. Ann. of Math. 123(1986), 265-301.

[LiRa 3] __________, Estimates for a class of integral operators and applications to the $\bar{\partial}$-Neumann problem, Invent. Math. 85(1986), 415-438.

[LiRa 4] __________, The kernel of the $\bar{\partial}$-Neumann operator on strictly pseudoconvex domains. (In preparation).

[Lig] Ligocka, E. The Hölder continuity of the Bergman projection and proper holomorphic mappings, Studia Math. 80(1984), 89-107.

[Pho] Phong, D.H., On integral representations for the Neumann operator, Proc. Nat. Acad. Sci. USA 76(1979), 1554-1558.

[PhSt] Phong, D.H., and Stein, E.M., Hilbert integrals, singular integrals, and Radon transforms I. Acta Math. (to appear).

[Ran 1] Range, R.M., An elementary integral solution operator for the Cauchy-Riemann equations on pseudoconvex domains in $\mathbb{C}^n$. Trans. Amer. Math. Soc. 274(1982), 809-816.

[Ran 2] __________, The $\bar{\partial}$-Neumann operator on the unit ball in $\mathbb{C}^n$, Math. Ann 266(1984), 449-456.

[Ran 3] ____________, Holomorphic Functions and Integral Representations in Several Complex Variables, Springer-Verlaq, New York, 1986.

[Siu] Siu, Y.-T., The $\bar{\partial}$-problem with uniform bounds on derivatives, Math. Ann. 207(1974), 163-176.

Department of Mathematics
State University of New York at Albany
Albany, NY 12222

TENTS AND INTERPOLATING SEQUENCES IN THE UNIT BALL*

Pascal J. Thomas

Abstract. A sufficient condition is given to make a sequence of points interpolating for $H^\infty(B^n)$. The methods are elementary: construction of the P. Beurling functions. This can be used to prove the sharpness of the exponent of Varopoulos's necessary condition for interpolation. The result is also compared with Berndtsson's recent sufficient condition, in conjunction with which it gives a slightly improved theorem.

Definitions and previous results.

If $\{a_k\}$ is a sequence of points in

$$B^n = \{z \in \mathbb{C}^n : |z| < 1\},$$

we say that $\{a_k\}$ is an (H^∞)-interpolating sequence if and only if for any $\{\alpha_k, k \in \mathbb{Z}_+\} \subset \Delta = B^1$, there exists $f \in H^\infty(B^n)$ (i.e., f is holomorphic and bounded on B^n) so that $f(a_k) = \alpha_k$, $k \in \mathbb{Z}_+$.

Carleson proved (see [2], [3], or [5]):

Theorem (1958). The following geometrical condition is necessary and sufficient for $\{a_k\} \subset \Delta$ to be an interpolating sequence:

There exists $\delta > 0$ so that for any $k \in \mathbb{Z}_+$,

$$(B_1) \qquad \prod_{j:j\neq k} \left| \frac{a_k - a_j}{1 - a_j \bar{a}_k} \right| \geq \delta .$$

Furthermore, the interpolating function f is given by a linear operator:

$$f(z) = \sum_{j\in Z_+} a_j F_j(z),$$

where $F_j(a_k) = \delta_{jk}$ (Kronecker symbol), and $\sum_j \left|F_j(z)\right| \leq M$, $z \in \Delta$ (see [2] or [3]). The F_j are P. Beurling functions.

More recently, Peter Jones [5] gave a new proof of Carleson's result about interpolating sequences in the disc, via explicit

---

*The present work, in a slightly altered form, constitutes part of the author's Ph.D. dissertation, written under the supervision of John B. Garnett at the University of California at Los Angeles.

formulas for P. Beurling functions. The same ideas were adapted by Bo Berndtsson [1] to prove that the following is a sufficient condition for interpolation:

Theorem (1984). If there is a $\delta > 0$ such that for all $k \in \mathbb{Z}_+$,

$$(B_n) \qquad \prod_{j:j\neq k} \left|\phi_{a_k}(a_j)\right| \geq \delta ,$$

then $\{a_k\} \subset B^n$ is an H^∞-interpolating sequence.

Here as in [7] we write ϕ_a for the following automorphism of B^n:

$$\phi_a(z) = \frac{a - P_a(z) - s_a Q_a(z)}{1 - z\cdot\bar{a}}, \quad a \neq 0;$$

$$\phi_0(z) = z,$$

where $z\cdot\bar{a} = \sum_{i=1}^{n} z_i\bar{a}_i$, $|z|^2 = z.\bar{z}$, $s_a = (1 - |a|^2)^{1/2}$, $P_a(z) = \frac{z\cdot\bar{a}}{|a|^2}a$ (the projection onto the complex line through a) and $Q_a(z) = z-P_a(z)$ (the complementary projection).

Note that $\phi_a(0) = a$, ϕ_a is an involution, and $|\phi_a(z)|^2 = 1 - \frac{(1 - |a|^2)(1 - |z|^2)}{|1 - z\cdot\bar{a}|^2}$.

Thus (B_n) is equivalent to the conjunction of the following two conditions;

$$(B_n') \qquad \exists\, M_1 > 0 \text{ such that } \sum_{j:j\neq k} \frac{(1 - |a_j|^2)(1 - |a_k|^2)}{|1 - a_j\cdot\bar{a}_k|^2} \leq M_1$$

and

$$(\text{Sep}) \quad \exists\, \delta_1 > 0 \text{ such that for } j \neq k, \ \frac{(1 - |a_j|^2)(1 - |a_k|^2)}{|1 - a_j\cdot\bar{a}_k|^2} \leq 1-\delta_1.$$

We shall call separated a sequence satisfying (Sep). Any interpolating sequence must be separated.

On the other hand, Varopoulos [10] has given a necessary condition for $\{a_k\}$ to be an H^∞-interpolating sequence (in a general context). For B^n, this condition reads:

Theorem (1972). If $\{a_k\} \subset B^n$ is an interpolating sequence, then

$$(V_n) \qquad \exists\, M_2 > 0 \text{ such that } \sum_{j:j\neq k} \left[\frac{(1 - |a_j|^2)(1 - |a_k|^2)}{|1 - a_j\cdot\bar{a}_k|^2}\right]^n \leq M_2$$

Of course, $(V_1) \Leftrightarrow (B_1')$, and so (V_1) and (Sep) are equivalent to (B_1), the necessary and sufficient condition for $\{a_k\}$ to be $H^\infty(\Delta)$-interpolating.

However, for $n \geq 2$, the two conditions (B_n') and (V_n) are quite far apart. (V_n) is far from sufficient.

Berndtsson [1] proved the more difficult fact that the exponent in Varopoulos' necessary condition is sharp. More precisely,

Theorem (1984). For any $\varepsilon > 0$, there exists an $H^\infty(B^n)$-interpolating sequence $\{a_k\}$ such that

$$\sum_j (1 - |a_j|^2)^{n-\varepsilon} = \infty .$$

Since

$$(1 - |a_j|^2)^{n-\varepsilon} \leq \left[\frac{(1 + |a_k|)^2}{1 - |a_k|^2}\right]^{n-\varepsilon} \left[\frac{(1 - |a_j|^2)(1 - |a_k|^2)}{|1 - a_j\cdot\bar{a}_k|^2}\right]^{n-\varepsilon},$$

the above result implies

$$\sum_{j:j\neq k} \left[\frac{(1 - |a_j|^2)(1 - |a_k|^2)}{|1 - a_j\cdot\bar{a}_k|^2}\right]^{n-\varepsilon} = \infty .$$

The expression $\dfrac{(1 - |a_j|^2)(1 - |a_k|^2)}{|1 - a_j\cdot\bar{a}_k|^2}$ is natural here because it is invariant under automorphisms (i.e., biholomorphic one-to-one and onto maps of B^n), and so is the problem (is $\{a_k\}$ interpolating?).

(V_n) can be reformulated. Consider, for any sequence $\{a_k\} \subset B^n$, the measure μ defined by

$$\mu = \sum_{k\in Z_+} (1 - |a_k|^2)^n \,\delta_{a_k},$$

where δ_{a_k} is the point mass at a_k. We say that a measure ν on B^n is a Carleson measure if and only if there exists $C > 0$ such that

$$\forall\, \zeta \in \partial B^n, \forall\, R > 0,\ \nu\Big(\{z \in B^n : |1 - z\cdot\bar{\zeta}| \leq R\}\Big) \leq CR^n .$$

(In this context, this definition was introduced by Hörmander [4]).

Then (V_n) is equivalent to μ being a Carleson measure, see e.g. Mantero [6].

Berndtsson pointed out in [1] that his example showing the sharpness of the exponent in Varopoulos' necessary condition (V_n) relied on the fact that the points of the sequence lie on "spread out" complex directions. To make such a spread quantitative, define the following regions:

$$T_t(a) = \left\{z \in B^n : |1 - z\cdot\bar{a}| \le t(1 - |a|^2)\right\}$$

where $t \ge 1$ is some constant. Then $a \in T_t(a)$, and by analogy with the real-variable case, $T_t(a)$ is called the tent at a. Those regions are commensurate to those used in the definition of Carleson measures. $\{z \in B^n : d_H(a,z) \le \delta\}$ is analogous to the top half of cubes in the real-variable case. Now consider the following condition on a sequence $\{a_k\} \subset B^n$:

(T_t) $$j \ne k \Rightarrow T_t(a_j) \cap T_t(a_k) = \emptyset$$

<u>Remark</u>. For any $t > 0$, (T_t) implies (V_n) and (Sep).

<u>Proof</u>. (Sep) was mentioned above, for $t \ge 1$. In general, observe that if the distances to the boundary of two points in the sequence have a ratio which is too close to 1, then the disjointness of the tents will force their projections onto the boundary to keep apart. More precisely, define the Koranyi balls in ∂B^n by

$$K(\zeta,R) = \left\{z \in \partial B^n : |1- z\cdot\bar{\zeta}| \le R\right\} .$$

Then $\bar{T}_t(a) \cap \partial B^n$ is commensurate to $K\left(\frac{a}{|a|}, t(1 - |a|^2)\right)$.

Let $K_j = K\left(\frac{a_j}{|a_j|}, ct(1 - |a_j|^2)\right)$; then for $c > 1$ well chosen, $j \ne k \Rightarrow K_j \cap K_k = \emptyset$.

To check (V_n), consider the measure μ. Given any $a \in B^n$,

$$\mu\left[T_1(a)\right] = \sum_{k:a_k\in T_1(a)} (1 - |a_k|^2)^n .$$

By the above,

$$\begin{aligned}\mu\left[T_1(a)\right] &\le C(t) \sum_{k:a_k\in T_1(a)} \sigma(K_j)\\ &\le C\sigma\left[K(\tfrac{a}{|a|}, 2(1 - |a|^2))\right]\\ &\le C(1 - |a|^2)^n, \text{ q.e.d.}\end{aligned}$$

Here σ stands for the normalized Lebesgue measure on ∂B^n.

We may now ask

<u>The Tents Question</u>. Is there some $t \ge 1$ such that any sequence $\{a_k\} \subset B^n$ satisfying (T_t) is an interpolating sequence?

A positive answer is known only for $n = 1$ (since in that case (V_1) and (Sep) together are equivalent to (B_1)).

The situation benomes much easier to deal with when a stronger condition is required. Define, for $\delta \ge 0$,

$$T_t^\delta(a) = \left\{z \in B^n : |1 - z\cdot\bar{a}| \le t(1 - |a|^2)^{1-\delta}\right\}$$

(For $\delta < 0$, disjointness of those new tents would not even imply separatedness of the sequence; for $\delta = 0$, we get the previously-defined tents).

The Main Theorem. If $\sigma = \{a_k\} \subset B^n$ is a sequence such that there exist $\delta > 0$ and $t \ge 1$ so that

$$(T_i^\delta) \qquad j \ne k \Rightarrow T_t^\delta(a_j) \cap T_i^\delta(a_k) = \emptyset ,$$

then σ is an $H^\infty(B^n)$-interpolating sequence.

Notice, before going on to the proof, that a shortcoming of this result is that it is not automorphism-invariant (the quantity $\frac{(1 - |z|^2)^{1-\delta}}{|1 - z\cdot\bar{w}|}$ is not preserved under automorphism of the ball, for any $\delta \ge 0$). The result is then formally strengthened in the following statement:

> If $\{a_k\} \subset B^n$ is such that there exists ϕ an automorphism of B^n such that $\{\phi(a_k)\}$ satisfies the condition (T_t^δ), then $\{a_k\}$ is an $H^\infty(B^n)$-interpolating sequence.

Proof of the Theorem. First notice that for all $A > 0$, there exists $\varepsilon > 0$ such that if $1 - |a|^2 \le \varepsilon$, then $(1 - |a|^2)^{1-\delta} \ge A(1-|a|^2)^{1-\delta/2}$. So we may restrict ourselves to the case of $T_{At}^\delta(a) = T^\delta(a)$, with At as large as necessary, by dropping a finite number of points from the sequence (it does not alter the final result, since that finite set is itself an interpolating sequence, separated from the rest of sequence, and the separated union of two interpolating sequences is interpolating as well, see Varopoulos [9]).

We will produce pseudo-P. Beurling functions for the sequence, i.e., functions $F_k \in H^\infty(B^n)$ satisfying the following properties:

$$\text{(i)} \qquad F_k(a_k) = 1$$

$$\text{(ii)} \qquad |F_k(a_j)| \le \tfrac{1}{2} \quad \text{for} \quad j \ne k$$

$$\text{(iii)} \qquad \sum_{k\in\mathbb{Z}_+} |F_k(z)| \le M < \infty .$$

Let $F_k(z) = \left(\frac{1 - |a_k|^2}{- z\cdot a^-_k|}\right)^n$.

(i) is obvious. Next

$$|F_k(a_j)| = \left(\frac{1 - |a_k|^2}{|1 - a_j\cdot\bar{a}_k|}\right)^n \le (tA)^{-n}(1 - |a_k|^2)^{\delta n}$$

since $a_j \notin T^{\delta}_{At}(a_k)$ for $j \ne k$. Choose A such that $(tA)^{-n} \le 1/2$, then $|F_k(a_j)| < 1/2$. Note that each F_k is bounded by 2^n. The proof of (iii) will require more work.

For a given $z \in B^n$, set

$$S_\ell = S_\ell(z) = \left\{a_k \in \sigma : 2^\ell \le \frac{|1 - a_k\cdot\bar{z}|}{(1 - |z|^2)^{1-\delta}} \le 2^{\ell+1}\right\}$$

for $\ell \in \mathbb{Z}$. However,

$$|1 - a_k\cdot\bar{z}| \ge 1 - |z| \ge \tfrac{1}{2}(1 - |z|^2)$$

thus

$$\log_2 \frac{1-a_k\cdot\bar{z}}{(1-|z|^2)^{1-\delta}} \ge -1 + \delta\log_2(1-|z|^2) = -1 - \delta\log_2 \frac{1}{1-|z|^2} = L_1(z).$$

Similarly,

$$\log_2 \frac{|1-a_k\cdot\bar{z}|}{(1-|z|^2)^{1-\delta}} \le 1 + (1-\delta)\log_2 \frac{1}{1-|z|^2} = L_2(z),$$

so the only relevant values are $L_1(z) \le 1 \le L_2(z)$.

Now at this point z

$$\sum_k |F_k(z)| = \sum_{\ell=L_1}^{L_2} \sum_{k:a_k\in S_\ell} \left(\frac{1 - |a_k|^2}{|1 - z\cdot\bar{a}_k|}\right)^n$$

$$\le \sum_{L_1}^{L_2} \left[2^\ell(1 - |z|^2)^{1-\delta}\right]^{-n} \sum_{k:a_k\in S_\ell} (1 - |a_k|^2)^n .$$

<u>Case 1</u>. There exists an ℓ, and $a_k \in S_\ell$ such that

$$(1 - |a_k|^2)^{1-\delta} \ge 4(At)^{-1}2^{\ell+1}(1 - |z|^2)^{1-\delta} .$$

Then let ℓ_0 be the largest such ℓ ($\ell \le L_2$). Suppose that there exists an $a_j \in S_\ell$ for $\ell \le \ell_0$, $j \ne k$. Then

$$|1 - a_j\cdot\bar{a}_k| \le 2\left(|1 - a_j\cdot\bar{z}| + |1 - z\cdot\bar{a}_k|\right)$$

$$\le 4.2^{\ell_0+1}(1 - |z|^2)^{1-\delta}$$

$$\le 8(1 - |a_k|^2)^{1-\delta} ,$$

so that $a_j \in T^\delta_{At}(a_k)$ for At sufficiently large, a contradiction.

So the sum can be rewritten as

$$\left(\frac{1 - |a_{k_0}|^2}{|1 - z\cdot\bar{a}_{k_0}|}\right)^n + \sum_{\ell\ge\ell_0+1}\left[2^\ell(1 - |z|^2)^{1-\delta}\right]^{-n}\sum_{k:a_k\in S_\ell}(1 - |a_k|^2)^n$$

where a_{k_0} is the unique point of the sequence in S_{ℓ_0}, and in the last sum

$$(1 - |a_k|^2)^{1-\delta} \le 4(At)^{-1}2^{\ell+1}(1 - |z|^2)^{1-\delta} .$$

Since the first term is bounded by 2^n, it will be enough to deal with

Case 2. $\forall\,\ell$, $\forall\, a_k \in S_\ell$, $(1 - |a_k|^2)^{1-\delta} \ge 4(At)^{-1}2^{\ell+1}(1 - |z|^2)^{1-\delta}$.

Now for $a_k \in S_\ell$, the regions $T^\delta(a_k) \cap \partial B^n$ are all contained in $K\left(\frac{z}{|z|}, c2^{\ell+1}(1 - |z|^2)^{1-\delta}\right)$ for some constant c, and they are all disjoint, thus $\sum_{k:a_k\in S_\ell}(1 - |a_k|^2)^{n(1-\delta)} \cong \sigma\left(\bigcup_{k:a_k\in S_\ell}T^\delta(a_k) \cap \partial B^n\right)$

where σ denotes Lebesgue measure on ∂B^n

$$\le C2^{\ell n}(1 - |z|^2)^{n(1-\delta)}$$

and

$$\sum_{k:a_k\in S_\ell}(1-|a_k|^2)^n \le C2^{\delta(\ell+1)n}(1-|z|^2)^{\delta n(1-\delta)}\sum_{k:a_k\in S_\ell}(1-|a_k|^2)^{n(1-\delta)}$$

$$\le C2^{\delta(1+\delta)\ell n}(1 - |z|^2)^{n(1-\delta^2)} .$$

So the original sum can be estimated by

$$\sum_{\ell=L_1} 2^{-\ell n}(1 - |z|^2)^{-n(1-\delta)}2^{(1+\delta)\ell n}(1 - |z|^2)^{n(1-\delta^2)}$$

$$= (1 - |z|^2)^{\delta n(1-\delta)}\sum_{L_1}^{L_2}2^{\delta\ell n} \le C(\delta)\ (1 - |z|^2)^{\delta n(1-\delta)}2^{\delta nL_2} = C(\delta) ,$$

by definition of L_2. (iii) is proved.

The theorem does cover cases which do not fall under Berndtsson's result; in fact, it allows to give a new proof of the sharpness of the exponent in Varopoulos' necessary condition:

Lemma. For all $\varepsilon \in (0,n)$, there exists a sequence $\{a_k\} \subset B^n$ and $\delta > 0$ such that $\{a_k\}$ satisfies (T_1^δ) and

$$\sum_k (1 - |a_k|^2)^{n-\varepsilon} = \infty .$$

Proof of the Lemma. Consider any infinite sequence of Koranyi balls in ∂B^n, $\{K_k\}$, so that the balls $\tilde{K}_k$ (with the same centers and quadrupled radii) are disjoint. Let $m_k = \sigma(K_k)$; $\sum_k m_k < \infty$; we may assume that m_k decreases to 0.

Then pick positive numbers ρ_k decreasing to 0 such that

$$0 < \rho_k^{1-\delta} \ll m_k^{1/n} .$$

Let $\{a_{k,j}^*\}_{1 \le j \le M_k} \subset K_k$ be the centers of a maximal collection of disjoint Koranyi balls of radius $C_0\rho_k^{1-\delta}$ included in K_k. C_0 is a numerical constant, to be specified. Since the union of fixed dilates of those balls will cover K_k,

$$M_k \cong \frac{m_k}{C_0^n \rho_k^{(1-\delta)n}} .$$

Set $a_{k,j} = (1 - \rho_k)^{1/2} a_{k,j}^*$, so that $1 - |a_{k,j}|^2 = \rho_k$.

Claim: $\{a_{k,j}\}$ satisfies (T_1^δ).

Proof of claim: Suppose that, for $(k,j) \neq (k',j')$, there exists a b so that

$$|1 - b\cdot\bar{a}_{k,j}| \le (1 - |a_{k,j}|^2)^{1-\delta}$$

$$\text{and} \quad |1 - b\cdot\bar{a}_{k',j'}| \le (1 - |a_{k',j'}|^2)^{1-\delta} .$$

Without loss of generality, we may assume $k \le k'$, so that, as usual, we derive

$$|1 - a_{k,j}\cdot\bar{a}_{k',j'}| \le 4\rho_k^{1-\delta} ,$$

and consequently

$$|1 - a_{k,j}^*\cdot\bar{a}_{k',j'}^*| \le 4\Big[|1 - a_{k,j}^*\cdot\bar{a}_{k',j'}| + |1 - a_{k,j}\cdot\bar{a}_{k',j'}| + |1 - a_{k,j}\cdot\bar{a}^*_{k',j'}|\Big]$$

$$\leq 4(\rho_k + 4\rho_k^{1-\delta} + \rho_{k'}) \leq 24\rho_k^{1-\delta} .$$

Now if $k \neq k'$, since $\rho_k^{1-\delta} \ll m_k^{1/n}$, this will violate the disjointness of $\tilde{K}_{k'}$; if $k = k'$, for C_0 large enough, it will violate the disjointness of the Koranyi balls of centers $a^*_{k,j}$ and $a^*_{k,j'}$ and radius $C_0\rho_k$, q.e.d.

We then have

$$\sum_{k,j} (1 - |a_{k,j}|^2)^{n-\varepsilon} = \sum_k \rho_k^{n-\varepsilon} M_k \cong \sum_k \rho_k^{n-\varepsilon} m_k \rho_k^{-n(1-\delta)} = \sum_k \rho_k^{\delta n-\varepsilon} m_k .$$

Choose ρ_k of the form $\rho_k = m_k^{\alpha}$, where $(1 - \delta)\alpha > 1/n$; then $\rho_k^{1-\delta} \ll m_k^{1/n}$. To make sure that the last series diverges, let

$$\alpha(\delta n - \varepsilon) + 1 = 0 ,$$

that is to say $\alpha = 1/(\varepsilon - \delta n)$. To get $\frac{1 - \delta}{\varepsilon - \delta n} > \frac{1}{n}$, we need to pick $\delta < \varepsilon/n$ and to have $\varepsilon - \delta n < n - \delta n$, which is always true for $\varepsilon < n$, q.e.d.

Of course, Berndtsson's sufficient condition also takes care of many sequences which do not satisfy (T^{δ}). Both results can be united in:

<u>Corollary</u>. Suppose that $\{a_{k,j}\} \subset B^n$ is a sequence satisfying (Sep) and that there exist numbers $\delta > 0$, $t_1, t_2 \geq 1$, and points $b_k \in B^n$ such that

$$k \neq k' \Rightarrow T^{\delta}_{t_2}(b_k) \cap T^{\delta}_{t_2}(b_{k'}) = \emptyset$$

$$\forall\, j \in \mathbb{Z}_+,\ a_{k,j} \in T_{t_1}(b_k) \subsetneq T^{\delta}_{t_2}(b_k)$$

and $$\sup_j \sum_{j':j'\neq j} \frac{(1 - |a_{k,j}|^2)(1 - |a_{k,j'}|^2)}{|1 - a_{k,j}\cdot\bar{a}_{k,j'}|^2} \leq M < \infty ;$$

then the sequence is $H^{\infty}(B^n)$-interpolating.

The hypotheses of this result are trivially verified when $(T^{\delta}_{t_2})$ is; and any sequence which satisfies (B'_n) can be split into a finite union of sequences of the above type, because we can find a finite number of tents to contain it.

<u>Proof</u>. $\{a_{k,j}\}_{j\in\mathbb{Z}_+}$ satisfies (B_n) for k fixed, uniformly in k. Let F_{kj} be P.Beurling functions for that sequence: $|F_{kj}(z)| \leq M_1$, $\forall\, z \in B^n$, where M_1 does not depend on j or k. Let

$$G_{kj}(z) = \left(\frac{1 - a_{k,j}\cdot\bar{b}_k}{1 - z\cdot\bar{b}_k}\right)^N F_{kj}(z) ,$$

$N \geq n$ to be determined. Proceed as for the proof of the main theorem.

(i) $$G_{kj}(a_{k,j}) = 1\cdot F_{kj}(a_{k,j}) = 1$$

(ii) $$|G_{kj}(a_{k,j'})| = 0 \quad \text{if} \quad j \neq j' ;$$

$$\text{for} \quad k \neq k', \ |G_{kj}(a_{k',j'})| \leq M_1\left(\frac{|1 - a_{k,j}\cdot\bar{b}_k|}{|1 - a_{k',j'}\cdot\bar{b}_k|}\right)^N$$

$a_{k',j'} \notin T^{\delta}_{t_2}(b_k) \Rightarrow |1 - a_{k',j'}\cdot\bar{b}_k| \geq t_2(1 - |b_k|^2)^{1-\delta}$

and $a_{k,j} \in T_{t_1}(b_k) \Rightarrow |1 - a_{k,j}\cdot\bar{b}_k| \leq t_1(1 - |b_k|^2)$.

So $|G_{kj}(a_{k',j'})| \leq M_1\left(\frac{t_1}{t_2}(1 - |b_k|^2)^{\delta}\right)^N$.

Now $T_{t_1}(b_k) \subsetneq T^{\delta}_{t_2}(b_k)$ implies that

$$\forall\ k, \frac{t_1}{t_2}(1 - |b_k|^2)^{\delta} < 1 ,$$

and since $|b_k| \longrightarrow 1$ as $k \longrightarrow \infty$, $\frac{t_1}{t_2}(1 - |b_k|^2)^{\delta} \longrightarrow 0$ as $k \longrightarrow \infty$, so there exists $\gamma < 1$ such that

$$\forall\ k, \frac{t_1}{t_2}(1 - |b_k|^2)^{\delta} \leq \gamma .$$

$$\text{Pick} \quad N \geq \max\left(n, \frac{\log 1/(2M_1)}{\log \gamma}\right),$$

then $|G_{kj}(a_{k',j'})| \leq 1/2$ whenever $(k,j) \neq (k',j')$.

(iii) $$\sum_{k,j} |G_{kj}(z)| = \sum_{k,j}\left(\frac{|1 - a_{k,j}\cdot\bar{b}_k|}{|1 - z\cdot\bar{b}_k|}\right)^N |F_{kj}(z)|$$

$$\leq t_1^N \sum_k \left(\frac{1 - |b_k|^2}{|1 - z\cdot\bar{b}_k|}\right)^N \sum_j |F_{kj}(z)|$$

$$\leq t_1^N 2^{N-n} M_1 \sum_k \left(\frac{1 - |b_k|^2}{|1 - z\cdot\bar{b}_k|}\right)^n$$

and the main step of the proof of the main theorem showed that this last sum is finite, q.e.d.

Remark. If, through each point of a sequence satisfying (T_t^δ), we run a complex hyperplane perpendicular to a complex line coming from the origin:

$$H_k := \left\{z \in B^n : z\cdot\bar{a}_k = |a_k|^2\right\},$$

then any family of uniformly bounded holomorphic functions $\{f_k\}$ defined on H_k can be interpolated by a function $f \in H^\infty(B^n)$. Since the F_k are constant on each hyperplane, simply take

$$f = \sum_k F_k \tilde{f}_k ,$$

where the $\tilde{f}_k$ are bounded extensions of the f_k. See [8] for another result of this type, with more details.

References

[1] Berndtsson, B., Interpolating sequences for H^∞ in the ball, Nederl. Akad. Wetensch, Indag. Math. 88 (1985).

[2] Carleson, L., An interpolation problem for bounded analytic functions, Amer. J. Math. 80 (1958), pp. 921-930.

[3] Garnett, J., Bounded Analytic Functions, Academic Press, 1981.

[4] Hörmander, L., L^p estimates for plurisubharmonic functions, Math. Scand. 20 (1967), pp. 65-78.

[5] Jones, P., L^∞-estimates for the $\bar{\partial}$-problem in a half-plane, Acta Math. 150 (1983), pp. 137-152.

[6] Mantero, A.M., Sur la condition de Carleson dans la boule unité de C^m, Boll. Un. Mat. Ital. (1983), 163-169.

[7] Rudin, W., Function Theory in the Unit Ball of $\mathbb{C}^n$, Springer-Verlag, 1980.

[8] Thomas, P.J., Interpolating sequences of complex hyperplanes in the unit ball of C^n, to appear in Ann. Inst. Fourier (Grenoble) 36 (1986), No. 4.

[9] Varopoulos, N.Th., Sur la réunion de deux ensembles d'interpolation d'une algèbre uniforme, C.R. Acad. Sci. Paris, Ser. A 272 (1970), pp. 950-952.

[10] Varopoulos, N.Th., Sur un problème d'interpolation, C.R. Acad. Sci. Paris Ser. A 274 (1972), pp. 1539-1542.

Department of Mathematics
University of California, Los Angeles

and

Department of Mathematics
Occidental College, Los Angeles

BALAYAGE AND POLYNOMIAL HULLS.

John Wermer

This paper is dedicated to Maurice Heins in gratitude and friendship on the occasion of his 70th birthday.

This paper is joint work with Herbert Alexander.

Let Y be a compact set in $\mathbb{C}^n$. By the polynomial hull of Y, denoted $\hat{Y}$, we mean the set of points y in $\mathbb{C}^n$ such that

$$|P(y)| \leq \max |P| \text{ over } Y$$

for every polynomial P on $\mathbb{C}^n$.

Given y in $\mathbb{C}^n$. Under what conditions does y belong to $\hat{Y}$? If n=1, the answer is simple: y is in $\hat{Y}$ if and only y lies in a bounded component of $\mathbb{C}\setminus Y$.

We shall study this question for $Y \subset \mathbb{C}^2$ under the assumption that Y lies over the unit circle $\Gamma: |\lambda|=1$, i.e. we assume that the λ-projection of Y is contained in Γ. We use λ, w to denote the coordinates in $\mathbb{C}^2$.

For λ in Γ, resp. λ in $|\lambda| < 1$, we put

$Y_\lambda = \{w \in \mathbb{C} \mid (\lambda, w) \in Y\}$, resp.

$\hat{Y}_\lambda = \{w \in \mathbb{C} \mid (\lambda, w) \in \hat{Y}\}$.

The following condition (*) is a sufficient condition for (λ_o, w_o) to belong to $\hat{Y}$.

(*) There exists a bounded analytic function f on $|\lambda| < 1$ with $f(\lambda_o) = w_o$ and $f(\lambda) \in Y_\lambda$ for a.a. λ in Γ.

Suppose (*) holds. Let P be a polynomial in λ,w. The function $P(\lambda, f(\lambda))$ is then analytic and bounded on $|\lambda| < 1$. Hence its value at λ_o is bounded in modulus by $|P(\lambda_1, f(\lambda_1))|$ for some λ_1 in Γ. Now since

$f(\lambda_1) \in Y_{\lambda_1}$, $|P(\lambda_1, f(\lambda_1))| \leq \max |P(\lambda_1, w)|$ over w in $Y_{\lambda_1} \leq \max |P|$ over Y. Hence $|P(\lambda_o, w_o)| = |P(\lambda_o, f(\lambda_o))| \leq \max |P|$ over Y. Thus (λ_o, w_o) is in $\hat{Y}$.

We now look at a second consequence of (*). With f as in (*), we put

$$F(\lambda, w) = (w - f(\lambda))^{-1}.$$

Provided that $R > \sup|f|$, we have that F is a bounded analytic function on the product domain: $|\lambda| < 1$, $|w| > R$ and F has boundary values at points (λ, w) with $|\lambda| = 1$ for a.a. λ. Further we have

(i) $$F(\lambda_o, w_o) = (w - w_0)^{-1}, \quad |w| > R, \text{ and}$$

if we define, for each λ in Γ the measure σ_λ as the unit point mass in the w-plane at the point $f(\lambda)$, then

(ii) $$F(\lambda, w) = - \int_{Y_\lambda} \frac{d\sigma_\lambda(\xi)}{\xi - w}, \ |w| > R, \text{ and } \operatorname{supp} \sigma_\lambda \subseteq Y_\lambda.$$

If Y is an arbitrary compact set in $\mathbb{C}^2$ lying over Γ there may not exist any bounded analytic function F on a domain $|\lambda| < 1$, $|w| > R$ which satisfies (ii).

<u>Example</u> 1: We take $Y = \{(\lambda, \bar{\lambda}) | \lambda \text{ in } \Gamma\}$, so that $Y_\lambda = \{\bar{\lambda}\}$ for each λ in Γ. Suppose F is bounded and analytic on $|\lambda| < 1$, $|w| > R$ and satisfies (ii). Then for λ in Γ, $|w| > R$,

$$F(\lambda, w) = \frac{1}{w - \bar{\lambda}} = \frac{\lambda}{w\lambda - 1}$$

and so

$$(1 - w\lambda)\ F\ (\lambda, w) = -\lambda, \ \lambda \text{ in } \Gamma.$$

Hence $(1 - w\lambda)F(\lambda, w) = -\lambda$ for $|\lambda| < 1$, $|w| > R$, which is false for $w = 1/\lambda$. So no such F exists.

Let now Y be an arbitrary compact set in $\mathbb{C}^2$ lying over Γ and fix (λ_o, w_o) with $|\lambda_o| < 1$. We assume without loss of generality that $Y \subset \{|w| < 1\}$.

Theorem 1: (λ_o, w_o) is in $\hat{Y}$ if and only if there exists a bounded analytic function F on the domain $\{|\lambda| < 1\} \times \{|w| > 1\}$ such that

(1) $F(\lambda_o, w) = (w - w_o)^{-1}$ and

for a.a. λ in Γ there exists a probability measure σ_λ supported on Y_λ such that

(2) $$F(\lambda, w) = - \int_{Y_\lambda} \frac{d\sigma_\lambda(\xi)}{\xi - w}, \quad |w| > 1.$$

Note: Since F is bounded and analytic on $\{|\lambda| < 1\} \times \{|w| > 1\}$, $\exists$ a set Λ on Γ on full Lebesgue measure such that F has a boundary value $F(\lambda, w)$ for λ in Λ, $|w| > 1$. We interpret the left hand side of (2) as that boundary value.

Proof: In one direction of the proof, we assume the existence of F satisfying (1) and (2). Fix a polynomial P in λ, w. Then $P(\lambda,w)F(\lambda,w)$ is analytic on $\{|\lambda| < 1\} \times \{|w| > 1\}$. Put

$$\phi(\lambda) = \frac{1}{2\pi i} \int_{|w|=2} P(\lambda,w)F(\lambda,w)dw, \quad |\lambda| < 1.$$

In view of (1), $\phi(\lambda_o) = P(\lambda_o, w_o)$. Also ϕ is analytic and bounded on $|\lambda| < 1$. For λ in Γ we define $\phi(\lambda)$ as the boundary value. By (2)

$$\phi(\lambda) = \frac{1}{2\pi i} \int_{|w|=2} P(\lambda,w) \left[\int_{Y_\lambda} \frac{d\sigma_\lambda(\xi)}{\xi - w} \right] dw$$

$$= \int_{Y_\lambda} d\sigma_\lambda(\zeta) \, \frac{1}{2\pi i} \int_{|w|=2} \frac{P(\lambda,w)dw}{w-\xi} = \int_{Y_\lambda} P(\lambda,\xi)d\sigma_\lambda(\xi)$$

Hence

$$|\phi(\lambda)| \le \int_{Y_\lambda} |P(\lambda,\xi)| d\sigma_\lambda(\xi) \le \max |P| \text{ over } Y.$$

Since this inequality holds for a.a. λ in Γ, the maximum principle gives that $|\phi(\lambda_o)| \le \max |P|$ over Y. Hence $|P(\lambda_o, w_o)| \le \max|P|$ over Y. So (λ_o, w_o) is in $\hat{Y}$.

To prove the necessity of (1) and (2) we shall use analytic vectors, defined as follows: an analytic vector on $|\lambda| < 1$ is a map: $\lambda \to V(\lambda)$ from $|\lambda| < 1$ to ℓ^∞ such that

$$V(\lambda) = \left\{ c_n(\lambda) \right\}_{n=0}^{\infty},$$

where $c_0 \equiv 1$ and $c_1, c_2, \ldots,$ are bounded analytic functions on $|\lambda| < 1$. Fix (λ_0, w_0) in $\hat{Y}$, with $|\lambda_0| < 1$.

<u>Assertion</u>. There exists an analytic vector V on $|\lambda| < 1$ so that

(3) $V(\lambda_0) = \left\{ w_0^n \right\}_0^\infty$, and

(4) For a.a. λ in Γ, $\exists$ probability measure σ_λ on Y_λ such that

$$V(\lambda) = \left\{ \int_{Y_\lambda} \xi^n \, d\sigma_\lambda(\xi) \right\}_0^\infty$$

We shall prove the Assertion by adapting an argument used in [1]. Since (λ_0, w_0) is in $\hat{Y}$, there exists a representing measure μ on Y for the point (λ_0, w_0) relative to the algebra of polynomials in λ, w. By disintegration of μ under the map λ: $Y \to \Gamma$, we get probability measures σ_λ on Y_λ such that for all f in C(Y)

$$\int_Y f d\mu = \int_\Gamma d\mu^* \int_{Y_\lambda} f \, d\sigma_\lambda$$

where μ^* is the projection of μ on Γ under λ. As in [1], we identify μ^* as the Poisson measure $K(\lambda_0, \lambda)\frac{d\theta}{2\pi}$ where

$$K(\lambda_0, \lambda) = \frac{1 - |\lambda_0|^2}{|\lambda - \lambda_0|^2}$$

For each k, k = 0,1,2,... we now put

$$W_k(\lambda) = \int_{Y_\lambda} w^k \, d\sigma_\lambda(w), \ \lambda \text{ in } \Gamma.$$

Then W_k is in $L^\infty(\Gamma)$. We claim: W_k is in $H^\infty(\Gamma)$.

To see this, fix g analytic in $|\lambda| < 1$, continuous in $|\lambda| \le 1$ with $g(\lambda_0) = 0$. Then

$$0 = \int_Y g(\lambda)w^k d\mu = \int_\Gamma K(\lambda_o,\lambda)\frac{d\theta}{2\pi}\left[\int_{Y_\lambda} g(\lambda)w^k d\sigma_\lambda\right]$$

$$= \int_\Gamma K(\lambda_o,\lambda)g(\lambda)W_k \frac{d\sigma}{2\pi} \quad .$$

It readily follows from this that W_k is in H^∞, as claimed. If we denote by $W_k(\lambda_o)$ the value at λ_o of the analytic extension of W_k to the unit disk, then

$$w_o^k = \int_Y w^k d\mu = \int_\Gamma K(\lambda_o,\lambda)\frac{d\theta}{2\pi}\left[\int_{Y_\lambda} w^k d\sigma_\lambda\right] = W_k(\lambda_o).$$

We now define for each λ in $|\lambda| < 1$ the vector $V(\lambda)$ by

$$V(\lambda) = \left\{ W_k(\lambda) \right\}_{k=0}^{\infty}$$

For a.a. λ in Γ, then,

$$V(\lambda) = \left\{ \int_{Y_\lambda} w^k \, d\sigma_\lambda(w) \right\}_{k=0}^{\infty} ,$$

So (4) holds. Further

$$V(\lambda_o) = \left\{ W_k(\lambda_o) \right\}_0^\infty = \left\{ w_o^k \right\}_o^\infty .$$

So (3) holds and the Assertion is proved.

Let now V be an analytic vector satisfying (3) and (4) and write

$$V(\lambda) = \left\{ c_m(\lambda) \right\}_0^\infty , \quad |\lambda| < 1.$$

Denote by R the maximum of $|w|$ over Y. Since $Y \subset \{|w| < 1\}$, $R < 1$. In view of (4) then, $|c_m(\lambda)| \le R^m$ for $|\lambda| = 1$, and hence $|c_m(\lambda)| < R^m$ for $|\lambda| < 1$, all m. We put

$$F(\lambda,w) = \sum_{m=o}^{\infty} \frac{c_m(\lambda)}{w^{m+1}} \quad , \; |\lambda| < 1, \; |w| > 1$$

Then the series converges absolutely for $|w| > 1$ and defines F as an analytic function on $|\lambda| < 1$, $|w| > 1$.

Since $c_m(\lambda_o) = w_o^m$, $m = 0,1,2,\ldots$, for $|w| > 1$

$$F(\lambda_o, w) = \sum_{m=o}^{\infty} \frac{w_o^{\,m}}{w^{m+1}} = (w - w_o)^{-1} \quad , \text{ i.e. } (1).$$

Fix λ in Γ. Then for $|w| > 1$,

$$F(\lambda, w) = \sum_{o}^{\infty} \frac{1}{w^{m+1}} \int_{Y_\lambda} \xi^m \, d\sigma_\lambda(\xi) = \int_{Y_\lambda} \left[\sum_{0}^{\infty} \frac{\xi^m}{w^{m+1}} \right] d\sigma_\lambda(\xi)$$

$$= - \int_{Y_\lambda} \frac{d\sigma_\lambda(\xi)}{\xi - w} .$$

So (2) holds and Theorem 1 is proved.

In the proof of Theorem 1 we constructed a function F satisfying (1) and (2) by starting with a representing measure for (λ_o, w_o). This relationship is reversible.

Given a compact set Y in $\mathbb{C}^2$ lying over Γ and fix a point (λ_o, w_o). We again take $Y \subseteq \left\{|w| < 1\right\}$. We define a set K of analytic functions as follows:

K consists of all analytic functions F on $\{|\lambda| < 1\} \times \{|w| > 1\}$ such that (1) and (2) are satisfied.

<u>Assertion</u>: <u>Fix</u> F <u>in</u> K. <u>Denote</u> <u>by</u> σ_λ <u>the</u> <u>probability</u> <u>measure</u> <u>occurring</u> <u>in</u> <u>the</u> <u>representation</u> (2) <u>for</u> F. <u>Define</u> <u>the</u> <u>measure</u> μ <u>on</u> Y <u>by</u> <u>the</u> <u>formula</u>

$$\int_Y f \, d\mu = \int_\Gamma K(\lambda_o, \lambda) \frac{d\theta}{2\pi} \int_{Y_\lambda} f(\lambda, \xi) \, d\sigma_\lambda(\xi) \quad , \ f \text{ in } C(Y).$$

<u>Then</u> μ <u>is</u> <u>a</u> <u>representing</u> <u>measure</u> <u>for</u> (λ_o, w_o).

<u>Proof</u>: Evidently μ is a probability measure on Y. Fix a non-negative integer ℓ. By (1) if $|w| > 1$, then

$$\frac{\lambda_o^\ell}{w - w_o} = \lambda_o^\ell F(\lambda_o, w) = \int_\Gamma \lambda^\ell F(\lambda, w) \, K(\lambda_o, \lambda) \frac{d\theta}{2\pi}$$

$$= \int_\Gamma K(\lambda_o, \lambda) \frac{d\theta}{2\pi} \int_{Y_\lambda} \frac{-\lambda^\ell d\sigma_\lambda(\xi)}{\xi - w}, \text{ so}$$

$$-\sum_{m=o}^{\infty} \frac{\lambda_o^{\ell}\, w_o^m}{w^{m+1}} = \int_{\Gamma} K(\lambda_o,\lambda)\, \frac{d\theta}{2\pi} \sum_{m=o}^{\infty} \left[-\int_{Y_\lambda} \frac{\lambda^{\ell}\xi^m}{w^{m+1}}\, d\sigma_\lambda(\xi) \right]$$

$$= -\sum_{m=o}^{\infty} \frac{1}{w^{m+1}} \int_{\Gamma} K(\lambda_o,\, \lambda) \left[\int_{Y_\lambda} \lambda^{\ell}\, \xi^m\, d\sigma_\lambda(\xi) \right] \frac{d\theta}{2\pi}$$

$$= -\sum_{m=o}^{\infty} \frac{1}{w^{m+1}} \int_{Y} \lambda^{\ell}\, \xi^m\, d\mu(\lambda,\xi).$$

Equating coefficients, we conclude

$$\lambda_o^{\ell} w_o^m = \int_Y \lambda^{\ell} \xi^m\, d\mu \qquad \ell, m \geq 0.$$

So μ is a representing measure for (λ_o, w_o) as claimed

We now compute F satisfying (1) and (2) for a two-sheeted set Y.

Example 2: Let $Y = \{(\lambda,w) | \lambda$ in Γ and $w^2 = \lambda\}$. So Y is a curve lying two-sheeted over Γ. For each λ in Γ,

$$Y_\lambda = \left\{ \sqrt{\lambda},\ -\sqrt{\lambda} \right\}.$$

Fix λ_o real and >0 in $|\lambda| < 1$. The point $(\lambda_o, \sqrt{\lambda_o})$ is in $\hat{Y}$. What is F for this point? (2) implies

$$(5) \quad F(\lambda,w) = \frac{c_1}{w-\sqrt{\lambda}} + \frac{c_2}{w+\sqrt{\lambda}}, \quad \lambda \in \Gamma,$$

where $c_1 \geq 0$, $c_2 \geq 0$, $c_1 + c_2 = 1$ and c_1, c_2 depend on λ

We try F in the form

$$F(\lambda,w) = \frac{w+a\lambda+a}{w^2-\lambda}$$, where a is a constant to be chosen.

We want $F(\lambda_o,w) = (w-\sqrt{\lambda_o})^{-1}$, so we choose a so that

$$(6) \quad a\lambda_o + a = \sqrt{\lambda_o}.$$

Then $F(\lambda_o,w) = \dfrac{w+\sqrt{\lambda_o}}{w^2 - \lambda_o} = \dfrac{1}{w-\sqrt{\lambda_o}}$, giving (1).

By (6) $a = \sqrt{\lambda_o} / (\lambda_o+1)$, so $|a| \leq 1/2$.

Now fix λ in Γ. $F(\lambda,w) = \dfrac{w+\sqrt{\lambda}\,(a\sqrt{\lambda} + \overline{a\sqrt{\lambda}})}{w^2 - \lambda}$. Put

$t = a\sqrt{\lambda} + \overline{a\sqrt{\lambda}}$. Then t is real and $-1 \le t \le 1$, so $c = (1+t)/2$ satisfies $0 \le c \le 1$. Also $2c - 1 = a\sqrt{\lambda} + \overline{a\sqrt{\lambda}}$, so $F(\lambda,w) =$

$= \dfrac{w+\sqrt{\lambda}\,(2c-1)}{w^2-\lambda} = \dfrac{c}{w - \sqrt{\lambda}} + \dfrac{1-c}{w + \sqrt{\lambda}}$, i.e. (5). So (2) holds, and F has the desired properties.

<u>Note</u>: In [2], paragraphs 1 and 2, Slodkowski has constructed a function b which is analytic on $U = \{(\lambda,w) \mid |\lambda| < 1,\ w \notin \hat{Y}_\lambda\}$ such that $b(\lambda_o,w) = (w_o - w)^{-1}$.

<u>Geometric Characterisation of $\hat{Y}$.</u>

Let Y be an arbitrary compact set in $\mathbb{C}^2$ lying over Γ. For each λ in Γ and each positive integer N we put

$K^{(N)}(\lambda)$ = the closed convex hull in $\mathbb{C}^{N+1}$ of the set of points $(1,w,w^2,\dots,w^N)$ with w in Y_λ.

The following theorem gives a geometric interpretation of Theorem 1.

<u>Theorem 2</u>: <u>Fix (λ_o,w_o) in $\mathbb{C}^2$, $|\lambda_o|<1$. Then $(\lambda_o,w_o)\in\hat{Y}$ if an only if for each positive inter N there exists an N-tuple of bounded analytic functions $\phi_1,\dots,\phi_N$ on $|\lambda|>1$ such that</u>

(7) $\phi_1(\lambda_o) = w_o, \phi_2(\lambda_o) = w_o^2, \dots, \phi_N(\lambda_o) = w_o^N$, <u>and</u>

(8) <u>For almost all λ in Γ, the point</u>

$$(1,\phi_1(\lambda),\phi_2(\lambda),\dots,\phi_N(\lambda)) \text{ is in } K^{(N)}(\lambda).$$

<u>Proof</u>: Fix λ in Γ. For each probability measure β on Y_λ we consider the point

$$(1, \int\xi\, d\beta(\xi), \int\xi^2 d\beta(\xi), \dots, \int\xi^N d\beta(\xi)) \text{ in } \mathbb{C}^{N+1}.$$

Each such point belongs to $K^{(N)}(\lambda)$. Also the totality of all such points is a closed convex set. Hence this totality equals $K^{(N)}(\lambda)$.

Fix now N and suppose $\phi_1, \dots, \phi_N$ exist satisfying (7) and (8). Let P be a polynomial in λ, w of degree N in w. Then $P(\lambda,w) = \sum_{j=o}^{N} a_j(\lambda)w^j$.

Put $\psi(\lambda) = \sum_{j=o}^{N} a_j(\lambda)\phi_j(\lambda)$, $|\lambda|<1$. Then ψ is a bounded analytic function on $|\lambda|<1$, and by (7), $\psi(\lambda_o) = P(\lambda_o,w_o)$. We denote by Γ_o the set of all λ in Γ such that (8) holds. Then Γ_o has full measure. Fix λ in Γ_o. For $j = 0,1,\dots,N$ let $\phi_j(\lambda) = \int \xi^j d\beta(\xi)$, where β is a suitable probability measure on Y_λ, and we have put $\phi_o \equiv 1$.

Hence

$$\psi(\lambda) = \sum_j a_j(\lambda) \int \xi^j d\beta(\xi) = \int (\sum_j a_j(\lambda)\xi^j) d\beta(\xi)$$

$$= \int P(\lambda,\xi)d\beta(\xi)$$

It follows that

$$|\psi(\lambda)| \le \max_{\xi\in Y_\lambda} |P(\lambda,\xi)| \le \|P\|_Y = \max_Y |P|.$$

Since this inequality holds for almost all λ in Γ, $|P(\lambda_o,w_0)| = |\psi(\lambda_o)| \le \|P\|_Y$. We thus have: if P is a polynomial of degree N in w, then

$$|P(\lambda_o,w_o)| \le \|P\|_Y.$$

If (7) and (8) hold for each N, then, $(\lambda_o,w_o)\in\hat{Y}$.

Conversely, suppose that $(\lambda_o,w_o)\in\hat{Y}$, $|\lambda_o|<1$. In the proof of theorem 1 above we constructed for a.a. λ in Γ a probability measure σ_λ, supported on Y, and we constructed a sequence of functions W_k, $k = 1,2,\dots$ in H^∞ such that

$$W_k(\lambda_o) = w_o^k, \; k = 1,2,\dots \text{ and}$$

$$W_k(\lambda) = \int \xi^k d\sigma_\lambda(\xi), \; k = 1,2,\dots$$

Fix N and put $\phi_k(\lambda) = W_k(\lambda)$, $1 \le k \le N$. Then (7) and (8) hold. This is true for each positive integer N. So the converse holds.

REFERENCES

[1] H. Alexander and J. Wermer, Polynomial Hulls with Convex Fibers, Math. Ann. 271, 99-109 (1985)

[2] Z. Slodkowski, Uniform algebras and analytic multifunctions, Atti della Accad. Nazion. dei Lincei,, vol. 75 (1983), 9-18

DEPARTMENT OF MATHEMATICS
BROWN UNIVERSITY
PROVIDENCE, R.I. 02912

CONVERGENCE OF FORMAL POWER SERIES AND ANALYTIC EXTENSION

J. Wiegerinck*

1. Introduction.

In this paper we study convergence properties of formal power series over $\mathbb{C}$ of n variables. Assume that there exists a big enough set of complex lines through the origin, such that the restriction of the series to any of these lines converges in some neighborhood of the origin. Then we prove in Theorem 1 convergence of the multiple series. Big enough has to be understood as having positive projective capacity.

Of course, the domain of convergence of such a series will be a Reinhardt domain. However, the holomorphic function which it represents may extend to a larger domain. For compact sets of lines, we give a description of the minimal domain to which the series extends holomorphically, Theorem 2.

Already Hartogs [3] gave a version of Theorem 1, namely the case where the restriction to every complex line through 0 converges. Abhyankar and Moh [1] and later Sathaye [11] studied a closely related problem. They considered convergence on complex hyperplanes of the form $H_\alpha = \{(\alpha z_2, z_2, \ldots, z_n)\}$, $\alpha \in \mathbb{C}$ and proved that if for α in a set of positive Hausdorff dimension, [1], or even of positive logarithmic capacity, [11], the restriction of a series to each H_α converges, then the series itself converges. Levenberg and Molzon [8] obtained a different proof of Theorem 1. They also showed how to reduce the case of complex subspaces to that of complex lines. In [15] Korevaar and the author obtained special cases of Theorems 1 and 2. Korevaar shows in [5] that Theorem 2 is implicit in Siciak's fundamental work [12]. In fact that approach allows to obtain a result like Theorem 2 in the non- compact case, but the proof is more involved.

Applications, among others to the edge-of-the-wedge theorem and to the support theorem for the Radon transformation are in [6] and

---

*Supported by the Netherland's organization for the advancement of pure research Z.W.O.

[14]. We include a stronger version of the latter result in section 4.

2. Definitions and facts.

We recall some basic properties of capacities in $\mathbb{C}^n$ and a well-known fact about Jensen measures.

A set $E \subset \mathbb{C}^n$ is called circular if $z \in E \Rightarrow e^{i\theta}z \in E$ for all $\theta \in (0,2\pi)$. In the sequel S denotes the boundary of the unit ball in $\mathbb{C}^n$, $\mathcal{C}$ will denote the collection of bounded circular subsets of $\mathbb{C}^n$, finally $\| \ \|_K$ denotes the sup norm on a compactum K.

Definition. For K compact $\in \mathcal{C}$ define

$$\text{Cap } K = \inf_{\substack{\text{homogeneous} \\ \text{polynomials } P}} \left(\frac{\|P\|_K}{\|P\|_S}\right)^{\frac{1}{\text{degree } P}},$$

for $\Omega \in \mathcal{C}$ open

$$\text{Cap } \Omega = \sup_{\substack{K \subset\subset \Omega \\ K \in \mathcal{C}}} \text{Cap } K ,$$

and for general $E \in \mathcal{C}$

$$\text{Cap } E = \inf_{\substack{\Omega \supset E \\ \Omega \in \mathcal{C} \text{ open}}} \text{Cap } \Omega .$$

We call Cap the projective capacity. Cap defines in fact an outer capacity on $\mathbb{P}^{n-1}$, the set of complex lines through the origin, via the obvious identification of subsets of $\mathbb{P}^{n-1}$ with circular subsets of S. The following properties of Cap are important and well-known, cf. [2, 9, 12].

$E_1 \subset E_2 \Rightarrow \text{Cap } E_1 \le \text{Cap } E_2$;

$\text{Cap } E = 0 \Leftrightarrow$ There exists a (logarithmically-homogeneous) plurisubharmonic function h such that $h \not\equiv -\infty$, $h|_E = -\infty$ (i.e., E is pluripolar);

$\text{Cap } E_i = 0,\ i = 1,2,\ldots \Rightarrow \text{Cap } (\bigcup_i E_i) = 0$;

$\text{Cap } K > 0 \Leftrightarrow$ the polynomially convex hull $\hat{K}$ of K contains a neighborhood of 0.

Examples.

1. $E \in \mathcal{C}$, E open $\Rightarrow \text{Cap } E > 0$.
2. $E \in \mathcal{C}$, E has positive Lebesgue measure $\Rightarrow \text{Cap } E > 0$.

For the applications the following ones are of interest. Let S^{n-1} denote the real unit sphere embedded in S in the usual way.

3. $E \in \mathcal{E}$, $E \cap S^{n-1}$ is open in $S^{n-1} \Rightarrow \operatorname{Cap} E > 0$.

4. $E \in \mathcal{E}$, $E \cap S^{n-1}$ has positive Lebesgue measure in $S^{n-1} \Rightarrow$ $\operatorname{Cap} E > 0$.

Let $P(X)$ denote the uniform algebra spanned by the polynomials on a compactum $X \subset \mathbb{C}^n$ and let $\hat{X}$ be the polynomially convex hull of X. Recall that a Jensen measure for $x \in \hat{X}$ is a positive measure μ on X such that

$$\log|f(x)| \le \int_X \log|f|d\mu, \quad f \in P(X) .$$

We need the following well-known fact, cf. [7, 13]. For every $x \in \hat{X}$ there exists a Jensen measure for x which is supported on the Šilov boundary of X (relative to $P(X)$).

3. Results.

We have to give one more definition, using standard multi-index notation. The restriction of a formal power series $F(z) = \sum_{|\alpha|\ge 0} c_\alpha z^\alpha \in \mathbb{C}\{z_1,\dots,z_n\}$ to the complex line L through $z_0 \in \mathbb{C}^n$ and the origin is the formal power series of one variable:

$$f(t) = \sum_{k=0}^{\infty} \left[\sum_{|\alpha|=k} c_\alpha z_0^\alpha \right] t^k .$$

We will also write this as $F(tz_0)$ or $F|_L$. Of course, $F(tz_0)$ may converge, while F does not.

Lemma. Let F be a formal power series. The set E of complex lines L through 0 such that $F|_L$ converges on a neighborhood of 0 is an F_σ in $\mathbb{P}^{n-1}$.

Proof. Let $F(z) = \sum_\alpha c_\alpha z^\alpha$ and let $P_k(z) = \sum_{|\alpha|=k} c_\alpha z^\alpha$. The subset of S which corresponds to E can be written as

$\{z \in S : \exists j \in \mathbb{N}$ such that $|P_k(z)|^{1/k} \le j$, for all $k = 1,2,\dots\}$.

This is clearly an F_σ, hence so is E.

Theorem 1. Let F be a formal power series. Suppose that there exists a set E of complex lines through 0 such that E is not contained in a pluripolar F_σ and for every $L \in E$, the restriction $F|_L$ converges on a neighborhood of 0 in $\mathbb{C}$. Then F converges on a neighborhood of 0 in $\mathbb{C}^n$.

Proof. In view of the lemma we can assume $\operatorname{Cap} E > 0$. Let $F(z) = \sum_\alpha c_\alpha z^\alpha$. We identify E with the corresponding subset of S. For $w \in E$ we have

$$F(tw) = \sum P_k(w)t^k, \quad \text{where} \quad P_k(w) = \sum_{|\alpha|=k} c_\alpha w^\alpha .$$

Since this series converges for some $t \neq 0$ there exists a positive constant C_w depending on w, such that for all integers $m \geq 1$ $|P_m(w)| < C_w^m$. In other words, the function

$$\phi(w) := \sup_{m=1,2,\ldots} |P_m(w)|^{1/m}, \quad w \in S$$

is finite for $w \in E$. Let $V_j = \{w \in S : \phi(w) \leq j\}$, a closed subset of S. Now $E \subset \bigcup_j V_j$ and by the properties of Cap, some V_k will have positive capacity. We infer

$$\|P_m\|_S^{1/m} \leq \frac{k}{\operatorname{Cap} V_k}, \quad \text{for all} \quad m \geq 1.$$

From this and the homogeneity of the P_m, we conclude that the series $\sum P_m(z)$ converges uniformly on compacta in the ball with radius $\frac{1}{k}\operatorname{Cap} V_k$. Clearly F will be the power series expansion of the limit, and will therefore converge.

Remark. It was shown by Siciak [12] and also be Levenberg and Molzon [8] that if $\operatorname{Cap} E = 0$ and E is an F_σ, then there exists a formal power series whose restriction to lines $L \in E$ converges, while the series itself is divergent. Therefore Theorem 1 is optimal.

Note that there exist sets of capacity 0 which are not contained in an F_σ of capacity 0. I learned the following example from U. Cegrell:

Example. Let Q be a countable dense subset of S. Since $\operatorname{Cap} Q = 0$ there exists a homogeneous plurisubharmonic function p which is identically $-\infty$ on Q. Let $E = \{z \in S : p(z) = -\infty\}$. Then E is a G_δ of capacity 0 and not contained in an F_σ of capacity 0. Indeed, let $\tilde{E}$ be an F_σ, $E \subset \tilde{E}$. Observe that $(S \backslash E) \cup \tilde{E}$ is a countable union of closed sets that contains S. Baire's theorem implies that one of these sets, say F, has non-empty interior (relative to S). Now $F \cap Q \neq \emptyset$ implies $F \subset S \backslash E$, hence $F \subset \tilde{E}$. The conclusion is that $\tilde{E}$ has non-empty interior and therefore positive capacity.

Theorem 2. Let K a circular compactum in $\mathbb{C}^n$, Cap $K > 0$. Suppose F is a formal power series such that for $w \in K$ the restriction $F(tw)$ converges for $|t| < 1$. Then the holomorphic function to which F converges on a neighborhood of 0 has an analytic continuation to Ω, the interior of the polynomially convex hull $\hat{K}$ of K.

Proof. Again write $F(z) = \sum_{k=0}^{\infty} P_k(z)$, where P_k is a homogeneous polynomial of degree k. We will show that the series $\sum P_k(z)$ converges uniformly on compacta in Ω. The set of complex lines through the origin that meet K has positive capacity. Hence Theorem 1 gives a positive A such that

$$(1) \qquad \overline{\lim_{k\to\infty}} \sup_{z\in S} |P_k(z)|^{1/k} \leq A .$$

By assumption, for every $z \in k$

$$(2) \qquad \overline{\lim_{k\to\infty}} |P_k(z)|^{1/k} \leq 1 .$$

Now let $z \in \hat{K}$ and let μ_z be a Jensen measure for z supported on K, so that

$$\frac{1}{k} \log|P_k(z)| \leq \int_K \frac{1}{k} \log|P_k| d\mu_z .$$

By (1), (2) and Fatou's lemma, we get for $z \in \hat{K}$

$$(3) \qquad \overline{\lim_{k\to\infty}} \frac{1}{k} \log|P_k(z)| \leq 0 .$$

Because of (1) Hartogs' lemma, cf. [3], applies to give (3) uniformly on compacta in Ω.

Since Ω is starlike relative to 0 and the P_k are homogeneous, the right-hand side of (3) may be replaced by a number $-\varepsilon < 0$ for any given compact subset of Ω. We conclude that the series $\sum P_k(z)$ converges uniformly on compacta in Ω.

In [15] Korevaar and the author were interested in a slightly different setup corresponding to the following

Corollary. Let Ω be a bounded circular domain in $\mathbb{C}^n$ with Šilov boundary K relative to $P(\bar{\Omega})$. Let H be defined by:

$$H = \{z : \exists t \in \mathbb{C},\ |t| < 1,\ w \in K \text{ such that } z = tw\}.$$

If f is a formal power series such that its restriction to every disc in H with center 0 is holomorphic, then f extends holomorphically to the interior of $\hat{K}$ (which contains Ω).

For the proof, observe that K is circular because Ω is and that $\mathrm{Cap}\, K > 0$, because $\hat{K} \supset \Omega$. Now apply Theorem 2.

4. An application to Radon transforms.

Let f be a rapidly decreasing, continuous function on $\mathbb{R}^n$. The Radon transform $\hat{f}$ is defined as follows

$$\hat{f}(w,t) = \int_{w\cdot s=t} f(s)dm(s),\ (w,t) \in S^{n-1} \times \mathbb{R} .$$

Here $\cdot$ denotes the Euclidean inner product on $\mathbb{R}^n$ and $m(s)$ the Lebesgue measure on the hyperplane $w\cdot s = t$.

We think of $\mathbb{R}^n$ as the real subspace of $\mathbb{C}^n$, so that the real unit sphere S^{n-1} is a subset of the complex unit sphere. For $\Omega \subset S^{n-1}$ it is convenient to define $\mathrm{Cap}\,\Omega$ as $\mathrm{Cap}\{e^{i\theta}w : \theta \in (0,2\pi),\ w \in \Omega\}$.

The following theorem strengthens the result of [14].

Theorem 3. Let f be a rapidly decreasing, continuous function on $\mathbb{R}^n$, whose Radon transform satisfies

$$(1) \qquad |\hat{f}(w,t)| < C_w e^{-\varepsilon_w t},\ \varepsilon_w > 0,\ (w,t) \in S^{n-1} \times \mathbb{R} .$$

Let $\Omega \subset S^{n-1}$, $\mathrm{Cap}\,\Omega > 0$ such that for $w \in \Omega$ $\hat{f}_w(t) = \hat{f}(w,t)$ has compact support. Then f has compact support.

Proof. Since f is continuous and rapidly decreasing, the Fourier transform $\mathcal{F}f$ is in $L^2(\mathbb{R}^n) \cap C^\infty(\mathbb{R}^n)$. By Fubini's theorem

$$(5) \qquad \mathcal{F}f(\lambda w) = \int_{\mathbb{R}^n} f(s)e^{-i\lambda w\cdot s}ds = \int_{-\infty}^{\infty} \hat{f}(w,t)e^{-i\lambda t}dt,\ \lambda \in \mathbb{R},\ w \in S^{n-1}.$$

It follows from (4) that for w fixed $\mathcal{F}f(\lambda w)$ can be extended holomorphically in λ to a neighborhood of the real λ-axis. Also, for $w \in \Omega$ $\mathcal{F}f(\lambda w)$ can be extended to an entire function of exponential type in the λ-variable. Hence by Theorem 1 the Taylor series of $\mathcal{F}f$ at 0 defines an entire function on $\mathbb{C}^n$, which is an extension of $\mathcal{F}f$ in view of the real analyticity of $\mathcal{F}f$ on the lines $\{\lambda w,\ \lambda \in \mathbb{R}\}$. In fact, the extension of $\mathcal{F}f$ will be of exponential type. This follows from the Sibony-Wong theorem [10], or directly as follows. Let $\mathcal{F}f(\lambda w) = \sum P_k(w)t^k$. Replacing Ω by a smaller subset of positive capacity, we can assume that for some $R > 0$ $\hat{f}(w,t) = 0$ if $|t| > R$, $w \in \Omega$. For $w \in \Omega$ the last integral in (5) is now over the interval $[-R,R]$ and we find $|P_k(w)| \le \frac{CR^{k+1}}{k!}$, $w \in \Omega$. This implies $|P_k(w)| \le C\left(\frac{R}{\mathrm{Cap}\ \Omega}\right)^{k+1} \cdot \frac{1}{k!}$ for $w \in S$. In turn, this gives $\mathcal{F}f$ is of exponential type. Now it follows from the Paley-Wiener theorem (or its precise form, the Plancherel-Pólya theorem) that f has compact support.

References

[1] Abhyankar, S.S. and T. Moh, A reduction theorem for divergent series, J. Reine Angew. Math. 241 (1970) 27-33.

[2] Alexander, H., Projective capacity. In P: Recent developments in several complex variables, 3-27. J.E. Fornaess, ed., Ann. of Math. studies 100, Princeton Univ. Press, 1981.

[3] Hartogs, F., Zur Theorie der analytischen Funktionen mehrerer unabhängiger Veränderlichen Math. Ann. 62 (1906) 1-88.

[4] Hörmander, L., An introduction to complex analysis in several variables. North Holland Publ. Co., Amsterdam, 1973.

[5] Korevaar, J., Polynomial approximation numbers, capacities and extended Green functions for $\mathbb{C}$ and $\mathbb{C}^n$. In: Proc. fifth Texas symposium on approximation theory (1986) Acad. Press (to appear).

[6] Korevaar, J. and J. Wiegerinck, A representation of mixed derivatives with an application to the edge-of-the-wedge theorem. Nederl. Akad. Wetensch. Proc. Ser. A 88 (1985) 77-86.

[7] Leibowitz, G.M., Lectures on complex function algebras. Scott, Foresman and Co., 1970.

[8] Levenberg, N. and R.E. Molzon, Convergence sets of formal power series. Preprint, Univ. of Kentucky, 1985.

[9] Levenberg, N. and B.A. Taylor, Comparison of capacities in $\mathbb{C}^n$. In: Analyse complexe, 162-172. E. Amar. ea., ed., LNM 1094 Springer, Berlin etc., 1984.

[10] Sibony, N. and P.M. Wong, Some results on global analytic sets. In: Sém. Lelong-Skoda 1978/1979, 221-237. LNM 822 Springer, Berlin etc., 1980.

[11] Sathaye, A., Convergence sets of divergent power series. J. Reine Angew. Math. 283 (1976) 86-98.

[12] Siciak, J., Extremal plurisubharmonic functions and capacities in $\mathbb{C}^n$. Sophia Kokyoroku in Math. 14, Sophia Univ. Tokyo, 1982.

[13] Stout, E.L., The theory of uniform algebras. Bogden & Quigley, Tarrytown-on-Hudson, NY, 1971.

[14] Wiegerinck, J., A support theorem for Radon transforms on $\mathbb{R}^n$. Nederl. Akad. Wetensch. Proc. Ser. A 88 (1985) 77-86.

[15] Wiegerinck, J. and J. Korevaar, A lemma on mixed derivatives and results on holomorphic extension. Nederl. Akad. Wetensch. Proc. Ser. A 88, (1985) 351-362.

Department of Mathematics
Princeton University
Fine Hall - Washington Road
Princeton, New Jersey 08544

LECTURE NOTES IN MATHEMATICS
Edited by A. Dold and B. Eckmann

Some general remarks on the publication of proceedings of congresses and symposia

Lecture Notes aim to report new developments - quickly, informally and at a high level. The following describes criteria and procedures which apply to proceedings volumes.

1. One (or more) expert participant(s) of the meeting should act as the responsible editor(s) of the proceedings. They select the papers which are suitable (cf. points 2, 3) for inclusion in the proceedings, and have them individually refereed (as for a journal). It should not be assumed that the published proceedings must reflect conference events faithfully and in their entirety. Contributions to the meeting which are not included in the proceedings can be listed by title. The series editors will normally not interfere with the editing of a particular proceedings volume - except in fairly obvious cases, or on technical matters, such as described in points 2, 3. The names of the responsible editors appear on the title page of the volume.

2. The proceedings should be reasonably homogeneous (concerned with a limited area). For instance, the proceedings of a congress on "Analysis" or "Mathematics in Wonderland" would normally not be sufficiently homogeneous.

   One or two longer survey articles on recent developments in the field are often very useful additions to such proceedings - even if they do not correspond to actual lectures at the congress. An extensive introduction on the subject of the congress would be desirable.

3. The contributions should be of a high mathematical standard and of current interest. Research articles should present new material and not duplicate other papers already published or due to be published. They should contain sufficient information and motivation and they should present proofs, or at least outlines of such, in sufficient detail to enable an expert to complete them. Thus resumes and mere announcements of papers appearing elsewhere cannot be included, although more detailed versions of a contribution may well be published in other places later.

   Surveys, if included, should cover a sufficiently broad topic, and should in general not simply review the author's own recent research. In the case of surveys, exceptionally, proofs of results may not be necessary.

   The editors of a volume are strongly advised to inform contributors about these points at an early stage.

.../...

4. Proceedings should appear soon after the meeeting. The publisher should, therefore, receive the complete manuscript within nine months of the date of the meeting at the latest.

5. Plans or proposals for proceedings volumes should be sent to one of the editors of the series or to Springer-Verlag Heidelberg. They should give sufficient information on the conference or symposium, and on the proposed proceedings. In particular, they should contain a list of the expected contributions with their prospective length. Abstracts or early versions (drafts) of some of the contributions are very helpful.

6. Lecture Notes are printed by photo-offset from camera-ready typed copy provided by the editors. For this purpose Springer-Verlag provides editors with technical instructions for the preparation of manuscripts and these should be distributed to all contributing authors. Springer-Verlag can also, on request, supply stationery on which the prescribed typing area is outlined. Some homogeneity in the presentation of the contributions is desirable.

   Careful preparation of manuscripts will help keep production time short and ensure a satisfactory appearance of the finished book. The actual production of a Lecture Notes volume normally takes 6 -8 weeks.

   Manuscripts should be at least 100 pages long. The final version should include a table of contents.

7. Editors receive a total of 50 free copies of their volume for distribution to the contributing authors, but no royalties. (Unfortunately, no reprints of individual contributions can be supplied.) They are entitled to purchase further copies of their book for their personal use at a discount of 33 1/3%, other Springer mathematics books at a discount of 20% directly from Springer-Verlag.

   Commitment to publish is made by letter of intent rather than by signing a formal contract. Springer-Verlag secures the copyright for each volume.

LECTURE NOTES

ESSENTIALS FOR THE PREPARATION OF CAMERA-READY MANUSCRIPTS

Springer

Springer-Verlag
Berlin Heidelberg New York
London Paris Tokyo

The preparation of manuscripts which are to be reproduced by photo-offset requires special care. Manuscripts which are submitted in technically unsuitable form will be returned to the author for retyping. There is normally no possibility of carrying out further corrections after a manuscript is given to production. Hence it is crucial that the following instructions be adhered to closely. If in doubt, please send us 1 - 2 sample pages for examination.

Typing area. On request, Springer-Verlag will supply special paper with the typing area outlined.

The CORRECT TYPING AREA is 18 x 26 1/2 cm (7,5 x 11 inches).

Make sure the TYPING AREA IS COMPLETELY FILLED. Set the margins so that they precisely match the outline and type right from the top to the bottom line. (Note that the page-number will lie outside this area). Lines of text should not end more than three spaces inside or outside the right margin (see example on page 4).

Type on one side of the paper only.

Type. Use an electric typewriter if at all possible. CLEAN THE TYPE before use and always use a BLACK ribbon (a carbon ribbon is best).

Choose a type size large enough to stand reduction to 75%.

Word Processors. Authors using word-processing or computer-typesetting facilities should follow these instructions with obvious modifications. Please note with respect to your printout that
i) the characters should be sharp and sufficiently black;
ii) if the size of your characters is significantly larger or smaller than normal typescript characters, you should adapt the length and breadth of the text area proportionally keeping the proportions 1:0.68.
iii) it is not necessary to use Springer's special typing paper. Any white paper of reasonable quality is acceptable.
IF IN DOUBT, PLEASE SEND US 1-2 SAMPLE PAGES FOR EXAMINATION. We will be glad to give advice.

Spacing and Headings (Monographs). Use ONE-AND-A-HALF line spacing in the text. Please leave sufficient space for the title to stand out clearly and do NOT use a new page for the beginning of subdivisions of chapters. Leave THREE LINES blank above and TWO below headings of such subdivisions.

Spacing and Headings (Proceedings). Use ONE-AND-A-HALF line spacing in the text. Start each paper on a NEW PAGE and leave sufficient space for the title to stand out clearly. However, do NOT use a new page for the beginning of subdivisions of a paper. Leave THREE LINES blank above and TWO below headings of such subdivisions. Make sure headings of equal importance are in the same form.

The first page of each contribution should be prepared in the same way. Therefore, we recommend that the editor prepares a sample page and passes it on to the authors together with these ESSENTIALS. Please take

.../...

the following as an example.

MATHEMATICAL STRUCTURE IN QUANTUM FIELD THEORY

John E. Robert
Fachbereich Physik, Universität Osnabrück
Postfach 44 69, D-4500 Osnabrück

Please leave THREE LINES blank below heading and address of the author. THEN START THE ACTUAL TEXT OF YOUR CONTRIBUTION.

Footnotes. These should be avoided. If they cannot be avoided, place them at the foot of the page, separated from the text by a line 4 cm long, and type them in SINGLE LINE SPACING to finish exactly on the outline.

Symbols. Anything which cannot be typed may be entered by hand in BLACK AND ONLY BLACK ink. (A fine-tipped rapidograph is suitable for this purpose; a good black ball-point will do, but a pencil will not). Do not draw straight lines by hand without a ruler (not even in fractions).

Equations and Computer Programs. Equations and computer programs should begin four spaces inside the left margin. Should the equations be numbered, then each number should be in brackets at the right-hand edge of the typing area.

Pagination. Number pages in the upper right-hand corner in LIGHT BLUE OR GREEN PENCIL ONLY. The final page numbers will be inserted by the printer.

There should normally be NO BLANK PAGES in the manuscript (between chapters or between contributions) unless the book is divided into Part A, Part B for example, which should then begin on a right-hand page.

It is much safer to number pages AFTER the text has been typed and corrected. Page 1 (Arabic) should be THE FIRST PAGE OF THE ACTUAL TEXT. The Roman pagination (table of contents, preface, abstract, acknowledgements, brief introductions, etc.) will be done by Springer-Verlag.

Corrections. When corrections have to be made, cut the new text to fit and PASTE it over the old. White correction fluid may also be used.

Never make corrections or insertions in the text by hand.

If the typescript has to be marked for any reason, e.g. for TEMPORARY page numbers or to mark corrections for the typist, this can be done VERY FAINTLY with BLUE or GREEN PENCIL but NO OTHER COLOR: these colors do not appear after reproduction.

Table of Contents. It is advisable to type the table of contents later, copying the titles from the text and inserting page numbers.

Literature References. These should be placed at the end of each paper or chapter, or at the end of the work, as desired. Type them with single line spacing and start each reference on a new line.
Please ensure that all references are COMPLETE and PRECISE.

Vol. 1117: D.J. Aldous, J.A. Ibragimov, J. Jacod, Ecole d'Été de Probabilités de Saint-Flour XIII – 1983. Édité par P.L. Hennequin. IX, 409 pages. 1985.

Vol. 1118: Grossissements de filtrations: exemples et applications. Seminaire, 1982/83. Edité par Th. Jeulin et M. Yor. V, 315 pages. 1985.

Vol. 1119: Recent Mathematical Methods in Dynamic Programming. Proceedings, 1984. Edited by I. Capuzzo Dolcetta, W.H. Fleming and T. Zolezzi. VI, 202 pages. 1985.

Vol. 1120: K. Jarosz, Perturbations of Banach Algebras. V, 118 pages. 1985.

Vol. 1121: Singularities and Constructive Methods for Their Treatment. Proceedings, 1983. Edited by P. Grisvard, W. Wendland and J.R. Whiteman. IX, 346 pages. 1985.

Vol. 1122: Number Theory. Proceedings, 1984. Edited by K. Alladi. VII, 217 pages. 1985.

Vol. 1123: Séminaire de Probabilités XIX 1983/84. Proceedings. Edité par J. Azéma et M. Yor. IV, 504 pages. 1985.

Vol. 1124: Algebraic Geometry, Sitges (Barcelona) 1983. Proceedings. Edited by E. Casas-Alvero, G.E. Welters and S. Xambó-Descamps. XI, 416 pages. 1985.

Vol. 1125: Dynamical Systems and Bifurcations. Proceedings, 1984. Edited by B.L.J. Braaksma, H.W. Broer and F. Takens. V, 129 pages. 1985.

Vol. 1126: Algebraic and Geometric Topology. Proceedings, 1983. Edited by A. Ranicki, N. Levitt and F. Quinn. V, 423 pages. 1985.

Vol. 1127: Numerical Methods in Fluid Dynamics. Seminar. Edited by F. Brezzi, VII, 333 pages. 1985.

Vol. 1128: J. Elschner, Singular Ordinary Differential Operators and Pseudodifferential Equations. 200 pages. 1985.

Vol. 1129: Numerical Analysis, Lancaster 1984. Proceedings. Edited by P.R. Turner. XIV, 179 pages. 1985.

Vol. 1130: Methods in Mathematical Logic. Proceedings, 1983. Edited by C.A. Di Prisco. VII, 407 pages. 1985.

Vol. 1131: K. Sundaresan, S. Swaminathan, Geometry and Nonlinear Analysis in Banach Spaces. III, 116 pages. 1985.

Vol. 1132: Operator Algebras and their Connections with Topology and Ergodic Theory. Proceedings, 1983. Edited by H. Araki, C.C. Moore, Ş. Strătilă and C. Voiculescu. VI, 594 pages. 1985.

Vol. 1133: K.C. Kiwiel, Methods of Descent for Nondifferentiable Optimization. VI, 362 pages. 1985.

Vol. 1134: G.P. Galdi, S. Rionero, Weighted Energy Methods in Fluid Dynamics and Elasticity. VII, 126 pages. 1985.

Vol. 1135: Number Theory, New York 1983–84. Seminar. Edited by D.V. Chudnovsky, G.V. Chudnovsky, H. Cohn and M.B. Nathanson. V, 283 pages. 1985.

Vol. 1136: Quantum Probability and Applications II. Proceedings, 1984. Edited by L. Accardi and W. von Waldenfels. VI, 534 pages. 1985.

Vol. 1137: Xiao G., Surfaces fibrées en courbes de genre deux. IX, 103 pages. 1985.

Vol. 1138: A. Ocneanu, Actions of Discrete Amenable Groups on von Neumann Algebras. V, 115 pages. 1985.

Vol. 1139: Differential Geometric Methods in Mathematical Physics. Proceedings, 1983. Edited by H. D. Doebner and J. D. Hennig. VI, 337 pages. 1985.

Vol. 1140: S. Donkin, Rational Representations of Algebraic Groups. VII, 254 pages. 1985.

Vol. 1141: Recursion Theory Week. Proceedings, 1984. Edited by H.-D. Ebbinghaus, G.H. Müller and G.E. Sacks. IX, 418 pages. 1985.

Vol. 1142: Orders and their Applications. Proceedings, 1984. Edited by I. Reiner and K. W. Roggenkamp. X, 306 pages. 1985.

Vol. 1143: A. Krieg, Modular Forms on Half-Spaces of Quaternions. XIII, 203 pages. 1985.

Vol. 1144: Knot Theory and Manifolds. Proceedings, 1983. Edited by D. Rolfsen. V, 163 pages. 1985.

Vol. 1145: G. Winkler, Choquet Order and Simplices. VI, 143 pages. 1985.

Vol. 1146: Séminaire d'Algèbre Paul Dubreil et Marie-Paule Malliavin. Proceedings, 1983–1984. Edité par M.-P. Malliavin. IV, 420 pages. 1985.

Vol. 1147: M. Wschebor, Surfaces Aléatoires. VII, 111 pages. 1985.

Vol. 1148: Mark A. Kon, Probability Distributions in Quantum Statistical Mechanics. V, 121 pages. 1985.

Vol. 1149: Universal Algebra and Lattice Theory. Proceedings, 1984. Edited by S. D. Comer. VI, 282 pages. 1985.

Vol. 1150: B. Kawohl, Rearrangements and Convexity of Level Sets in PDE. V, 136 pages. 1985.

Vol 1151: Ordinary and Partial Differential Equations. Proceedings, 1984. Edited by B.D. Sleeman and R.J. Jarvis. XIV, 357 pages. 1985.

Vol. 1152: H. Widom, Asymptotic Expansions for Pseudodifferential Operators on Bounded Domains. V, 150 pages. 1985.

Vol. 1153: Probability in Banach Spaces V. Proceedings, 1984. Edited by A. Beck, R. Dudley, M. Hahn, J. Kuelbs and M. Marcus. VI, 457 pages. 1985.

Vol. 1154: D.S. Naidu, A.K. Rao, Singular Pertubation Analysis of Discrete Control Systems. IX, 195 pages. 1985.

Vol. 1155: Stability Problems for Stochastic Models. Proceedings, 1984. Edited by V.V. Kalashnikov and V.M. Zolotarev. VI, 447 pages. 1985.

Vol. 1156: Global Differential Geometry and Global Analysis 1984. Proceedings, 1984. Edited by D. Ferus, R.B. Gardner, S. Helgason and U. Simon. V, 339 pages. 1985.

Vol. 1157: H. Levine, Classifying Immersions into $\mathbb{R}^4$ over Stable Maps of 3-Manifolds into $\mathbb{R}^2$. V, 163 pages. 1985.

Vol. 1158: Stochastic Processes – Mathematics and Physics. Proceedings, 1984. Edited by S. Albeverio, Ph. Blanchard and L. Streit. VI, 230 pages. 1986.

Vol. 1159: Schrödinger Operators, Como 1984. Seminar. Edited by S. Graffi. VIII, 272 pages. 1986.

Vol. 1160: J.-C. van der Meer, The Hamiltonian Hopf Bifurcation. VI, 115 pages. 1985.

Vol. 1161: Harmonic Mappings and Minimal Immersions, Montecatini 1984. Seminar. Edited by E. Giusti. VII, 285 pages. 1985.

Vol. 1162: S.J.L. van Eijndhoven, J. de Graaf, Trajectory Spaces, Generalized Functions and Unbounded Operators. IV, 272 pages. 1985.

Vol. 1163: Iteration Theory and its Functional Equations. Proceedings, 1984. Edited by R. Liedl, L. Reich and Gy. Targonski. VIII, 231 pages. 1985.

Vol. 1164: M. Meschiari, J.H. Rawnsley, S. Salamon, Geometry Seminar "Luigi Bianchi" II – 1984. Edited by E. Vesentini. VI, 224 pages. 1985.

Vol. 1165: Seminar on Deformations. Proceedings, 1982/84. Edited by J. Ławrynowicz. IX, 331 pages. 1985.

Vol. 1166: Banach Spaces. Proceedings, 1984. Edited by N. Kalton and E. Saab. VI, 199 pages. 1985.

Vol. 1167: Geometry and Topology. Proceedings, 1983–84. Edited by J. Alexander and J. Harer. VI, 292 pages. 1985.

Vol. 1168: S.S. Agaian, Hadamard Matrices and their Applications. III, 227 pages. 1985.

Vol. 1169: W.A. Light, E.W. Cheney, Approximation Theory in Tensor Product Spaces. VII, 157 pages. 1985.

Vol. 1170: B.S. Thomson, Real Functions. VII, 229 pages. 1985.

Vol. 1171: Polynômes Orthogonaux et Applications. Proceedings, 1984. Edité par C. Brezinski, A. Draux, A.P. Magnus, P. Maroni et A. Ronveaux. XXXVII, 584 pages. 1985.

Vol. 1172: Algebraic Topology, Göttingen 1984. Proceedings. Edited by L. Smith. VI, 209 pages. 1985.

Vol. 1173: H. D...
329 pages. 198...

Vol. 1174: Categories in Continuum Physics, Buffalo 1982. ...
Edited by F.W. Lawvere and S.H. Schanuel. V, 126 pages. ...

Vol. 1175: K. Mathiak, Valuations of Skew Fields and ...
Hjelmslev Spaces. VII, 116 pages. 1986.

Vol. 1176: R.R. Bruner, J.P. May, J.E. McClure, M. ...
H_∞ Ring Spectra and their Applications. VII, 388 pages. ...

Vol. 1177: Representation Theory I. Finite Dimens...
Proceedings, 1984. Edited by V. Dlab, P. Gabriel an...
340 pages. 1986.

Vol. 1178: Representation Theory II. Groups an...
dings, 1984. Edited by V. Dlab, P. Gabriel and G...
pages. 1986.

Vol. 1179: Shi J.-Y. The Kazhdan-Lusztig Cells in ...
Groups. X, 307 pages. 1986.

Vol. 1180: R. Carmona, H. Kesten, J.B. Walsh, École d'Été de Probabilités de Saint-Flour XIV – 1984. Édité par P.L. Hennequin. X, 438 pages. 1986.

Vol. 1181: Buildings and the Geometry of Diagrams, Como 1984. Seminar. Edited by L. Rosati. VII, 277 pages. 1986.

Vol. 1182: S. Shelah, Around Classification Theory of Models. VII, 279 pages. 1986.

Vol. 1183: Algebra, Algebraic Topology and their Interactions. Proceedings, 1983. Edited by J.-E. Roos. XI, 396 pages. 1986.

Vol. 1184: W. Arendt, A. Grabosch, G. Greiner, U. Groh, H.P. Lotz, U. Moustakas, R. Nagel, F. Neubrander, U. Schlotterbeck, One-parameter Semigroups of Positive Operators. Edited by R. Nagel. X, 460 pages. 1986.

Vol. 1185: Group Theory, Beijing 1984. Proceedings. Edited by Tuan H.F. V, 403 pages. 1986.

Vol. 1186: Lyapunov Exponents. Proceedings, 1984. Edited by L. Arnold and V. Wihstutz. VI, 374 pages. 1986.

Vol. 1187: Y. Diers, Categories of Boolean Sheaves of Simple Algebras. VI, 168 pages. 1986.

Vol. 1188: Fonctions de Plusieurs Variables Complexes V. Séminaire, 1979–85. Edité par François Norguet. VI, 306 pages. 1986.

Vol. 1189: J. Lukeš, J. Malý, L. Zajíček, Fine Topology Methods in Real Analysis and Potential Theory. X, 472 pages. 1986.

Vol. 1190: Optimization and Related Fields. Proceedings, 1984. Edited by R. Conti, E. De Giorgi and F. Giannessi. VIII, 419 pages. 1986.

Vol. 1191: A.R. Its, V.Yu. Novokshenov, The Isomonodromic Deformation Method in the Theory of Painlevé Equations. IV, 313 pages. 1986.

Vol. 1192: Equadiff 6. Proceedings, 1985. Edited by J. Vosmansky and M. Zlámal. XXIII, 404 pages. 1986.

Vol. 1193: Geometrical and Statistical Aspects of Probability in Banach Spaces. Proceedings, 1985. Edited by X. Fernique, B. Heinkel, M.B. Marcus and P.A. Meyer. IV, 128 pages. 1986.

Vol. 1194: Complex Analysis and Algebraic Geometry. Proceedings, 1985. Edited by H. Grauert. VI, 235 pages. 1986.

Vol. 1195: J.M. Barbosa, A.G. Colares, Minimal Surfaces in $\mathbb{R}^3$. X, 124 pages. 1986.

Vol. 1196: E. Casas-Alvero, S. Xambó-Descamps, The Enumerative Theory of Conics after Halphen. IX, 130 pages. 1986.

Vol. 1197: Ring Theory. Proceedings, 1985. Edited by F.M.J. van Oystaeyen. V, 231 pages. 1986.

Vol. 1198: Séminaire d'Analyse, P. Lelong – P. Dolbeault – H. Skoda. Seminar 1983/84. X, 260 pages. 1986.

Vol. 1199: Analytic Theory of Continued Fractions II. Proceedings, 1985. Edited by W.J. Thron. VI, 299 pages. 1986.

Vol. 1200: V.D. Milman, G. Schechtman, Asymptotic Theory of Finite Dimensional Normed Spaces. With an Appendix by M. Gromov. VIII, 156 pages. 1986.

Vol. ...
Edited by A. ...
1986.

Vol. 1210: Probability Measures ...
Edited by H. Heyer. X, 386 pages. 1986.

Vol. 1211: M.B. Sevryuk, Reversible Systems. V, 319 p...

Vol. 1212: Stochastic Spatial Processes. Proceedings, 1984. Ed...
by P. Tautu. VIII, 311 pages. 1986.

Vol. 1213: L.G. Lewis, Jr., J.P. May, M. Steinberger, Equivariant Stable Homotopy Theory. IX, 538 pages. 1986.

Vol. 1214: Global Analysis – Studies and Applications II. Edited by Yu.G. Borisovich and Yu.E. Gliklikh. V, 275 pages. 1986.

Vol. 1215: Lectures in Probability and Statistics. Edited by G. del Pino and R. Rebolledo. V, 491 pages. 1986.

Vol. 1216: J. Kogan, Bifurcation of Extremals in Optimal Control. VIII, 106 pages. 1986.

Vol. 1217: Transformation Groups. Proceedings, 1985. Edited by S. Jackowski and K. Pawalowski. X, 396 pages. 1986.

Vol. 1218: Schrödinger Operators, Aarhus 1985. Seminar. Edited by E. Balslev. V, 222 pages. 1986.

Vol. 1219: R. Weissauer, Stabile Modulformen und Eisensteinreihen. III, 147 Seiten. 1986.

Vol. 1220: Séminaire d'Algèbre Paul Dubreil et Marie-Paule Malliavin. Proceedings, 1985. Edité par M.-P. Malliavin. IV, 200 pages. 1986.

Vol. 1221: Probability and Banach Spaces. Proceedings, 1985. Edited by J. Bastero and M. San Miguel. XI, 222 pages. 1986.

Vol. 1222: A. Katok, J.-M. Strelcyn, with the collaboration of F. Ledrappier and F. Przytycki, Invariant Manifolds, Entropy and Billiards; Smooth Maps with Singularities. VIII, 283 pages. 1986.

Vol. 1223: Differential Equations in Banach Spaces. Proceedings, 1985. Edited by A. Favini and E. Obrecht. VIII, 299 pages. 1986.

Vol. 1224: Nonlinear Diffusion Problems, Montecatini Terme 1985. Seminar. Edited by A. Fasano and M. Primicerio. VIII, 188 pages. 1986.

Vol. 1225: Inverse Problems, Montecatini Terme 1986. Seminar. Edited by G. Talenti. VIII, 204 pages. 1986.

Vol. 1226: A. Buium, Differential Function Fields and Moduli of Algebraic Varieties. IX, 146 pages. 1986.

Vol. 1227: H. Helson, The Spectral Theorem. VI, 104 pages. 1986.

Vol. 1228: Multigrid Methods II. Proceedings, 1985. Edited by W. Hackbusch and U. Trottenberg. VI, 336 pages. 1986.

Vol. 1229: O. Bratteli, Derivations, Dissipations and Group Actions on C*-algebras. IV, 277 pages. 1986.

Vol. 1230: Numerical Analysis. Proceedings, 1984. Edited by J.-P. Hennart. X, 234 pages. 1986.

Vol. 1231: E.-U. Gekeler, Drinfeld Modular Curves. XIV, 107 pages. 1986.